THE EURASIAN HUCHEN, *HUCHO HUCHO*

Perspectives in Vertebrate Science

Volume 5

Series Editor
EUGENE K. BALON

The Eurasian Huchen, *Hucho hucho*

Largest Salmon of the World

By

J. HOLČÍK, K. HENSEL, J. NIESLANIK and L. SKÁCEL

1988
Dr. W. JUNK PUBLISHERS
A MEMBER OF THE KLUWER ACADEMIC PUBLISHERS GROUP
DORDRECHT / BOSTON / LANCASTER

Distributors

for the United States and Canada: Kluwer Academic Publishers, P.O. Box 358, Accord Station, Hingham, MA 02018-0358, USA
for Albania, Bulgaria, China, Cuba, Czechoslovakia, German Democratic Republic, Hungary, Democratic People's Republic of Korea, Mongolia, Poland, Rumania, U.S.S.R., Democratic People's Republic of Vietnam, and Yugoslavia: VEDA, Klemensova 19, 814 30 Bratislava, CSSR
for all other countries: Kluwer Academic Publishers Group, Distribution Center, P.O. Box 322, 3300 AH Dordrecht, The Netherlands

Library of Congress Cataloging in Publication Data

The Eurasian huchen, *Hucho hucho.*

(Perspectives in vertebrate science; v. 5)
Bibliography: p.
Includes indexes.
1. Huchen. 2. Fishes–Europe. 3. Fishes–Asia.
I. Holčík, Juraj. II. Series.
QL638.S2E86 1988 597'.55 87-3792

Scientific Editor: Academician Ľudovít Weismann

ISBN 90-6193-643-8 (this volume)
ISBN 90-6193-897-X (series)

Joint edition published by
Dr. W. Junk Publishers,
P.O. Box 163, 3300 AD Dordrecht, The Netherlands
and VEDA, Klemensova 19, 814 30 Bratislava, CSSR

PRINTED IN CZECHOSLOVAKIA

"Once I caught a huchen just at noon. I overcome her and hauled her to the shallows. She lay there, quiet and shining, and as she breathed she shone out and dimmed time and again. The valley was majestic, the water roared, and I heard a voice saying: 'What a beauty you are!' Not a soul was there. It was me saying the words, overwhelmed by the beauty. Carefully removing the hook, I turned and released the huchen. She swam slowly through the shallow water among the boulders, looking back as though to see if I was in earnest, then came on deeper water, speeding up a little, and, feeling the depth below, she bolted, disappeared like lightning. To this day she tells this tale of me to her little ones."

Jan Werich, National Artist
(Mladý svět 1980, **22:** *51)*

Preface

The need to gather available data on the Eurasien huchen – an important salmonid species – has been forced by a plain and, unfortunately, common fact of our times: the numbers and distribution of this biggest of salmonids have begun to decline and its range has begun to shrink. A seminar on the huchen – the European form of the species *Hucho hucho* – held in Žilina in February 1973 as a result of a suggestion of the Section for the Conservation of Fauna of the Slovak Zoological Society, indicated very clearly the sad situation. Data on the biology of the huchen are regrettably scarce despite several recent papers (Ivaška 1951, Svetina 1962, Prawocheński and Kolder 1968) with the aim of filling this gap. Supposing that without a thorough knowledge it is practically impossible to conserve any plant or animal species, the participants of the seminar concluded that the existing knowledge on the huchen should be compiled in an exhaustive monograph. The first such outline originated in 1977 under the authorship of J. Holčík, K. Hensel and L. Skácel, and was submitted as a research report to some of the central authorities. Even during the compilation of the report it became evident, however, that there is no difference between the huchen and its relative, the taimen. Consequently, we immediately began revising our first report, which took over three years. We have endeavoured to present not only a simple compilation, but also a critical analysis and synthesis as a basis for both subsequent scientific investigations and practical conservation measures. For this reason, several problems had to be tackled during the preparation of the monograph, in the form of separate scientific papers. In spite of the fact that our knowledge of the huchen has thus advanced considerably, it would be preposterous to believe that all problems associated with this species have been solved. On the contrary, we can only confirm Kanep's (1976) recent statement that the genus *Hucho* is still among the least known forms of this great and important salmonoid group.

Our monograph reflects the present state of knowledge of both the huchen and the taimen. But both forms were, and unfortunately still are, primarily the concern of fishermen, especially anglers, and only much less of ichtyologists and zoologists. That is why individual chapters of the present monograph are so unbalanced: for example, data on the anatomy of the huchen are still fragmentary and lag behind those on economic importance and farming.

In this team work individual authors participated as follows: K. Hensel wrote Chapters 1, 4 and 7, compiled the maps showing the distribution (Figs. 17–21) and re-drew Figs. 7–12, J. Nieslanik and L. Skácel wrote the fundamentals to Part V and supplied some of the photographs and other documentation. J. Holčík wrote the whole Parts III and IV, Chapters 3, 6, 8 and 22; he cooperated with K. Hensel in working out Chapters 2 and 5, supplemented and compiled Part V, re-drew or compiled most of the illustrations, supervised the work of the team and finally edited the manuscript.

It is more than a pleasant duty of the authors, in particular the senior one, to extend cordial thanks to a wide circle of collaborators without whose help this work could not have been completed. They contributed to it by providing literature inaccessible to the authors, by submitting unpublished data, by providing some of the necessary computer time, by their valuable advice and discussions, or in various other ways. Their names are listed below in an alphabetical order, regardless of their titles, functions, or nationalities:

P. Áč, Faculty of Arts, Comenius University, Bratislava;

M. Andraščík, Bratislava;

I. Bastl, Institute of Fishery Research and Hydrobiology, Bratislava;

J. Bél, Seenforschungslaboratorium der EAWAG/ETH, Kastanienbaum, Switzerland;

J. F. Bergeron, Service de l'aménagement et de l'exploitation de la faune, Montréal, Canada;

P. Blahák, Institute of Science, Slovak National Museum, Bratislava;

M. Bohl, Bayerische Landesanstalt für Wasserforschung, Wielenbach, FRG;

P. Brunovský, Institute of Applied Mathematics and Computation, Comenius University, Bratislava;

K. Buss, Pennsylvania Fish Commission, Boalsburg, Pennsylvania, USA;

T. Cavender, Museum of Zoology, The Ohio State University, Columbus, Ohio, USA;

Xin-Luo Chu, Kunming Institute of Zoology, Academia Sinica, Kunming, Yunnan, People's Republic of China;

I. A. Chereshnev, Institute of Biology and Soil, Far East Centre, USSR Academy of Sciences, Vladivostok, USSR;

J. Černý, Institute of Fishery Research and Hydrobiology, Bratislava;

E. A. Dorofeeva, Institute of Zoology, USSR Academy of Sciences, Leningrad, USSR;

V. Dyk, Brno;

R. Frank, Baia Mare, Rumanian Socialist Republic;

O. F. Gritsenko, All-union Research Institute of Marine Fisheries and Oceanography, Moscow, USSR;

L. Grman, Editorial Board of the magazine Poľovníctvo a rybárstvo, Bratislava;

J. Guziur, Department of Fisheries, Academy of Agriculture and Technology, Olsztyn, People's Republic of Poland;

Jan Henricson, Fishery Board of Sweden, Härnősand, Sweden;

P. Holec, Department of Geology and Paleontology, Faculty of Science, Comenius University, Bratislava;

L. Jedlička, Department of Zoology, Faculty of Science, Comenius University, Bratislava;

M. Jungwirth, Limnologisches Institut, Österreichische Akademie der Wissenschaften, Wien, Österreich;

E. Kainz, Bundesinstitut für Gewässerforschung und Fischereiwirtschaft, Scharfling, Österreich;

A. Kaliský, Liptovský Hrádok;

L. Kattoš, Local constituent, Slovak Anglers' Union, Prievidza;

V. Křička, Local constituent, Slovak Anglers' Union, Bardejov;

V. Lacko, Research Institute of Chemical Fibres, Svit;

P. Laurent, Station d'hydrobiologie lacustre, Institut National de la Recherche Agronomique, Thonon-les-Baines, France;

J. Lepiksaar, Naturhistoriska Museet, Göteborg, Sweden;

J. Lobón-Cerviá, Centro de Zoologia Aplicada, Instituto Nacional para la Conservación de la Naturaleza, Madrid, Spain;

A. Makara, Rožňava;

P. Mamatej, Local constituent, Slovak Anglers' Union, Martin;

N. V. Martin, Fisheries Branch, Ontario Ministry of Natural Resources, Maple, Ontario, Canada;

C. Mondejar-Reyna, Seccion de Hidrología, Instîtuto Nacional para la Conservación de la Naturaleza, Madrid, Spain;

F. Moravec, Institute of Parasitology, Czechoslovak Academy of Sciences, České Budějovice;

N.-A. Nilsson, Institute of Freshwater Research, Drottningholm, Sweden;

I. Novák, Local constituent, Slovak Anglers' Union, Martin;

L. Nyman, Institute of Freshwater Research, Drottningholm, Sweden;

O. Oliva, Department of Systematic Zoology, Faculty of Science, Charles University, Prague;

M. Papadopol, Faculty of Biology, University of Bucharest, Bucharest, Rumanian Socialist Republic;

M. Peňáz, Institute of Systematic and Ecological Biology, Czechoslovak Academy of Sciences, Brno;

V. Pikna, Waterworks Investment, Bratislava;

J. Poupě, Central Committee, Czech Anglers' Union, Prague;

J. Příhoda, Central Committee, Slovak Anglers' Union, Žilina;

S. U. Qadri, Faculty of Science, University of Ottawa, Ottawa, Ontario, Canada;

P. Ráb, Institute of Animal Physiology and Genetics, Czechoslovak Academy of Sciences, Liběchov;

J. Sedlár, Department of Poultry and Zoology, University of Agriculture, Nitra;

S. Stokłosova, Institute of Zoology, Jagellonian University, Cracow, People's Republic of Poland;

J.-A. Timmermans, Station de Recherches des Eaux et Forêts, Groenendaal, Belgium;

P. Vivier, Paris, France;

H. Willoughby, U. S. Fish and Wildlife Service, Denver, Colorado, USA;

A. Witkowski, Natural History Museum, University of Wrocław, Wrocław, People's Republic of Poland;

Xien-Wen Wu, Institute of Hydrobiology, Academia Sinica, Shanghai, Hubei, People's Republic of China;

J. Zubrický, Regional Committee, Slovak Anglers' Union, Košice;

M. Zhivkov, Institute of Zoology, Bulgarian Academy of Sciences, Sofia, People's Republic of Bulgaria.

We are much obliged to Mrs. M. Baradlaiová for accurately drawing Figs. 2–5, 13–15 and 23, Mr. J. Dubeň for kindly providing most of the photographs, and Mrs. M. Brinzová for typing the difficult manuscript. We are also deeply indebted to Dr. Ing. R. Obrtel not only for the painstaking English translation, but also for some critical comments, notes and linguistic assistance.

The English edition of this monograph represents a substantially updated, revised and corrected original Slovak version issued by VEDA, Publishing House of the Slovak Academy of Sciences, Bratislava, 1984. Papers appearing after December 1985 are not included in this book.

Contents

PART IV – UTILIZATION

PART V – FARMING

List of Symbols and Abbreviations

ad-C	distance between adipose fin and caudal fin
A	anal fin
A-C	distance between base of anal and base of caudal fin
BR	branchiostegal rays
C (C_1, C_2, C_3)	caudal fin (upper lobe, medial part and lower lobe of caudal fin, respectively)
D	dorsal fin
Fl	fork length (Smitt length)
h	minimum body depth (least depth of caudal peduncle)
hA	depth of anal fin
hD	depth of dorsal fin
hc	head depth (at nape)
hmx	depth of upper jaw
H	body depth (maximum body depth)
io	interorbital distance (skull width)
K	Fulton's coefficient of condition
lc	length of head
l.l.	lateral line (pored scales)
lmd	lower jaw length
lmx	upper jaw length
lpc	length of caudal peduncle
lA	length of anal fin base
lC (lC_1, lC_2, lC_3)	length of caudal fin (length of upper lobe, medial part and lower lobe of caudal fin, respectively)
lD	length of dorsal fin base
lP	length of pectoral fin
lV	length of ventral (pelvic) fin
Oh	horizontal diameter of eye
pop	preopercular distance
poD	postdorsal distance
poO	postorbital distance

prO	preorbital distance (snout length)
pA	preanal distance
pD	predorsal distance
pV	preventral distance
P	pectoral fin
PC	pyloric caeca
P-V	distance between pectoral and ventral fin
rA	rays in anal fin
rD	rays in dorsal fin
rP	rays in pectoral fin
rV	rays in ventral (pelvic) fin
Sl	standard length
Sp.br.	branchial spines (gill rakers)
Squ.	transverse rows of scales
Squ. inf.	scales below lateral line
Squ. sup.	scales above lateral line
Tl	total length
TU	temperature unit (day-degree)
V	ventral (pelvic) fin
V-A	distance between ventral and anal fin
Vert.	vertebrae
w	individual weight

Part I

TAXONOMY, SYSTEMATICS AND EVOLUTION

1. Nomenclature

VALID SCIENTIFIC NAME

Hucho hucho (Linnaeus, 1758)
Original combination: *Salmo hucho* Linnaeus, 1758 (Systema Naturae, **10:** 309)

SYNONYMS

Huch Germanorum Gessner, 1560
Trutta piscinaria Gessner, 1575
Trutta fluviatilis altera Marsilius, 1726
Salmo hucho Linnaeus, 1758
Salmo taimen Pallas, 1773
Salmo Schiefermülleri Bloch, 1782
Salmo fluviatilis Pallas, [1814]
Salmo (Hucho) hucho Günther, 1866
Salmo (Salvelinus) lossos Günther, 1866
Salmo (Epitomynis) hucho Schulze, 1890
Salvelinus (Hucho) taimen Berg, 1909
Hucho taimen Berg, 1916
Hucho hucho Berg, 1948

2. Taxonomy and Systematics

RELATIONS

Suprageneric Relations

Phyllum: Vertebrata
Subphyllum: Gnathostomata
Supraclassis: Pisces
Classis: Osteichthyes
Subclassis: Actinopterygii
Infraclassis: Teleostei
Cohors: Euteleostei
Divisio: Neognathi
Supraordo: Protacanthopterygii
Ordo: Salmoniformes
Subordo: Salmonoidei
Suprafamilia: Salmonoidea
Familia: Salmonidae
Subfamilia: Salmoninae
Tribus: Huchonini

The classification is based on Greenwood et al. (1966, 1967), Rosen (1974) and Holčík (1982).

Generic Relations

Hucho Günther, 1866 (Cat. Fish. Brit. Mus., **6:** 125)
Typical species: *Salmo hucho* Linnaeus, 1758

The following is the diagnosis of the genus *Hucho* Günther (according to Berg 1948, Lindberg and Legeza 1965, Shaposhnikova 1968, 1975 and Holčík 1982):

The teeth on os vomere form a continuous, horseshoe-like band with the palatal teeth. The shaft of os vomere lacks teeth (Fig. 12). Ossa basibranchialia are mostly without teeth. Ossa postorbitalia occupy on average not more than 45% of distance from posterior margin of the orbit to preopercular bone. The distance between tip of snout and posterior margin of the upper jaw (os maxillare) is usually more than 45% of head length. Skull broad (Pl. 1.2), os supraethmoideum relatively short, with very varied posterior margin (Fig. 11). Head compressed (Pl. 2.1). Mouth large. Mandible articulating with skull behind the vertical line touching the posterior margin of eye. Maxilla of adults reaching to posterior margin of eye. Body cylindrical, prolonged, covered with relatively small scales forming 100–290 transversal rows. Scales of lateral line exhibit peculiar structure: they lack circular ridges and resemble a narrow and thick osseous plate with canal on medial side imperfectly closed, opening on the surface on extreme edge of the caudal part of the scale (Fig. 1). Adipose fin large: the length of its base is over 50% of that of anal fin on average (Pl. 1.1). Coloration lacks pale spots but large, round, crescent-shaped and X-shaped black markings are present. D II–VII 8–14, A I–V 7–11.

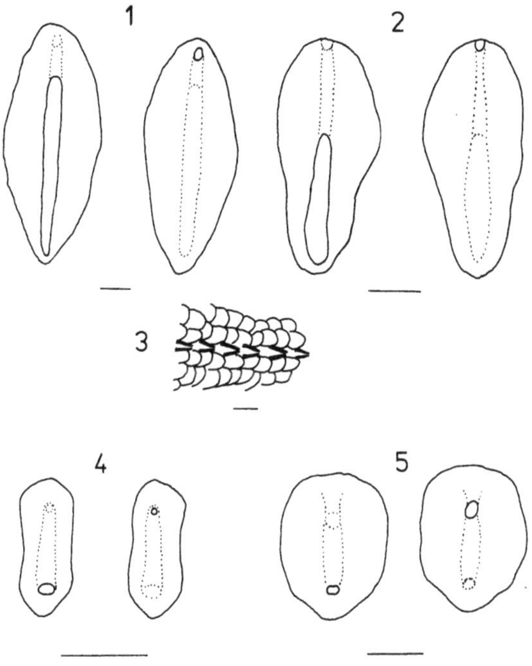

Fig. 1. Lateral line scales of *Hucho hucho* (1 – *Fl* 926 mm) and *Brachymystax lenok* (2 – *Fl* 363 mm) in comparison with those of *Salvelinus fontinalis* (4 – *Fl* 148 mm) and *Salmo trutta* m. *fario* (5 – *Fl* 294 mm). Ventral (left) and dorsal (right) view. Ridges in 4 and 5 omitted. No. 3 shows scalation of cleared and stained specimen of *Hucho hucho* (*Fl.* 148 mm) – lateral line scales ossified only partly. Bar = 1 mm (from Holčík 1982).

The huchens were originally included in the genera *Salmo* or *Salvelinus*. Recently, all authors have agreed on a separate position and valid definition of the genus *Hucho*, except for Spillman (1961), who continues to include the huchen in the genus *Salvelinus* Richardson.

The difference between huchen and salmon (genus *Salmo*) was first pointed out by Günther (1866) who established a separate subgenus *Hucho* for the huchen. The name was first used by Jordan and Snyder (1902) as generic in connection with *Hucho perryi*. Norden (1961) believes that the genus *Hucho* Günther is intermediate between the genera *Salmo* Linnaeus and *Brachymystax* Günther. Detailed osteological studies enabled Shaposhnikova (1967, 1968, 1975) to prove that the genus *Hucho* is very closely related to the genus *Brachymystax*. Both these genera have analogously arranged teeth on os vomere and on the palatal bones, and also the shape and position of other skull bones are similar. The close relation of the two genera is also indicated by the conspicuous similarity of their karyotypes (Dorofeeva 1977, Ráb 1981, Ráb and Liehman 1982), by the shape and size of their adipose fins, and by the identical structure of the lateral line scales (Holčík 1982) (Fig. 1, Pl. 3.2). It is also evidenced by the findings of natural hybrids between *Hucho h. taimen* and

Brachymystax lenok in the Amur river system (Hsieh et al. 1959). The genera *Hucho* and *Brachymystax* differ mainly in the size of the jaws of adults; in the genus *Hucho* they are considerably bigger. On the contrary, the coloration is more varied in the genus *Brachymystax*.

On the other hand, there is no doubt that even the genera *Hucho* and *Salvelinus* have many characters in common and are close to each other not only in morphological and anatomical characters, but also in the similarity of their proteins, cytomorphology and the degree of homology of unique sequence DNA (Vladykov 1954, Behnke 1968, Tuturov and Omel'chenko 1976, Dorofeeva 1977, Cavender 1980, Holčík 1982). In this connection the North American lake trout, *Salvelinus (Cristivomer) namaycush* (Walbaum 1792), is usually mentioned. According to the chromosomal characters, the genus *Hucho* occupies an intermediate position between the genera *Brachymystax* and *Salvelinus* (Table 1) and appears to be somewhat closer to the former.

TABLE 1

Chromosomal characters of the Salmoninae (compiled from Kang and Park 1973, Kirpichnikov 1974, Dorofeeva 1977, Sofradžija 1979, Vasil'ev 1980, Ráb and Liehman 1982, Anbinder et al. 1982, Viktorovskii et al. 1985)

Genus	2*n*	*M*	*A*	*NF*
Brachymystax	90—92	10—13	82	102—116
Hucho	62—84	9—19	66	100—116
Salvelinus	68—86	14—22	56—72	96—100
Salmo	54—84	12—50	6—70	72—108
Salmothymus	82	12	70	94
Oncorhynchus	52—74	20—52	0—56	94—106

In spite of considerable progress of research in the past decade, the knowledge of the genus *Hucho* is still insufficient. The appurtenance of this genus in the family Salmonidae is beyond any doubt. Until recently, there has been no doubt either of the genus belonging to the subfamily Salmoninae. Holčík (1982) believes, however, that the genera *Hucho* and *Brachymystax* constitute either the separate subfamily Huchoninae, once established by Jordan and McGregor (1925), or the separate tribus Huchonini. Accepting the opinion of most present authors (Berg 1940, 1955, Behnke 1968, Rosen 1974, Svetovidov et al. 1975, 1976, Dorofeeva 1977, Kendall and Behnke 1984) that salmons, whitefishes and graylings constitute only the respective subfamilies Salmoninae, Coregoninae and Thymallinae, then the huchen and the lenok *(Brachymystax lenok)* can only be separated as a tribus. However, if one should accept the opinion of Reshetnikov (1980, not altogether unsubstantiated either) that the above groups constitute valid families (Salmonidae, Coregonidae, Thymallidae) within the superfamily Salmonoidea, then the genera *Hucho* and *Brachymystax* can be considered to constitute the separate subfamily Huchoninae.

The genus *Hucho* includes four species:

1 – *Hucho hucho* (Linnaeus, 1758) (Figs. 2 and 3, Pl. 1 and 5)

First gill arch with 9–19 gill rakers (spinae branchiales), lateral line (linea lateralis) with 107–194 perforated scales, with 25–38 scale rows above and 25–39 rows below the lateral line. Body with 141–288 scale rows (squamae transversales). Vertebrae 60–72 in number, of which 42–49 are trunk and 20–25 caudal. Pyloric caeca (appendices pyloricae) 150–284 in number, 9–13 branchiostegal rays. D II–VI 8–14, A I–VI 7–11. A big species, attaining over 2000 mm in length and over 100 kg in body weight.

The distribution of this species is disjunct: rivers of the Danube river system in the west and rivers emptying into the Northern Arctic Ocean in the east, and from the Pechora river in the west to the Yana river in the east. It also occurs in the Kama river drainage area (Volga river system) and in those of the rivers Uda, Tugur and Amur. An exclusively freshwater species.

Fig. 2. The huchen, *Hucho hucho hucho* (Linnaeus, 1758). The Danube river near Radvaň nad Dunajom (river kilometre 1749), 10 May 1967. Total length 607 mm (orig. M. Baradlaiová).

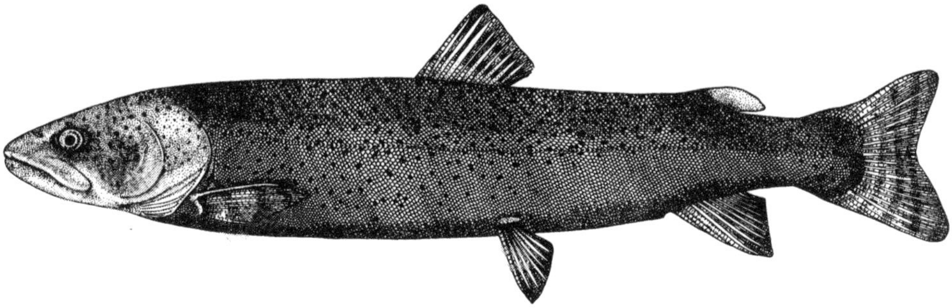

Fig. 3. The taimen, *Hucho hucho taimen* (Pallas, 1773). The Kama river near Perm. Total length about 870 mm (after Berg 1948–1949).

2 – *Hucho ishikawai* Mori, 1928

First gill arch with about 14 gill rakers, lateral line with approximately 126 perforated scales, body with about 150 longitudinal scale rows. D IV 7, A III 7.

An endemic freshwater species populating the upper reaches of the Yalu river (Korea).

3 – *Hucho bleekeri* Kimura, 1934 (Fig. 4)

First gill arch with 12–14 gill rakers. Lateral line with about 150 (probably transverse rows) of scales. D III 7, A III 8.

A freshwater species from the upper reaches of the Chang-jiang (Yangtze) river in the Sichuan Province. Attains 4 kg in body weight and 720 mm in length.

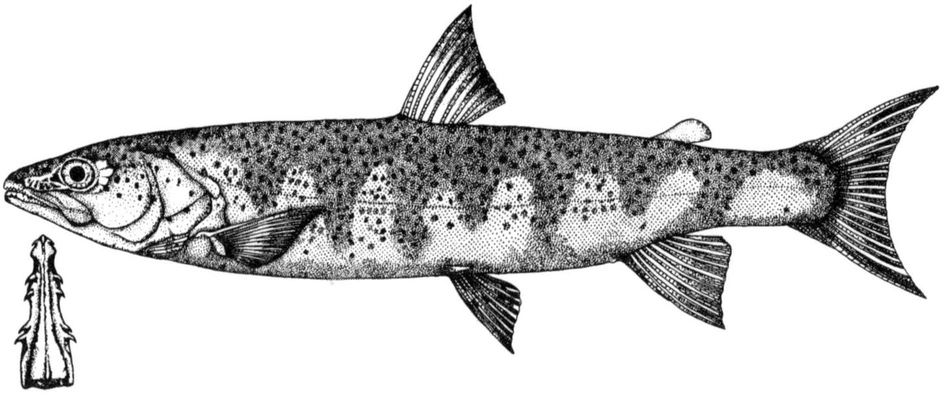

Fig. 4. *Hucho bleekeri* Kimura, 1934. A mountain brook near Guanxian, Sichuan Province. Total length 280 mm (after Kimura 1934). Inset, also after Kimura, shows os linguale plattum which Kimura erroneuosly denoted and Berg erroneously interpreted as the vomer. The dark transverse stripes on the flanks of the fish indicate that it is a sexually immature individual. This fact, together with the length of the fish and the conspicuously large adipose fin, evidence that it belongs to the genus *Hucho*.

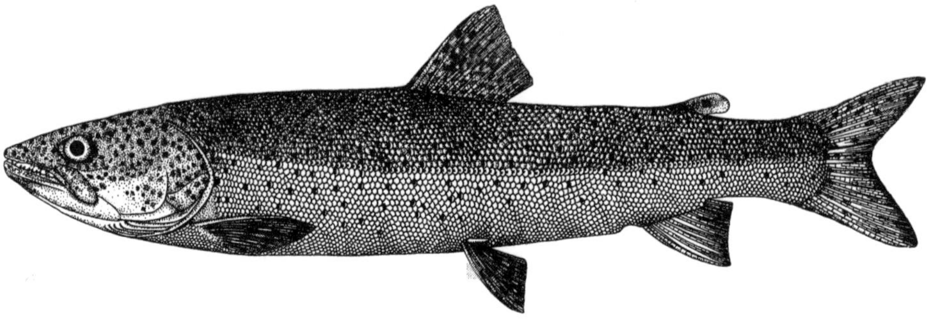

Fig. 5. *Hucho perryi* (Brevoort, 1856). Vladimir Bay (Maritime Region, Sea of Japan), 30 June 1913. Length (*Fl*?) 462 mm (after Berg 1948–1949).

4 – *Hucho perryi* (Brevoort, 1856) (Fig. 5)

First gill arch with 10–21 gill rakers. Lateral line with 87–121 perforated scales, with 17 scale rows above and 16 rows below the lateral line. Body with 100–119 transverse scale rows. Vertebrae 55–62 in number, 35–39 being trunk and 20–22 caudal. Pyloric caeca 184–254 in number, 9–13 branchiostegal rays. D II–IV 8–10, A I–III 7–11. The species attains up to 60 kg in body weight and over 1500 mm in length.

Contrary to the other members of the genus, *H. perryi* is diadromous. Adults mostly stay in river estuaries or in the sea off the coast, migrating up river only for spawning and/or overwintering. The species lives in the Sea of Japan (from the Tatar Strait in the north to Vladivostok and Cap Rokko in the south) and in the adjacent part of the Pacific Ocean and the Sea of Okhotsk (approximately to the town of Morioka on the Honshu island in the south and to Nyiskii Bay on the Sakhalin in the north.

In the spawning period, it migrates from the sea into rivers draining the eastern slopes of the Sikhote-Alin' Mts. and into the rivers and lakes of the islands of Honshu, Hokkaido, Kunashir and Sakhalin. Sakhalin appears to be the centre of the range of the species.

The diagnoses of the individual species and their distribution have been compiled from the data of the following authors: Heckel and Kner (1858), Siebold (1863), Mori (1928), Kimura (1934, 1966), Berg (1948–1949), Pravdin (1949), Shmidt (1950), Nikol'skii (1956), Lindberg and Legeza (1965), Bukirev (1967), Kirillov (1962, 1972, 1976), Shaposhnikova (1968), Misharin and Shutilo (1971), Gritsenko et al. (1974), Wu (1979), Wu and Chen (1979), Hensel (1976, 1980), Hensel and Pivnička (1980), Kuo (1981), Hensel and Holčík (1983) and Bushuev (1983).

It should be added that the inclusion of the species *H. bleekeri* Kimura, 1934 in the genus *Hucho* has been questionable until recently. Describing the species, Kimura himself considers its incidental appurtenance to the genus *Salmo*. Berg (1948–1949), on the basis of (allegedly) Kimura's description and illustration of its os vomere, doubts that it is a representative of *Hucho* or *Salmo*. Berg is mistaken, however: Kimura does not illustrate the os vomere but the os linguale plattum (Fig. 4)! Nor does the supplement to the description of this species, based on a revision of the type material, recently published by Kuo (1981) make any contribution to knowledge of the species, as it is terse and unclear, and lacks illustrations.[1] Nevertheless, it is certain that it is in fact a member of the genus *Hucho:* the shape of os linguale plattum, as illustrated by Kimura, the size of the adipose fin and, above all, the structure of the lateral line scales (several scales were available to us from a specimen 640 mm *Sl* and 4 kg in body weight, kindly sent by Dr. Xin-Luo Chu, Kunming) are in good agreement with these features in the huchen. The small number of pyloric caeca (60) is not decisive for the time being, since Kimura mentions that the appendages were damaged. According to Wu and Chen (1979), *H. bleekeri* reaches 720 mm in length and 4 kg in body weight. Thus it is among the smallest species of the genus. It occurs in the rapids of mountain gorges in the upper reaches of the rivers Min and Dadu (the Yangtze river drainage area).

The situation of *Hucho ishikawai* is likewise uncertain. To our knowledge,

[1] The following is a verbatim translation of Kuo's (1981) supplement by Dr. Xin-Luo Chu (Kunming) (in a letter of 23 Feb. 1981): "There are two strong teeth at the anterior end of palatine instead of four small teeth (as stated by Kimura, note J. H.). A conspicuous process at the base of each tooth. Palatine proper and its both edges without any teeth instead of 4 teeth as original description recorded."

a detailed description of this species has not yet been presented, the only description available being the original one by Mori (1928) which is rather incomplete. Vladykov (1963) presumed *H. ishikawai* to be conspecific with *H. perryi*, differing only in the larger number of transverse scale rows, which alone does not suffice for the separation of a species. We are inclined to believe, however, that it is closer to *H. hucho*, as indicated not only by the almost identical number of transverse scale rows and the number of perforated scales in the lateral line, but also by the fact that it populates the upper reaches of the Yalu river (which is longer than any river on Sakhalin or Hokkaido islands) and obviously is not diadromous.

Vladykov (1963) presumed that *H. perryi* forms a separate subgenus *Parahucho*, which differs from the remaining huchen especially in the presence of basibranchial teeth, a supplementary medial tooth row on os linguale plattum, the presence of a hypethmoideum, a smaller number of vertebrae, a smaller number of scales in the lateral line, and differences in coloration. Shaposhnikova (1968) confirmed this assumption by osteological analysis, and finally Vladykov and Gruchy (1972) additionally described the subgenus *Parahucho*. Thus the genus *Hucho* is divided into two subgenera, viz.:

Hucho Günther; basibranchialia without a toothed plate, os linguale plattum with teeth on its edge only. The outer margin of frontale lacking a projection above os sphenoticum. Hypethmoideum absent. Vertebrae around 70 in number. The number of transverse scale rows is distinctly larger than that of perforated scales in the lateral line.

Parahucho Vladykov; basibranchialia with a plate covered with teeth, os linguale plattum with a medial row of small teeth besides the marginal ones. The outer margin of os frontale forms a projection above os sphenoticum. Hypethmoideum occurs in big individuals. Vertebrae around 60 in number.

Viktorovskii et al. (1965) believe that *H. perryi* does not belong to the genus *Hucho* but is a member of the separate genus *Parahucho*. In their opinion, *H. perryi* is less closely related to the genus *Hucho* than this genus to *Brachymystax*. It is not clear on what grounds they draw this conclusion, since both the anatomical characters including the karyotype (Tables 1 and 5), and the ecology of *H. perryi* clearly indicate its appurtenance to the genus *Hucho*.

Specific Relations

Linnaeus (1758) dit not state the type specimen or the terra typica of *Salmo hucho*. According to the reference to Artedi, as stated by Berg (1948–1949), however, it is the Danube near Vienna. So far, nobody has designated a neotype of the species. The characters in which *Hucho hucho* differs from the remaining members of the genus are given in the preceding section.

TABLE 2

Comparison of meristic characters of *Hucho hucho* from different river systems

	Slovak rivers[1] (Váh, Orava, Turiec)				Danube[2] (*n* = 2)	Tisza[3] (*n* = ?)	Cheremosh[4]			
	n	$\bar{x}$	$s_{\bar{x}}$	*s*	Range		*n*	$\bar{x}$	$s_{\bar{x}}$	*s*
rD	23	10.65	0.17	0.83	11—11	9—11	—	—	—	—
rA	23	9.26	0.09	0.45	9—9	8—9	—	—	—	—
rP	—	—	—	—	—	—	—	—	—	—
rV	—	—	—	—	—	—	—	—	—	—
l.l.	—	—	—	—	152—156	—	23	132.83	1.38	6.63
Squ.	19	185.16	1.34	5.85	—	207—233	23	253.90	4.84	23.25
Squ.sup.	20	31.85	0.55	2.46	32—32	30—34	23	34.35	0.47	2.27
Squ.inf.	20	29.20	0.52	2.31	26—28	25—33	23	29.48	0.63	3.01
Sp. br.	23	14.12	0.19	0.92	15—15	17—19	26	13.08	0.24	1.21
BR	—	—	—	—	—	—	—	—	—	—
Vert.	—	—	—	—	—	—	—	—	—	—
PC	9	191.33	8.63	25.89						

[1]Hensel and Pivnička (1980); [2]Balon (1968), Blahák (1972); [3]Vladykov (1931); [4]Kulmatycki (1931c) (*Squ.* includes also pored scales); our data for *PC* (9 specimens from the Turiec river).

SYSTEMATIC STATUS

The systematic status of *Hucho hucho* has been established on the basis of external morphological characters, later supplemented by osteological and cytomorphological ones. While on the one hand there is a clear difference between *Hucho hucho* and the remaining members of the genus, the differences between the central European and the eastern European-Siberian-Amur river populations of *Hucho hucho* are dubious. Until the publication of our revision (Hensel and Holčík 1983), the former have been denoted as *Hucho hucho* (Linnaeus, 1758), the latter as *Hucho taimen* (Pallas, 1773). As seen in Tables 2 and 3 and especially in Table 4, which summarize the variation ranges of taxonomically important characters, there are practically no differences worth mentioning. The osteological differences are minimal and insufficient for a differentiation between *H. hucho* and *H. taimen.* Already Bajkov (1924) and later also Vladykov (1963), Behnke (1968) and Shaposhnikova (1968) presumed that at most subspecific differences exist between the two forms. This presumption is corroborated by the data on chromosomal characteristics of the two forms, which are practically identical (Ráb and Liehman 1982, Viktorovskii et al. 1985). Moreover, there are practically no differences even in the ecology of the two

TABLE 2

(Continued)

	Danube[5] basin (n = ?)	Vishera[6]				Ob'[7] (n = 3)	Shishkhid[8]			
	Range	n	$\bar{x}$	$s_{\bar{x}}$	s	Range	n	$\bar{x}$	$s_{\bar{x}}$	s
rD	8—11	64	11.28	0.09	0.69	10—10	71	10.10	0.08	0.65
rA	7—9	59	9.29	0.06	0.49	9—9	71	8.87	0.06	0.51
rP	16—16	48	13.25	—	—	14—16	—	—	—	—
rV	8—9	47	8.4	—	—	9—9	—	—	—	—
l.l.	—	22	137.95	0.90	4.20	—	10	134.50	2.18	6.88
Squ.	180—200	19	195.50	0.50	2.17	194—210	55	189.16	1.17	8.60
Squ.sup.	18—20	—	—	—	—	—	65	30.92	0.28	2.26
Squ.inf.	20—24	—	—	—	—	—	65	30.09	0.36	2.93
Sp.br.	10—18	59	11.58	0.16	1.12	11—11	71	10.65	0.12	0.99
BR	10—11	—	—	—	—	11—11	—	—	—	—
Vert.	—	29	65.48	0.36	1.90	—	—	—	—	—
PC	200—200	35	237.10	3.96	23.40	—	—	—	—	—

[5]Berg (1948), Bănărescu (1964); [6]Bukirev (1967) (data for transverse rows calculated from those of Bukirev, for pored scales from those supplied by O.F.Gritsenko); [7]Varpakhovskii (1899) and Berezovskii (1924); [8]Hensel and Pivnička (1980).

forms, as recently pointed out by Hensel and Holčík (1983). Besides, hybrids between *H. hucho* and *H. taimen* are fertile, whereas those between *H. taimen* and *H. perryi* are not (Wendland 1965). All this leads to the conclusion that the two forms must be conspecific. By priority, this single species must be named *Hucho hucho* (Linnaeus, 1758), the name *H. taimen* (Pallas, 1773) being a junior synonym.

SUBSPECIES

1. *Hucho hucho hucho* (Linnaeus, 1758), the huchen (for the terra typica, etc., see Specific Relations).
2. *Hucho hucho taimen* (Pallas, 1773), the taimen (terra typica: "Ascendit flumina Siberiae facile omnia Oceanum glaciale influentia, et praeruptos maxima alpestrium torrentes petit"; original combination: *Salmo taimen* Pallas, Reise **1:** 216, 716; 1773).

The two subspecies cannot be separated on the basis of morphometric and meristic characters (see Table 4). Berg (1948–1949) states that the differences in the number of gill rakers are the "key characters" separating the huchen and the taimen: 16 in *H. hucho* and (9) 11–12 (14–16 with the rudimentary ones) in *H. taimen*. The author also mentions the difference in the number of longitudinal and transverse scale rows: *H. hucho,* 180 $\frac{18-20}{20-24}$ 200; *H. taimen,* 193 $\frac{26-30}{26-30}$ 242. However, sufficiently

TABLE 2

(Continued)

	Enisei[9] (n = 37)	Angara[10] (n = 100)			Frolikha[11] (n = 47)			Khatanga[12] (n = 9)
	Range	$\bar{x}$	$s_{\bar{x}}$	s	$\bar{x}$	$s_{\bar{x}}$	s	Range
rD	9—12	10.69	0.02	0.19	9.87	0.05	0.33	9—11
rA	8—9	8.98	0.01	0.13	8.64	0.07	0.48	7—9
rP	14—16	14.43	0.02	0.21	14.62	0.06	0.40	16—16
rV	9—9	8.83	0.01	0.12	8.98	0.02	0.14	10—10
l.l.	107—160	—	—	—	—	—	—	135—194
Squ.	209—210	191.65	0.70	6.99	186.40	1.02	6.92	—
Squ.sup.	—	—	—	—	—	—	—	25—31
Squ.inf.	—	—	—	—	—	—	—	25—37
Sp.br.	9—12	12.71	0.09	0.89	12.02	0.02	0.70	12—13
BR	10—11	—	—	—	—	—	—	—
Vert.	—	—	—	—	—	—	—	—
PC	—	—	—	—	—	—	—	—

[9]Isachenko (1912); Podlesnyi (1958); [10,11]Misharin and Shutilo (1971); [12]Luk'yanchikov (1967).

numerous data, particularly on the huchen, were not available to Berg. Besides, the numbers of gill rakers stated by different authors depend on their different ways of counting them, i.e. whether or not a particular author included (and mentioned having done so) the rudimentary rakers in the raker formula. Hensel and Pivnička (1980), who applied the same method in counting these rakers in the huchen from the Váh river system and in the taimen from the Enisei river system, found no difference in this character. In *H. hucho* they counted 10–14 (avg. 11.9) and in *H. taimen* 9–13 (avg. 10.7) normal gill rakers. Almost all individuals examined had from 1 to 5 rudimentary rakers (avg. 2.2 in *H. hucho* and 2.4 in *H. taimen*). The authors found no differences even in the number of longitudinal and transverse scale rows between the huchen and taimen examined. They found an average of 185.2 transverse and 31.9/30.9 longitudinal scale rows in *H. h. hucho,* and 189.2 and 29.2/30.1 in *H. h. taimen*, respectively. The coefficients of difference computed from the data fail to reveal any differences.

Shaposhnikova (1968) pointed out certain differences in the mean values of certain morphometric criteria, such as the length of mandible, length of snout, and eye diameter. Hensel and Pivnička (1980) and Hensel and Holčík (1983) demonstrated, not only on exanples of the huchen from the Váh river and the taimen from the Enisei river systems (in which the coefficients of difference of the characters mentioned dit not exceed 0.9), but also by comparing data from practically the whole ranges of the two subspecies, *that H. h. hucho* and *H. h. taimen* cannot be separated morphologically.

Osteological differences between the huchen and the taimen are likewise small (Figs. 7–12). According to Shaposhnikova (1968), the greatest difference exists in

TABLE 2

(Continued)

	Anabar[13] (n = 26)			Olenek[14]				Vitim[15] (n = 19)		Vilyui[16] (n = 20)
	$\bar{x}$	$s_{\bar{x}}$	s	n	$\bar{x}$	$s_{\bar{x}}$	s	$\bar{x}$	Range	Range
rD	11.31	0.19	0.95	24	10.21	0.13	0.64	10.5	10—11	9—12
rA	9.77	0.11	0.58	24	8.54	0.14	0.70	8.8	8—9	8—10
rP	15.03	0.14	0.75	24	15.04	0.13	0.67	14.7	14—15	—
rV	9.04	0.08	0.44	24	9.33	0.25	1.23	8.8	8—9	—
l.l.	—	—	—	—	—	—	—	—	—	—
Squ.	(180—224)	—	—	—	—	—	—	209.3	192—288	190—216
Squ.sup.	—	—	—	—	—	—	—	—	—	—
Squ.inf.	—	—	—	—	—	—	—	—	—	—
Sp.br.	12.08	0.18	0.96	24	13.04	0.29	1.42	13.0	12—16	10—14
BR	—	—	—	—	—	—	—	12.1	11—13	11—13
Vert.	66.80	0.28	1.38	13	67.76	0.36	1.31	—	—	64—69
PC	—	—	—	—	—	—	—	—	193—260	160—240

[13]Kirillov (1976); [14]Kirillov (1972); [15]Kalashnikov (1978); [16]Kirillov (1962).

the shape of os supraethmoideum and in that of the anterior margin of ossa frontalia. According to Hensel and Holčík (1983), however, no such differences exist.

The only major difference so far exists in coloration (Pl. 23). In the taimen *(H. h. taimen)*, in the spawning season, but often also beyond it, the posterior and ventral parts of the body are raspberry red or bright orange, either all over or in the form of large blotches only. The corresponding coloration of the huchen *(H. h. hucho)* is copper red, and occurs on the sides and the dorsal surface of the body. Occasionally, there are individuals with larger oval blotches on the belly, too. It is remarkable that in the taimen the red colour reaches to the anal and caudal fins (or at least parts of them), which has never been observed in the nominate subspecies. Exceptions are provided by certain populations of *H. h. taimen* in streams draining the Ural Mts., which form a kind of transition to the nominate form, as they lack the red coloration of their anal and caudal fins (Fig. 15).

3. *Origin, Evolution and Phylogeny*[2]

Several hypotheses exist as to the origin and evolution of the genus *Hucho,* almost all of which pertain to *H. hucho* and deal above all with the possible causes of the disjunct area of its distribution.

According to Bajkov (1924), the genus *Hucho* is of marine origin. It arose by

TABLE 2

(Continued)

	Lena[17] (n = 29)	Lena[18]				Yana[19] Lena Olenek (n = 86)	Uda[20] (n = 2)	Amur[21] (n = 5)
	Range	n	x	$s_{\bar{x}}$	s	Range	Range	Range
rD	9—11	58	10.39	0.08	0.64	9—12	11—11	9—11
rA	8—10	57	8.74	0.09	0.71	7—10	9—9	8—9
rP	14—17	55	14.76	0.05	0.53	14—16	—	—
rV	8—10	57	9.05	0.05	0.44	9—10	—	—
l.l.	—	—	—	—	—	—	—	—
Squ.	201—242	—	—	—	—	190—242	184—200	190—220
Squ.sup.	—	—	—	—	—	—	—	—
Squ.inf.	—	—	—	—	—	—	—	—
Sp.br.	9—16	53	12.85	0.19	1.40	10—16	13—15	12—14
BR	11—13	—	—	—	—	11—13	11—12	12—13
Vert.	64—72	37	67.73	0.33	2.03	64—72	—	—
PC	172—250	—	—	—	—	160—250	—	—

[17]Borisov (1928), Sych-Averintseva (1933), Karantonis et al. (1956); [18,19]Kirillov (1972); [20]Lindberg and Dul'keit (1929); [21]Nikol'skii (1956).

separation from the genus *Salvelinus* in Siberia at the beginning of the Glacial Period when the migratory *Salvelinus* spp. were gradually cut off from the sea and had to adapt to permanent life in fresh water. This was the origin of the ancestral form of the huchen. There was a different situation in the Far East where, due to the warmer climate, the Sakhalin species, *Hucho perryi,* was not denied access to the sea by glaciation and has been preserved in the original diadromous form. According to Bajkov (l.c.), the huchen disappeared from the region between the Don and Dniester rivers. The original form, the taimen *(Hucho h. taimen)* of the present, remained in the east whereas in the west only a relict was preserved which gradually changed into the present-day huchen.[3]

Bajkov emphasizes that the genus *Hucho* dit not originate by separation from the genus *Salmo* and points out, at the same time, that except for differences in coloration and size the huchen is identical to the taimen.

Hankó (1932) considers the huchen to be an autochthonous relict species previously inhabiting the Pannonian Sea.

Krauss (1933) believes that the huchen arose through transformation from the salmon *(Salmo salar)* which in the Tertiary got through waters connecting the Atlantic Ocean and the Pannonian Sea into the drainage area of the Black Sea where subsequent glaciation isolated it from the oceanic salmon and caused its gradual

[2] This chapter was also published separately in a more complex form in 1982 (Holčík, J., 1982: Review and evolution of *Hucho* (Salmonidae). Acta Sci. Natur. Brno **16** (3): 1–29.

[3] Bajkov (l.c.) considered the two forms valid species, denoted as *Hucho taimen* and *Hucho hucho*.

TABLE 3

Morphometrical characteristics of *Hucho hucho* from different river systems

	Váh, Orava, Turiec[1] (*n* = 23)			Danube[2] (*n* = 2)
	$\bar{x}$	$s_{\bar{x}}$	s	Range
Sl (mm)	528.48	26.70	128.2	517—556
Fl (mm)	575.52	29.00	139.1	548—607
In % of Fl				
lc	24.21	0.15	0.69	23.7—24.7
pD	46.57	0.21	0.98	44.6—46.3
pV	56.25	0.26	1.26	53.4—55.6
pA	73.17	0.90	4.32	70.5—77.1
poD	34.73	0.18	0.86	—
H	17.16	0.35	1.67	17.2—23.7
lpc	11.92	0.13	0.63	9.4—12.0
h	7.10	0.06	0.31	6.6—7.2
P—V	34.30	0.32	1.56	33.6—34.1
V—A	19.01	0.16	0.76	19.0—20.4
A—C	6.63	0.11	0.54	—
ad—C	6.39	0.11	0.47	—
lD	10.83	0.19	0.93	11.5—11.8
lA	8.16	0.08	0.37	7.5—8.4
lC_1	13.67	0.20	0.95	10.9—13.4
lC_2	7.92	0.13	0.64	—
lC_3	—	—	—	10.2—14.3
lP	12.57	0.13	0.62	11.7—12.7
lV	10.40	0.08	0.45	9.5—10.2
hD	8.58	0.17	0.81	9.5—11.0
hA	10.35	0.12	0.57	9.5—11.4
In % of lc				
prO	28.83	0.21	1.02	27.0—30.4
Oh	12.24	0.26	1.25	11.8—13.8
poO	60.27	0.38	1.81	62.9—63.9
hc	53.75	0.57	2.72	56.8—57.9
pop	78.37	0.29	1.41	—
io	28.43	0.28	1.32	31.1—39.6
lmx	39.31	0.35	1.64	—
hmx	10.80	0.20	0.96	—
lmd	62.17	0.53	2.53	—

[1]Hensel and Pivnička (1980); [2]Balon (1968), Blahák (1972).

transformation into the huchen. His opinion has been wholly adopted by Vokač (1973).

Berg (1948–1949) on the one hand considers that a previously continuous range populated by *Hucho taimen* was later divided into two separate parts, with the relict form *Hucho hucho* having evolved in the west. On the other hand, however, Berg

TABLE 3

(Continued)

	Tisza[3] (n = 18)	Vişeul, Ruscova[4] (n = 4)	Vishera[5]				Ob'[6] (n = 3)
	Range	Range	n	$\bar{x}$	$s_{\bar{x}}$	s	$\bar{x}$
Sl (mm)	—	—	—	—	—	—	—
Fl (mm)	450—900	575—710	64	515.63	18.24	146.00	411.0
In % of Fl							
lc	22.7—24.9	22.8—24.3	64	21.92	0.13	1.03	23.2
pD	—	—	64	44.25	0.23	1.91	44.4
pV	—	—	64	54.37	0.49	3.92	55.3
pA	—	—	64	73.61	0.38	3.05	—
poD	—	—	64	37.80	0.22	1.73	—
H	15.1—18.5	17.6—20.1	64	16.33	0.22	1.74	16.1
lpc	19.1—22.8	18.4—21.4	64	13.11	0.13	1.08	13.6
h	6.1—6.8	6.3—7.3	64	6.17	0.09	0.70	6.5
P—V	—	34.5—35.4	64	32.97	0.31	2.51	33.7
V—A	—	20.4—20.7	64	19.19	0.14	1.09	20.1
A—C	—	—	—	—	—	—	—
ad—C	—	—	—	—	—	—	—
lD	10.3—11.1	10.7—11.5	64	10.52	0.09	0.74	10.0
lA	7.3—8.2	7.9—8.6	64	7.22	0.09	0.69	7.7
lC_1	—	—	40	11.95	0.17	1.09	—
lC_2	—	—	40	5.38	0.13	0.81	—
lC_3	—	—	40	11.51	0.25	1.60	—
lP	—	11.5—12.8	—	—	—	—	12.6
lV	—	8.9—10.0	—	—	—	—	9.8
hD	9.9—11.7	10.3—11.0	64	10.94	0.14	1.08	10.6
hA	10.0—11.1	11.1—12.3	64	10.55	0.13	1.04	12.5
In % of lc							
prO	25.2—32.4	26.2—28.2	64	27.20	0.39	3.09	27.8
Oh	12.1—16.4	12.6—12.9	64	13.80	0.23	1.81	14.2
poO	—	—	64	60.11	0.41	3.32	—
hc	—	—	64	54.41	0.54	4.32	—
pop	—	—	—	—	—	—	—
io	—	—	—	—	—	—	29.1
lmx	34.8—39.8	37.9—41.2	64	37.98	0.25	1.98	38.7
hmx	—	9.0—10.6	64	10.42	0.11	0.85	—
lmd	54.3—62.9	59.4—63.5	64	59.16	0.35	2.78	—

[3]Vladykov (1931); [4]Frank (1971); [5]Bukirev (1967); [6]Varpakhovskii (1899); Berezovskii (1924).

admits the possibility that the Danube river system was populated by huchen through immigration of the taimen from Siberia. While Berg considers the genus *Hucho* as arctic, he does not exclude the possibility that there were originally Tertiary fishes which inhabited warmer regions and later became adapted to life in the temperate and colder climate. Berg presumes that the present occurrence of the

TABLE 3

(Continued)

	Enisei[7] (n = 33)		Shishkhid[8] (n = 71)			Angara[9] (n = 100)		Frolikha[10]		
	$\bar{x}$	$s_{\bar{x}}$	$\bar{x}$	$s_{\bar{x}}$	s	$\bar{x}$	$s_{\bar{x}}$	n	$\bar{x}$	$s_{\bar{x}}$
Sl (mm)	—	—	554.20	21.60	182.10	—	—	—	—	—
Fl (mm)	—	—	599.00	25.50	198.10	691.00	1.60	54	652.00	4.35
In % of Fl										
lc	22.71	0.15	24.29	0.11	0.95	22.82	0.07	47	22.15	0.15
pD	46.40	0.24	47.14	0.16	1.33	45.17	0.12	45	45.42	0.18
pV	56.89	0.28	56.98	0.17	1.44	55.40	0.12	47	55.97	0.21
pA	75.05	0.34	74.82	0.16	1.36	72.38	0.12	47	74.32	0.20
poD	—	—	35.50	0.14	1.20	—	—	—	—	—
H	17.52	0.22	18.47	0.11	0.94	17.84	0.11	47	18.35	0.22
lpc	12.67	0.21	11.60	0.10	0.87	13.36	0.07	47	12.88	0.10
h	6.07	0.15	6.66	0.04	0.35	6.62	0.02	47	6.03	0.06
P—V	33.40	0.48	33.67	0.13	1.13	31.69	0.13	47	34.57	0.22
V—A	18.27	0.12	18.65	0.10	0.83	16.71	0.08	47	19.09	0.16
A—C	—	—	6.18	0.88	0.64	—	—	—	—	—
ad—C	—	—	6.76	0.06	0.55	—	—	—	—	—
lD	10.20	0.25	10.73	0.06	0.56	10.58	0.05	47	10.23	0.07
lA	7.11	0.15	7.49	0.05	0.40	7.61	0.05	47	7.07	0.07
lC_1	14.56	0.20	14.93	0.09	0.75	13.72	0.10	47	14.04	0.17
lC_2	5.41	0.14	8.00	0.08	0.63	5.30	0.09	47	5.23	0.03
lC_3	—	—	—	—	—	—	—	—	—	—
lP	12.40	0.19	13.23	0.07	0.59	10.92	0.06	47	11.91	0.08
lV	9.65	0.19	10.14	0.05	0.43	8.78	0.08	47	9.14	0.07
hD	10.56	0.23	10.93	0.10	0.88	11.09	0.06	47	9.03	0.11
hA	9.96	0.26	10.79	0.07	0.61	11.26	0.05	47	10.44	0.08
In % of lc										
prO	27.79	0.25	27.63	0.15	1.30	27.06	0.13	—	—	—
Oh	9.82	0.21	12.52	0.20	1.70	14,26	0.15	47	11.77	0.27
poO	62.23	0.19	62.63	0.24	2.06	61.02	0.17	—	—	—
hc	54.79	0.18	53.50	0.37	3.08	52.23	0.31	47	54.92	0.45
pop	—	—	77.08	0.22	1.82	—	—	—	—	—
io	26.47	0.30	27.11	0.17	1.45	27.38	0.13	47	26.30	0.27
lmx	38.30	0.26	37.40	0.19	1.58	36.34	0.20	47	37.24	0.29
hmx	9.82	0.17	10.18	0.10	0.84	10.38	0.21	47	9.92	0.11
lmd	60.59	0.19	60.07	0.25	2.13	58.76	0.20	47	59.00	0.34

[7, 9, 10]Misharin and Shutilo (1971); [8]Hensel and Pivnička (1980)

taimen in the Kama river can be explained by its penetration from the Pechora river, since the headwaters of the two systems have common sources even at present.

According to Svetina (1966), *Hucho perryi* is the ancestral form of the recent genus *Hucho*, whose original range was in the Angaris, a Tertiary continent lying in today's northeastern Asia. Diluvial glaciation advanced gradually southwards;

TABLE 3

(Continued)

	Khatanga[11] (n = 9)		Anabar[12] (n = 26)			Olenek[13] (n = 21)		
	$\bar{x}$	Range	$\bar{x}$ (range)	$s_{\bar{x}}$	s	$\bar{x}$ (range)	$s_{\bar{x}}$	s
Sl (mm)	—	—	—	—	—	—	—	—
Fl (mm)	—	—	(405—703)	—	—	(683—1500)	—	—
In % of Fl								
lc	24.2	23.1—25.4	22.35	0.13	0.66	23.95	0.20	0.93
pD	36.1	26.8—45.1	—	—	—	45.67	0.24	1.13
pV	—	—	—	—	—	55.28	0.46	2.07
pA	—	—	—	—	—	74.50	0.31	1.46
poD	—	—	—	—	—	38.26	0.32	1.47
H	—	—	—	—	—	14.86	0.31	1.46
lpc	13.2	12.9—13.7	—	—	—	12.80	0.16	0.74
h	—	—	5.45	0.04	0.23	5.02	0.06	0.27
P—V	—	—	32.58	0.27	1.39	33.26	0.18	0.83
V—A	—	—	—	—	—	20.09	0.35	1.63
A—C	—	—	—	—	—	—	—	—
ad—C	—	—	—	—	—	—	—	—
lD	—	—	—	—	—	10.57	0.15	0.71
lA	—	—	—	—	—	7.46	0.12	0.57
lC_1	—	—	—	—	—	—	—	—
lC_2	—	—	—	—	—	—	—	—
lC_3	—	—	—	—	—	—	—	—
lP	—	—	—	—	—	—	—	—
lV	—	—	—	—	—	—	—	—
hD	—	—	9.57	0.11	0.57	8.55	0.11	0.53
hA	—	—	9.58	0.10	0.52	9.38	0.11	0.50
In % of lc								
prO	29.9	25.0—33.8	—	—	—	25.66	0.53	2.42
Oh	8.6	8.2— 8.9	12.56	0.20	1.02	9.89	0.17	0.78
poO	—	—	59.74	0.31	1.57	59.78	0.29	1.36
hc	—	—	—	—	—	50.15	0.71	3.26
pop	—	—	—	—	—	—	—	—
io	—	—	23.67	0.17	0.88	25.81	0.26	1.20
lmx	—	—	—	—	—	40.25	0.46	2.08
hmx	—	—	—	—	—	9.83	0.28	1.30
lmd	59.7	53.9—65.6	—	—	—	62.98	0.53	2.46

[11]Luk'yanchikov (1967); [12]Kirillov (1976); [13]Kirillov (1972).

Hucho perryi and *Hucho ishikawai* survived in the eastern refuge, while a third species, not named by Svetina, survived in the refuge of the present Pontic-Caspian-Aralian province. As glaciation retreated, the species proceeded westwards along the Danube and simultaneously northwards along the Siberian rivers. In the region between the Ural Mts. and the Carpathians this species later became extinct as the

TABLE 3

(Continued)

	Lena[14]				Vitim[15] (n = 19)		Uda[16] (n = 2)	Amur[17] (n = 5)	
	n	$\bar{x}$ (range)	$s_{\bar{x}}$	s	$\bar{x}$	Range	$\bar{x}$	$\bar{x}$	Range
Sl (mm)	—	—	—	—	—	—	—	—	150—840
Fl (mm)	58	(452—750)	—	—	712	—	252	—	—
In % of Fl									
lc	58	21.79	0.14	1.14	22.6	21.3—24.4	24.3	25.8	23.6—27.1
pD	46	45.94	0.25	1.75	46.2	43.8—48.4	44.2	48.1	46.0—49.5
pV	46	55.31	0.26	1.77	57.1	53.2—78.6	54.8	—	—
pA	46	74.22	0.18	1.28	77.0	74.0—78.6	—	—	—
poD	46	36.79	0.35	2.41	40.0	38.8—41.2	—	—	—
H	58	15.35	0.15	1.21	17.3	15.2—18.9	15.3	15.4	14.0—16.6
lpc	58	13.11	0.05	0.39	13.9	12.8—15.4	14.6	12.9	11.8—14.2
h	58	5.64	0.04	0.36	6.1	5.8—6.6	6.1	5.4	5.4—7.1
P—V	45	32.41	0.16	1.09	34.0	31.0—35.6	31.1	32.1	31.0—33.0
V—A	46	19.37	0.15	1.07	20.4	19.0—21.5	19.5	—	—
A—C	—	—	—	—	—	—	—	—	—
ad—C	—	—	—	—	—	—	—	—	—
lD	58	10.68	0.11	0.87	10.8	10.1—11.5	11.1	10.9	9.8—12.4
lA	58	7.54	0.06	0.48	7.4	6.8—8.0	7.7	7.2	6.2—8.4
lC_1	—	—	—	—	—	—	—	—	—
lC_2	—	—	—	—	—	—	—	—	—
lC_3	—	—	—	—	—	—	—	—	—
lP	—	—	—	—	12.4	11.1—13.1	12.4	13.3	12.4—14.2
lV	—	—	—	—	9.9	9.2—10.5	9.8	10.5	9.6—11.6
hD	58	9.86	0.14	1.09	9.6	7.9—11.3	12.4	9.4	9.3—15.5
hA	58	10.43	0.13	1.03	10.7	9.0—11.6	11.0	10.6	10.6—13.6
In % of lc									
prO	58	25.24	0.61	4.68	28.2	26.8—29.3	—	26.7	—
Oh	58	12.06	0.23	1.75	13.1	11.0—14.7	—	10.6	—
poO	46	62.95	0.57	3.90	—	—	—	57.9	—
hc	46	54.00	1.24	8.45	49.5	43.9—53.9	—	—	—
pop	—	—	—	—	—	—	—	—	—
io	58	27.81	0.28	2.18	—	—	—	24.0	—
lmx	58	40.66	0.60	4.60	37.2	33.3—39.5	—	—	—
hmx	57	9.82	0.12	0.92	10.3	9.8—11.1	—	—	—
lmd	58	61.40	0.57	4.40	59.8	52.0—64.6	—	—	—

[14]Kirilov (1972); [15]Kalashnikov (1978); [16]Lindberg and Dul'keit (1929); [17]Nikol'skii (1956).

climate became warmer. The late occurrence of the huchen in Europe is indicated, according to Svetina, by its absence from the Vardar river, and the eastern origin of the species is evidenced by the population in the Prut river which, while belonging to *Hucho hucho,* is said to be close to *Hucho taimen* in certain morphological

TABLE 4

Comparsion of variation ranges of meristic characters in the two subspecies of *Hucho hucho* (modified from Hensel and Holčík 1983)

Character	*Hucho hucho hucho*	*Hucho hucho taimen*
rD	8—12	8—14
rA	7—10	7—11
l.l.	115—160	107—194
Squ.	175—240	141—288
$l.l._1$	28—38	25—33
$l.l._2$	25—35	26—39
Sp.br.[1]	10—14 (19)	9—16 (18)
BR	10—11	9—13
Vert.	67—68	60—72
PC	138—232[2]	150—284

[1]The value in brackets includes rudimentary rakers. [2]Our data.

characters. However, it is unknown from what evidence Svetina drew such conclusions, since it is evident from the data of Kulmatycki (1931c) that the huchen from the Cheremosh river (Prut river system) cannot be separated from other huchen populations and, as already mentioned above, there are practically no differences between the huchen and the taimen.

Finally, Balon (1967) and Sigunov (1972) consider the occurrence of the huchen in Europe as a Quaternary event and explain it by immigration of the taimen from Siberia, mediated by the waters of the edges of thawing glaciers.

At present, thanks to advanced knowledge especially of the related genus *Salvelinus* (Balon 1980), a more detailed picture can be compiled of the time of origin and evolution of the genus *Hucho,* even though still on a hypothetical level. As a matter of fact, no fossil evidence of members of the genus *Hucho* has been found as yet. Only subfossil skeletal remains of *Hucho hucho* are known. They belong predominantly to the taimen from epi-Palaeolithic (Mesolithic), Neolithic, Bronze Age and early Iron Age strata in the primaeval settlements of man in the drainage areas of the Lena, Aldan, Angara, Baikal and Kama rivers, i.e. from upper Pleistocene and Holocene, dating from 14,000 to 200 years B.C. (Tsepkin 1976, 1979, 1980a). All remains of the nominate form discovered so far come from the Danube river system and are likewise very young: Brillenhöhle near Ulm–Upper Pleistocene; Inzigskofen near Sigmaringen and Lautereck between Sigmaringen and Ulm – Mesolithic or Neolithic; Nuremberg Castle – 16th century A.D. (Lepiksaar 1968, 1973; 1977 in litt.). Hence a hypothesis on the origin and evolution of the genus *Hucho* must be based on such data as (1) morphological structures and early ontogenic characteristics of the recent forms in comparison with those of other Salmoninae, (2) the ecology of the recent species, (3) a knowledge of the origin and evolution of the family Salmonidae and the subfamily Salmoninae, (4) the present geographical distribution of the recent members of the genus *Hucho*, (5) the affinity of the genus *Hucho* to other related genera within the subfamily, (6) the degree of differentiation of individual species, and (7) the geomorphological evolution of the Earth.

(1) Almost all recent authors are unanimous in considering the genus *Hucho* to be rather primitive and archaic in both phenotype and genotype (Norden 1961, Rounsefell 1962, Behnke 1968, Shaposhnikova 1975, Tuturov and Omel'chenko 1976, Dorofeeva 1977, Cavender 1980, Ráb 1981, Ráb and Liehman 1982). In particular, this pertains to the morphology and organization of the head and caudal skeleton, dentition of vomer, number of vertebrae, number of scales in the lateral line, structure of gill rakers, number of pyloric caeca, and character of the karyotype. Hence, they place the genus *Hucho* somewhere at the beginning, near the hypothetical ancestral form or consider it part of the beginning of the *Brachymystax*–*Hucho*–*Salvelinus* evolutionary lineage (Behnke 1968, Kendall and Behnke 1984). We may add that in our opinion the primitive and ancient character of the genus *Hucho* and its proximity to the ancestral form are confirmed by further important, but little mentioned or previously unknown characters. *Hucho hucho* and *H. perryi* are the biggest salmonid species, the former attaining 105 kg (Sigunov 1972), the latter 60 kg (Khatkevich 1973) in body weight. The remaining salmonid species attain only smaller dimensions, often being dwarfed by those of *Hucho* spp. (Table 19). Late sexual maturation is another such character. *Hucho hucho* attains sexual maturity only after having reached at least some 2 kg body weight and the age of 4–8 years (Table 6); *H. perryi* attains sexual maturity when over 7–8 years of age (Gritsenko and Churikov 1977, Bushuev 1983). In connection with this, the maximum life-span of *Hucho* spp. is also shifted considerably upwards (Chapter 10). The primitive character of *Hucho* spp. is also indicated by the large number of somites in its embryo: 72 in *H. hucho*, compared to 63 in *Brachymystax lenok*, 54–59 in the genus *Salmo*, and 60–62 in the genus *Oncorhynchus* (Peňáz and Příhoda 1981, Smol'yanov 1961; Pavlov 1979, 1980). The lateral line scales and the conspicuously large adipose fin may also be considered archaic characters (Holčík 1982). It should be added that compared to other members of the Salmoninae, huchen young are less advanced at the end of the embryonic period, the latter being very short (Chapter 9). Thus the genus *Hucho* appears on the whole hyperomorphic and altricial, whereas the genera *Salmo*, *Oncorhynchus* and *Salvelinus* are hypomorphic and precocial, showing various degrees of neotenization. Viktorovskii et al. (1985) are the only present-day authors known to us who assume a contrary evolutionary order and believe that the genera *Hucho* and *Brachymystax* evolved later than the genera *Salmo* and *Salvelinus*. Moreover, they believe that the genus *Hucho* did not originate until an Upper Caenozoic epoch, the Pliocene. They do not, however, substantiate this opinion in any way whatever; we believe that the arguments stated above are sufficient to refute it.

(2) The ecology of *Hucho hucho* and *H. perryi* (that of the other species being unknown) indicates that the species are primarily oxyphilous and cold-water, mostly

potamic, but otherwise without any specific habitat requirements. They are predatory, their diet consisting largely of fish. *H. perryi* also consumes considerable proportions of invertebrates. For reproduction the two species are confined to mountain streams with a clear gravelly or rocky bottom. Spawning takes place in spring.

(3) While the recent salmonids were previously predominantly believed to have evolved from a marine ancestral form (Zenkevich 1933, Shmidt 1947, 1950), the idea of their freshwater origin is gaining ground at present (Tchernavin 1939, Berg 1948, Hoar 1958, Yakovlev 1961, Privol'nev 1967, Nikol'skii 1971, Viktorovskii 1976). Most authors, however, still consider the salmonids a relatively young but rapidly evolving group, the origin of which is mostly placed in the Tertiary (Zenkevich l.c., Tchernavin l.c., Nikol'skii 1956, Yakovlev 1961, Kirpichnikov 1974, Viktorovskii 1976, Viktorovskii et al. 1985, Dorofeeva 1977). According to the most recent data, however, the Salmonidae is a far more ancient family: the giant *Smilodonichthys rastrosus* (1900 mm), described by Cavender and Miller (1972), comes from the Pliocene; the earliest known charr comes from the Miocene (Cavender 1980); and the earliest known member of the Salmonidae, *Eosalmo driftwoodensis,* was found in the Middle Eocene (Wilson 1977). It follows that, as mentioned by Cavender (1980; 296), "in reality salmonids must have developed during the Mesozoic and salmonines were already distributed over the Northern Hemisphere land masses by early Tertiary time". The existence of the genus *Salvelinus* in the Miocene and its relation to the recent representatives of the genus evidence in turn, as pointed out by Cavender (1980), that the divergence and distribution of the major evolutionary lines of charrs had been completed by that epoch and that consequently the existing distribution of the genus *Salvelinus* must have been completed millions of years before the first Quaternary glaciation. According to Gritsenko (1975), the origin of the genus *Salvelinus* should most probably be placed in the Eocene. The author connects it with the transgression of the World's Ocean which may have caused a disintegration of the previously continuous range of the common ancestral form of the genera *Salvelinus* and *Hucho*.

An important part in the evolution of salmonids was played by evolutionary polyploidy, as already pointed out by Svärdson (1945). The importance of the phenomenon was summarized in detail by Ohno (1970) and in the salmonids in particular by Rees (1964), Svetovidov et al. (1975, 1976), Dorofeeva (1977) and Vasil'ev (1977). It is assumed that in an ancestral salmonid form with 50–60 chromosomes, a total tetraploidization of the genome took place through polyploidy (Ohno 1970). During subsequent evolution, redundant genetic information was then gradually reduced by diploidization. In the salmonids this process involves a major transformation of the karyotype structure, with the number of chromosomes ($2n$) decreasing, but the number of chromosome arms *(NF)* remaining fairly even in all species. For a time after the tetraploidization of their genome, polyploid organisms are much more flexible and variable than diploid ones, but their subsequent divergence is slower due to causes still unknown. One may thus assume that the Salmonidae as such diverged more slowly from their common branch or ancestral form, whereby they also retained many archaic characters (Dorofeeva l.c., Vasil'ev l.c.). As indicated by karyological data, the evolution of the Salmonidae proceeded along two lines. One of them was through internal karyotype transformations – pericentric inversions and to a smaller extent also Robertsonian translocations. This line is represented by freshwater species or by those species which do not migrate far into the sea. The other line is connected with a greater reduction of the chromosome number and especially with Robertsonian fusions, and one can even observe an early stage of evolution of heteromorphic sex chromosomes. This line is represented by diadromous salmonids.

Let us add that the reduction in the chromosome number is in very good agreement with the oligomerization of certain organs (e.g., somites, number of vertebrae, number of scale rows, number of lateral line scales, number of fin rays, number of pyloric caeca) (Dogel' 1954), although it is obviously governed by other factors. In any case, the character of the karyotype is an important morphological feature (Benazzi 1973); a larger number of chromosomes indicates greater evolutionary age and hence originality, and provides a strong argument for the freshwater origin of the genus *Hucho* and of the Salmonidae in general.

(4) The distribution of the recent *Hucho* spp. is disjunct and consistently allopatric (Fig. 17).

(5) All contemporary authors agree that there is a close affinity between the genera *Brachymystax* and *Hucho* on the one hand, and between *Hucho* and *Salvelinus* on the other. Most authors assume, directly or indirectly, that *Hucho* is intermediate between the evolutionarily more primitive, monotypic *Brachymystax* and the more specialized, polytypic *Salvelinus* (Norden 1961, Lindsey 1964, Shaposhnikova 1975, Behnke 1968, Gritsenko 1975, Dorofeeva 1977, Cavender 1980, Ráb and Liehman 1982). Cavender (l.c.) and Behnke (l.c.) even place the genus *Hucho* at the beginning of the evolutionary series, and Behnke (l.c.) considers the possibility that *Brachymystax* is in fact only a subgenus of *Hucho*. Considering the chromosomal characteristics as the most important character (Tables 1 and 5), then *Brachymystax* is more ancient than *Hucho,* but the two genera certainly present an evolutionarily archaic type (they still have an identical *NF*). On the other hand, however, the differences existing between *Brachymystax* and *Hucho* provide grounds for considering the genus *Hucho* as a separate branch in salmonid evolution (Dorofeeva et al. 1980). In their unique and unsubstantiated opinion, Viktorovskii et al. (1985) exclude *H. perryi* from the genus *Hucho* and place it at the beginning of the evolutionary lineage *(Parahucho – Hucho – Brachymystax).*

(6) It follows from the diagnoses and descriptions of the individual forms of the genus *Hucho* that their differentiation is on a low level: only *H. hucho, H. perryi* and *H. bleekeri* can be considered valid species. *H. ishikawai* is probably conspecific with *H. hucho. H. perryi* appears to be most specialized, considering its diadromous life habits and the small number of chromosomes (Table 5).

(7) The most important factors affecting the distribution of freshwater fishes may include the orogenic processes of the formation of relief of the Earth's surface, variations in the level of the World's Ocean which enable the appearance or disappearance of extensive, united and entire river systems, as well as freshwater or brackish lakes in some depressions, and climatic changes. There is a lot of evidence now, that all continents once constituted a single, continuous land mass, Pangaea, which was later divided into Laurasia (the present Eurasia and North America) and Gondwana (the Antarctica, the Indian subcontinent, Australia, South America, and Africa). The splitting of Pangaea took place some 180 million years ago, i.e. in the Mesozoic, and the new continents were then subject to massive orogenic processes, especially in the Tertiary. The climatic changes and the transgressions of the World's Ocean took place mainly in the Pliocene and the Pleistocene, their frequency and amplitude being often so great (in geological terms) that they were catastrophic. Data on these phenomena were summarized by King (1967), Tarling and Tarling (1972) and Lindberg (1972).

Synthesizing these items of knowledge and facts, the following hypothetical origin and evolution of the genus *Hucho* may be outlined (Fig. 6).

Beyond doubt, the genera *Brachymystax* and *Hucho* are very ancient. Their ancestral forms became separated from the common ancestral salmonid form very early, probably towards the end of the Mesozoic (Cretaceous?) after an assumed polyploidization of salmonoid fishes. One may suppose that this ancestral form

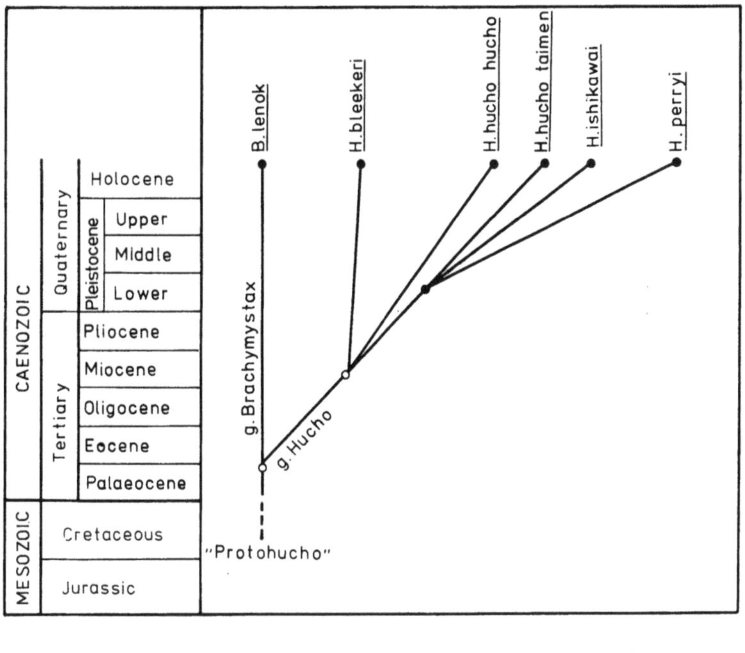

Fig. 6. Scheme showing assumed phylogeny of the genus *Hucho* (the size of the compartments for successive epochs does not correspond to their duration) (orig. J. Holčík).

– *Protohucho* – existed in the eastern part of Laurasia, east of the present Danube and north of the present 30°–40° northern latitude. Having become separated from *Protohucho,* the genus *Hucho* gradually populated a more extensive range, as it was more progressive than *Brachymystax.* Its continuous range was then divided in consequence of orogenic processes on the one hand, and of transgressions of the World's Ocean on the other.

Hucho bleekeri was probably the first to become separated from the common ancestor, very probably in the Miocene, when the fundamental parts of the Himalayan range were thrown up. It is probable that at the time also the previously continuous range of *Hucho hucho* was divided into the western and the eastern part. It is well known that a large part of Europe has been continuously sinking since the Upper Triassic, the greatest depressions being those of the Northern German Basin, the Baltic Basin and three depressions within the present European part of the USSR, between the northern and southern uplift zones, where the drop was by as much as 3000 m in places.

Since it is a polyploid form with a conservative tendency, and since the environmental conditions in the regions in which the present derivatives of the ancestral form are encountered have not changed much, we are dealing, in the cases of *H. hucho* and *H. taimen,* not with valid species but at most with subspecies. It is likely to assume that *H. hucho* probably never occurred north of the Carpathian Mountains and north and north-west of the Alps or, in the contrary case, became entirely extinct before the last but two (sensu Lindberg 1972) or Piacenza transgression, i.e. about 4 million years ago. If during the following Villafranchian regression the huchen repopulated the rivers of the present Russian Lowland (which is improbable), it must have definitively become extinct there during the next, lower Pleistocene (Sicilian) transgression. In our opinion, the present occurrence of the taimen in the upper part of the Volga river system is secondary and is connected with the still existing connection between the headwaters of the Volga and Pechora rivers (Sabaneev 1911, Berg 1948–1949), or it penetrated there once from the Ob' river system. In the Pechora itself, affected by the transgression in a large section, the huchen found refuge in the upper reaches, and a similar situation apparently existed in the Prut river as well. Beyond doubt, the largest continuous range of the huchen at the time of those events was the Danube river system in the region down to the Iron Gate gorge. During the great Quaternary fluctuations of the World's Ocean, this region was not flooded by the sea in spite of the fact that a large part of it lay below the 200 m isohypse, but that it was a vast freshwater lake, as demonstrated by Lindberg (1972).

Hucho perryi appears to be the youngest derivative, having arisen through separation from *H. hucho taimen.* The beginning of its origin can be dated back to the Sicilian transgression at latest. Up to that time, the rivers of Sakhalin and those entering the Gulf of Peter the Great, together with the Amur, formed a uniform and entire river system of the Paleoamur. Transgression compelled the taimen to retreat upstream into the rivers. While this was no major problem in the Amur, in the rivers of the Gulf of Peter the Great, Sakhalin and the Kurile Islands the taimen had to retreat practically up to their headwaters. But lack of space and undoubtly also poor trophic conditions forced the big fish to stay near the mouths of rivers flooded by the seas. Thus the foraging fishes often came into contact with salt water, which meant that the originally freshwater species became gradually adapted and changed into a partly diadromous form, finally to transform into the recent *H. perryi.* That this is a relatively young species is also indicated by the fact that it never withdraws far from the shore, continuously staying in mostly brackish waters of river estuaries. Moreover, the Sakhalin taimen often stays in rivers even after the spawning time, entering them again in early winter, this time to overwinter, migrating upstream several tens of kilometres. Besides, young fish drift downstream to the sea rather late – in the Tym river, for example, not until 5–7 years old and not until having attained 40–50 cm in fork length (Gritsenko and Churikov 1977). This fact clearly indicates that compared to other species of the genera *Salmo, Salvelinus* and *Oncorhynchus, H. perryi* has become neotenized only slightly or not at all. As a matter of fact, smoltification most frequently occurs at the age of 1–3 years in the genera mentioned above. All this evidences the relatively recent change to diadromous life habits in *H. perryi.* The presence of toothed basibranchialia in *H. perryi,* which is generally considered a very primitive character (Nelson 1969, Vladykov 1970, Cavender and Miller 1972), can be explained by this structure's probable importance in preying upon smaller-sized prey (besides small fish species, also benthic and airborne insects, Gritsenko et al. 1974), whereas it has lost importance and gradually disappeared in *H. hucho* which feeds on invertebrates only when very young.

The recent papers by Anbinder et al. (1982) and Vasil'ev (1983) on the karyotype of *H. perryi* (Table 5), fully confirm the above hypothesis. The small number of chromosomes in this species indicates that (1) a change in lifestyle habits results in a reduction of the chromosome complement, and (2) it is an evolutionarily rapid process (in terms of geological age).

Beyond doubt, even *H. ishikawai* represents a very young derivative (if not identical with *H. hucho taimen* and, hence, conspecific with *H. hucho*). It has apparently penetrated into the Yalu river only quite recently. The headwaters of the Yalu in fact are close to those of the Sungari – thus the species *H. hucho* got there through the Sungari from the Amur at approximately the time when the sea transgression isolated the populations of the present *H. perryi.* Penetration from the Sea of Japan is excluded, as the Yalu, together with the Liao-ho and other rivers of the southeastern part of Korea had emptied themselves into the Hoang-ho, the mouth of which lay considerably more to the east.

Our hypothesis thus assumes speciation having taken place as a result of geographic isolation – due to disintegration of the range populated by the ancestral form. Migrations from one river system into another were of minor importance in this process, as they were of merely local extent (e.g., certain Siberian rivers, the Volga–Pechora, or the Enisei–Amur connections); they played a more important

TABLE 5

Chromosomal characteristics of *Hucho hucho* in comparison with *Hucho perryi* and *Brachymystax lenok*

Species (locality)	*2n*	*NF*	Types of chromosomes	Author
Brachymystax lenok				
(eastern Korea)	90	116	13 *M, SM* + 32 *A—T*	Kang and Park (1973)
(unknown)	92	102	?	Dorofeeva (1977)
(lower Amur)	92	110	8 *M* + 8 *SM* + 30 *A*	Viktorovskii et al. (1985)
Hucho hucho hucho				
(Yugoslavia)	82	106	12 *M, SM* + 29 *A—T*	Sofradžija (1979)
(Yugoslavia — Sava river)	82	114	13 *M* + 3 *SM* + 6 *ST* + 19 *A*	Ráb and Liehman (1982)
(Slovakia — Turiec river)	82	112	13 *M* + 2 *SM* + 6 *ST* + 19 *A*	Ráb and Liehman (1982)
Hucho hucho taimen				
(unknown)	84	102	9 *M, SM* + 33 *A—T*	Dorofeeva (1977)
(lower Amur)	84	112—116	10 *M* + 6 *SM* + 26 *A*	Viktorovskii et al. (1985)
Hucho perryi				
(Sakhalin — Belaya river)	62	100	19 *M, SM* + 12 *ST, A*	Anbinder et al. (1982)
(unknown)	62	100	19 *M, SM* + 4 *ST* + 8A	Vasil'ev (1983)

part only at the time of the World's Ocean regressions when *H. hucho* may have invaded, through renewed connections between rivers, those streams in which the freshwater fauna was destroyed by sea waters. It is definitely necessary to reject the hypothesis which assumes that the recent *H. h. hucho* got into the Danube from Siberia in the Pleistocene. Considering such a long migration, ecology must be taken into account: the species was certainly unable to cover the distance between the Ural and the Danube within a short period of time and without having reproduced. The conditions prevailing in that area during the period from the presumed origin of the genus *Hucho* up to present preclude the reproduction of the huchen (during the transgressions, the Russian Lowland was largely flooded by the sea; glaciation in the form of floating icebergs occurred in the north as far as the Valdai Hills; at the time of the regressions, the huchen would have had to cover the distances between the mouths of individual river systems by migrating in salt water, which is impossible). Thus it is obvious that the recent huchen represents an ancient relic form, confined in the Danube depression, where it occurs up to present on its northern, southwestern and eastern slopes and in the Prut river system. In spite of the fact that, due to orogenic processes, climatic and hydrological conditions this form is practically divided into four more or less isolated populations (viz. Alpine-Balkanic, northern Carpathian, eastern Carpathian, and southern Carpathian, see Fig. 18), its evolution has proceeded so slowly that there are no differences between these populations as yet. The range of the huchen, like that of the other members of the genus, has periodically increased or decreased in area but mostly in the vertical direction only, according to what climatic and hydrological conditions obtained in any one period. Since it is a form which has obviously retained its original relation to its environment (assumed for the common ancestral form), its evolution has been very slow, and hence the Danubian huchen does not show, up to present, any such changes that would justify acknowledging its valid specific status.

Besides geographic isolation, ecological specialization was involved in the origin of *H. perryi*. This is the reason why this species, in spite of its indubitable phyletic youth, shows the greatest difference from the other members of the genus.

4. *Vernacular Names*

In this part, we give all hitherto known vernacular names of all members of the genus *Hucho*, excerpted from the available literature (Antipa 1909, Berg 1909, 1948–1949, Bajkov 1924, Bănărescu 1964, Baskakov 1967, Blanc et al. 1971, Borne 1942, Dyk 1956, Drenski 1948, Ferianc 1948, Gove 1966, Gozmány 1979, Herman 1887, Hirtz 1956, Ivaška 1951, Karbe 1970, Kašpar 1886, Kimura 1934, 1966, Kirillov 1972, Klemm 1973, Kohaut 1905, Kornhuber 1860, Kowalska et al. 1973, Lindberg et al. 1980, Lindberg and Chzhu 1964, Lindberg and Gerd 1972,

Mahen 1926, Majewski 1889, Martinka 1931, Mori 1928, Munda 1926, Nilsson and Svärdson 1963, O-n 1955, Preudhomme 1953, Rass 1971, Ricker 1962, Sigunov 1972, Taler 1953, Tchang and Liu 1957, Tret'yakov 1947, Vasiliu 1959, Vladykov 1926, 1963, Wheeler 1961, Zaplata and Taler 1932). The officially acknowledged or most frequently used names in any of the languages are shown in **bold-face letters.**

Hucho hucho hucho (Linnaeus, 1758)

Bulgarian: dunavska săomka, **dunavska păstărva**[4];
Czech: dunajský losos, hlavatice, hlavatka, **hlavatka podunajská,** huch, losos dunajský, losos jezerní, siven dunajský;
Danish: donaulaks;
English: Danube trout, Danube salmon, Danubian salmon, huch, **huchen,** hucho, huck, river salmon;
French: grand omble du Danube, heuch, heuchon, huch, hucho, **huchon,** saumon du bassin de Danube, saumon du Danube, saumon huch;
German: Dickkopf, Donaulachs, Donau-Lachs, Donausalm, Donau-Salm, Donauwels (incorrectly), Gauchforelle, Groszkopf, Hauch, Hauchforelle, Hausen (incorrectly), Heuch, Huch, Huch der Donau, **Huchen,** Huchforelle, Hueche, Hüch, Hüchl, Lachsforelle, Lachsforen, Rotfisch, Rothfisch, Rothuchen;
Hungarian: **dunai galóca,** dunai galócza, dunalazac (incorrectly), fejes, gadóca, gadoci, gadócza, galóca, galócza, huhó szemling, huko, hukó, hukó huhkó, retke;
Polish: **głowacica,** hołowatycia, hucho, łosoś rudopstrąg, pstrąg brudny, rudopstrąg;
Rumanian: lostită, lostocă, lostosă, lostoste, lostoza, lostoză, lostrița, lostriță, lostrucă, lostruță, lostucă, puica, puică;
Russian: dunajskij losos', **dunajskij tajmen',** golovatka, losos' dunajskij;
Serbo-Croatian: glavačica, glavatica, glavatica mladica, glavatić, gonjavac, huj, ljepotica, losos dunavski, mlačak, mladevčić, **mladica,** mladica glavatica, mladuljak, mladunjak, mlađerčić, pastrva glavašica, pastrva riječna, sulac, sulać, sulač, sulak, sulc, sulček, sulčik, sulec, sulica, suljač, suljak, štulac, šulica, šumaj, uj;
Slovak: gadovica, hadovica, hlavatka, hlavatka vážska, hlavátka, hlavátka obyčajná, **hlavátka podunajská,** hlavatica, hlawatka, losos jäzerný, tuba, tubica (only subadult specimens);
Slovenian: glavatica, selač, solač, soláč, sulač, sulc, sulec;
Swedish: bäckröding, donaulax, **donaulaxen,** huchen;
Ukrainian: dunajs'kyj losos', holavl', holovač, **holovatycja,** holovec', holovel', holoven', holovyč, losos', dunajs'kyj, suraviak.

[4] The Cyrillic characters in this part have been transliterated according to the ISO Recommendation R 9, "International system for the transliteration of Slavic Cyrillic characters", 1968.

Hucho hucho taimen (Pallas, 1773)

Bashkir: kizil-balyk, kysil-balyk, nehai, nengaha;
Buryat: toloch, tulu, šumu;
Chinese: tàimén zhéloyú[5];
English: Siberian taimen, taimen;
Estonian: ihereus;
Evenki (Tungus): dželi, džili, eli, toloh, tulu;
Finnish: taimen;
German: Taimen, Taimen-Huchen;
Gilyak (Nivkh): čili, goj;
Hungarian: tajmen;
Japanese: taimen, Amuru itô;
Karakalpak: tajmen', tajmen' balyk;
Khalkha-Mongolian: tul;
Khanty (Ostyak): adi, alen, aln, alna, aľna, alten, althen, as-tu-sart, ass-tu-sart, atne, atue ting, ting, tiu, tyhng;
Komi: len';
Korean: cachi, cengcange[5];
Mansi (Vogul): alen;
Nanai (Goldi): dzeli, dželi, dželʹ;
Nenets (Yurak, Samoyed): nengchaj-nenchaj, njangochej, njangechej, tej;
Orochi: džili, džjali, sachana, šva;
Polish: tajmień;
Russian: čusovskoj len', golovan, krasnaja ryba, krasnaja ščuka, krasulja, len, leněk, lenok, len', lin', ljachsforel', loch, **obyknovennyj tajmen',** pestruška, sibirskaja tajmen', strežnevoj lin', strežnevoj loch, tajmen, tajmen', talmen', tal'ma, tal'men', trošnica;
Slovak: **hlavátka sibírska,** tajmeň;
Tatar: bel, bell, bil-balyk, bill-balyk, rghai balyk;
Ulchi: dželʹ;
Vietnamese: cá hồi taimen;
Yakut: belejach, bil balik, bil balyk, bilʹ, čončkur, mindimen, sjulʹges.

Hucho ishikawai Mori, 1928

Chinese: Shízhōnshì zhéloyú;
English: Korean taimen, chachi;
German: Korea-Huchen;
Japanese: tyati, iwana, Kôrai itô;
Khalkha-Mongolian: Solongosyn tul;

[5] Chinese names are given in the official pinyin-zimu transliteration, Japanese ones in the kunreisiki romazi transliteration, and Korean ones in that of the Yale University.

Korean: Cosen cachi, Cosen cengcange;
Russian: Korejskij tajmen';
Slovak: hlavátka kórejská;
Vietnamese: cá hõi Triên-thiên.

Hucho bleekeri Kimura, 1934

Chinese: hǔyú, hǔjiāyú, māoyú, māorjú;
Japanese: okoze;
Slovak: hlavátka čínska.

Hucho perryi (Brevoort, 1856)

Chinese: Yuǎn-dōng zhéloyú;
English: Sakhalin taimen, goy;
German: Sachalin-Huchen, Pazifik-Tschewiza;
Gilyak (Nivkh): goj;
Japanese: itô;
Khalkha-Mongolian: Sachaliny tul;
Korean: Ssahaluyn cachi;
Lithuanian: Sachalino šlakȳs;
Orochi: chué;
Polish: czewica;
Russian: čevica, dal'nevostočnaja čevica, goj, golovastik, **Sachalinskij tajmen';**
Slovak: hlavátka sachalinská;
Vietnamese: cá hõi Xakhalin.

5. *Morphology*

DESCRIPTION

Many data are available on the morphometric and meristic characters of *Hucho hucho* (Tables 2–4).

Of the anatomical characters, only the skeleton is well known, thanks to Shaposhnikova (1968).

H. hucho has a relatively low neurocranium. Os supraethmoideum is short and broad, with lateral projections. Its posterior end is divided into a large number of projections which are either wedged into a horizontal fissure or between the projections of the anterior margins of ossa frontalia. In *H. perryi* the posterior part of os supraethmoideum is uneven, often pointed, always resting on the frontal bones. *H. hucho* never has a hypethmoid (Fig. 11). Ossa frontalia are variable in shape,

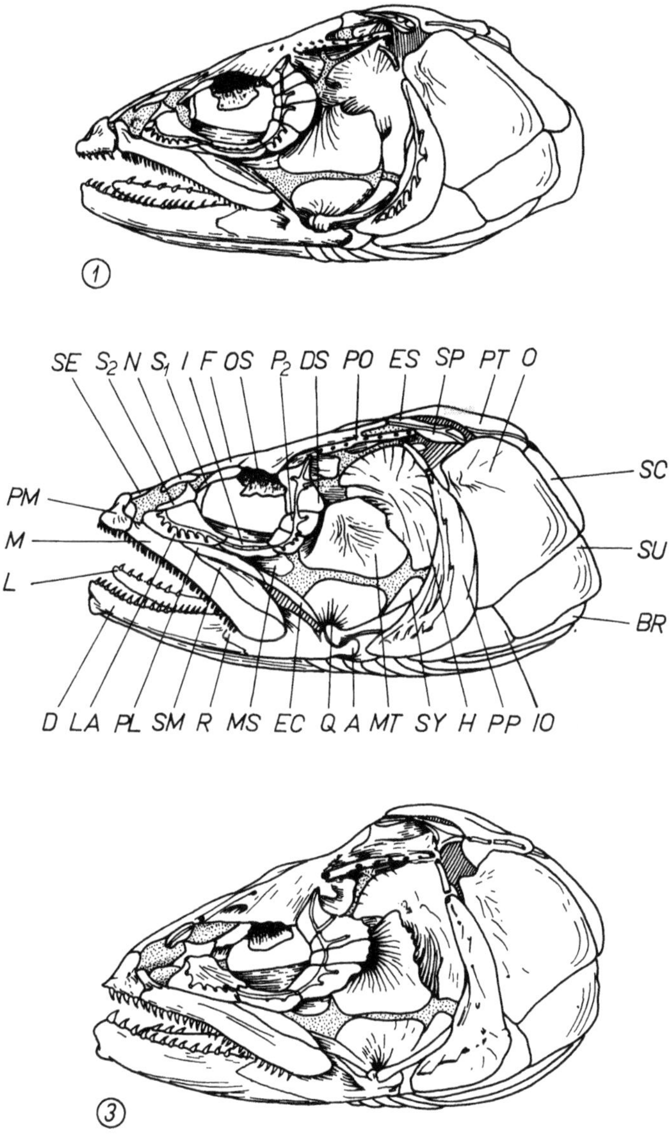

Fig. 7. Lateral view of the skull of (1) the huchen, (2) the taimen and (3) *Hucho perryi.* A, os angulare; BO, os basioccipitale; BR, os radii branchiostegi; C, os cleithrum; D, os dentale; DS, os dermosphenoticum; EC, os ectopterygoideum; EP, os epioticum; ES, os extrascapulare; F, os frontale; H, os hyomandibulare; I, os infraorbitale; IO, os interoperculum; L, os linguale plattum; LA, os lacrimale; M, os maxillare; MS, os mesopterygoideum; MT, os metapterygoideum; N, os nasale; O, os operculum; OO, os opisthoticum; OS, os orbitosphenoideum; P, os postorbitalium; PA, os parietale; PE, os pterosphenoideum; PF, os praefrontale; PL, os palatinum; PM, os praemaxillare; PO, os pteroticum; PP, os praeoperculum; PR, os prooticum; PS, os parasphenoideum; PT, os posttemporale; Q, os quadratum; R, os retroarticulare; S, os supraorbitalium (1–2); SC, os supracleithrum; SE, os supraethmoideum; SH, os sphenoticum; SM, os supramaxillare; SO, os supraoccipitale; SP, os suprapraeoperculum; SU, os suboperculum; SY, os symplecticum; V, os vomere (adapted from Shaposhnikova 1968).

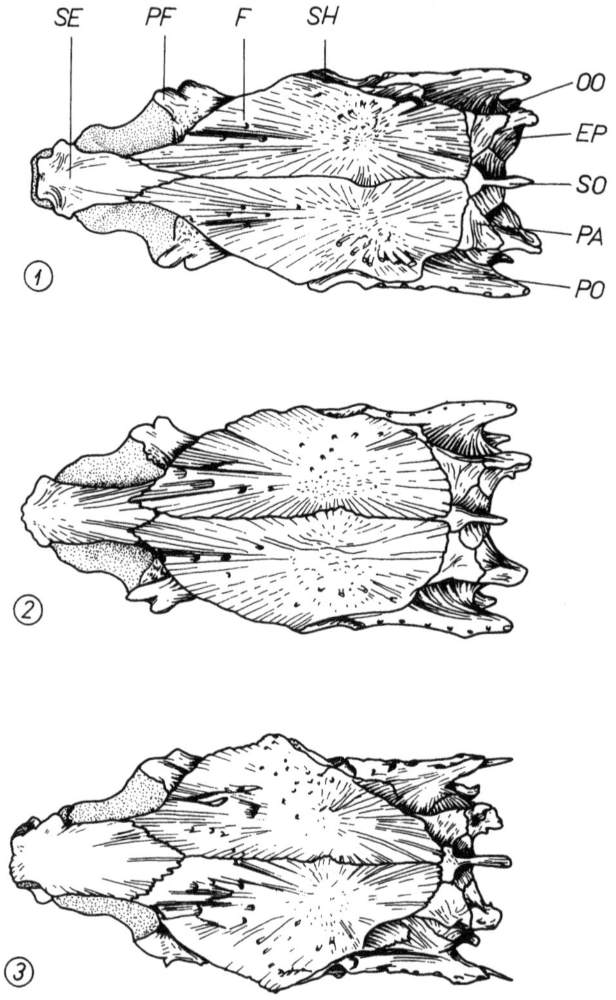

Fig. 8. Dorsal view of the neurocranium of (1) the huchen, (2) the taimen and (3) *Hucho perryi*. Lettering as in Fig. 7 (adapted from Shaposhnikova 1968).

particularly their posterior margin. They reach quite a long way posteriorly, almost completely overlapping ossa parietalia (Figs. 7 and 8). Os vomere comparatively short, with a single row of teeth on its head, but the posterior projection of the os vomere, the so-called shaft (manubrium), always lacks teeth (Fig. 12). There are teeth even on ossa palatina, forming a continuous arch with those on os vomere. Os parasphenoideum is slightly enlarged in the region of ossa parafrontalia, but tapers anteriorly and forms, on its ventral surface, a groove into which the dorsal edge of the vomer's manubrium fits. The caudal end of os parasphenoideum is either rounded or slightly excised (Fig. 9). Its profile is almost straight. Os basisphenoideum is small. Its anterior margin is enlarged and sometimes even bipartite. Os orbitosphenoideum

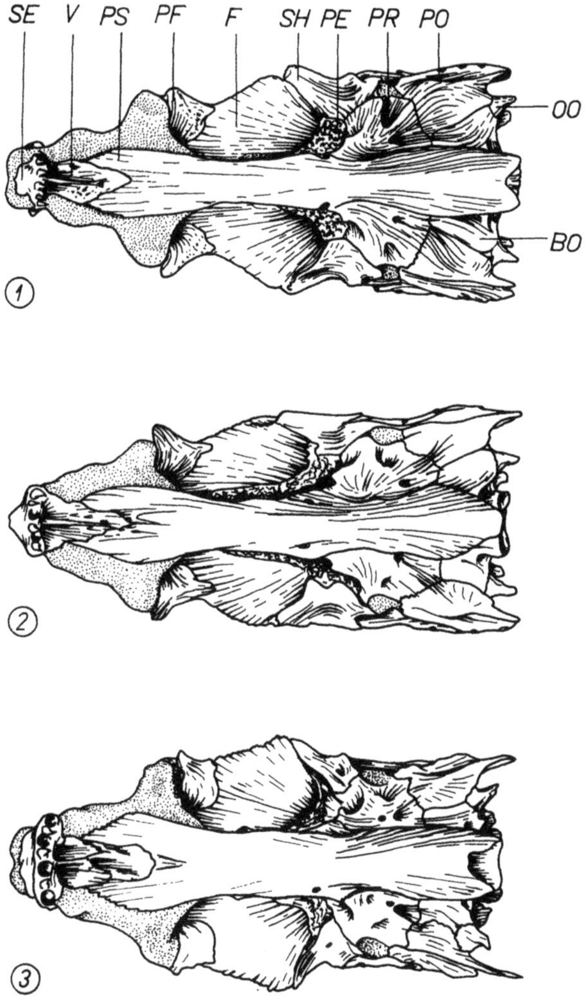

Fig. 9. Ventral view of the neurocranium of (1) the huchen, (2) the taimen and (3) *Hucho perryi*. Lettering as in Fig. 7 (adapted from Shaposhnikova 1968).

is well developed, with a slightly pointed anterior projection. The number of ossa postorbitalia is variable. They are medium broad, caudally reaching just about the first third part of the distance between the posterior margin of the orbit and os praeoperculum; in *H. perryi* they cover almost half this distance (Fig. 7). The caudal end of ossa maxillaria of adult fishes reaches below the caudal margin of the orbit. The jaws are relatively strong and bear strong, conical teeth. The number of the teeth changes with advancing growth of the fish, but is constant after a certain body length has been attained. Besides well-developed teeth, firmly stuck in the jaws, there are smaller teeth hidden beneath the skin. The jaw teeth are exchanged in the course of the year. There are teeth even on the "tongue" (os linguale plattum), always

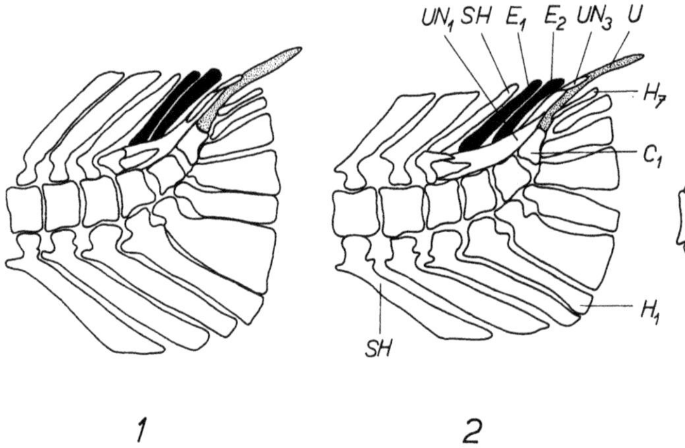

Fig. 10. Caudal skeleton of (1) the huchen, (2) the taimen and (3) *Hucho perryi*. C, os centrum; E, os epurale; H, os hypurale; U, urohyale; UN, os uroneurale; SH, spina haemalis; SN, spina neuralis (adapted from Shaposhnikova 1968).

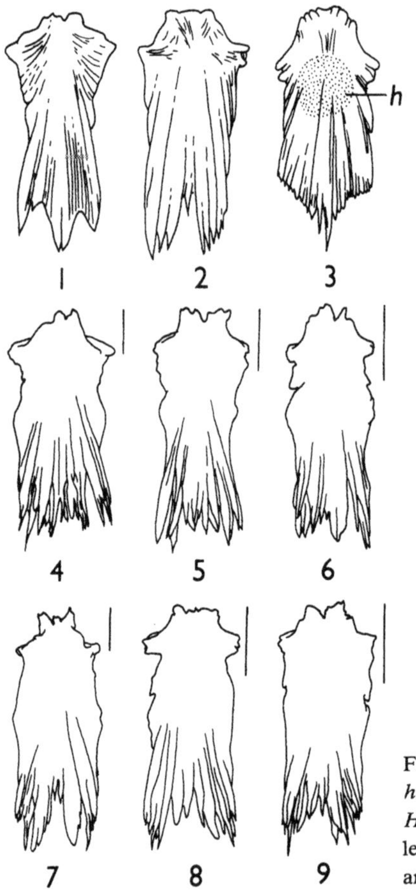

Fig. 11. Os supraethmoideum of (1, 4, 5, 6) *Hucho hucho hucho,* (2, 7, 8, 9) *H. hucho taimen* and (3) *H. perryi;* h, os hypethmoideum; abscissae indicate length of 10 mm. After Shaposhnikova 1968 (1–3) and Hensel and Holčík 1983 (4–9).

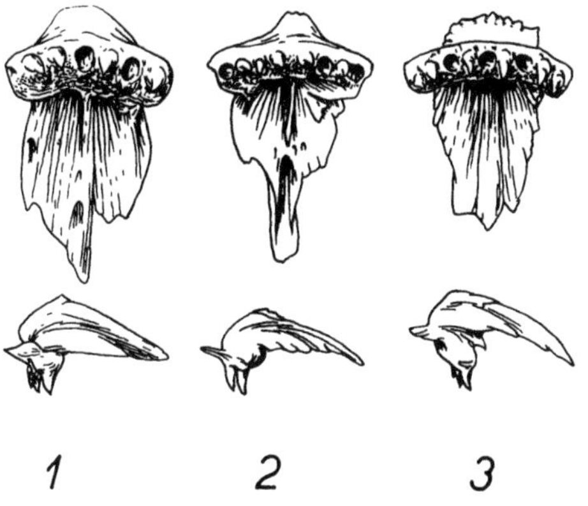

Fig. 12. Ventral and lateral view of the vomer (os vomere) of (1) the huchen, (2) the taimen and (3) *Hucho perryi* (after Shaposhnikova 1968).

arranged in two marginal rows. The medial row of minute teeth, known to occur in *H. perryi,* is absent in *H. hucho,* the same as a toothed plate on ossa basibranchialia. Spinae branchiales are sharp and stout, their number on the first gill arch varying between 9 and 16. In addition to normal gill rakers also rudimentary ones occur; on the first gill arch there may be up to five. They do not occur in all individuals, however. The number of vertebrae in *H. hucho* is considerably larger than in *H. perryi,* particularly of the trunk ones. Consequently, there are more ossa praedorsalia. The caudal skeleton contains three pairs of uroneural bones, two ossa epuralia and seven ossa hypuralia. Three ossa epuralia, typical of *H. perryi,* occur in *H. hucho* only exceptionally (Fig. 10).

The number of transverse scale rows in *H. hucho* considerably exceeds that of the pored lateral line scales. In *H. perryi* the two numbers are much the same. *H. hucho* has 141−288 transverse scale rows on its body and 107−194 pored lateral line scales. There are 25−38 scale rows above and 25−39 below the lateral line.

Bohl (1982) found that huchen from Bavarian streams and pond cultures bigger than 70 mm have the cranial part of the gas bladder turned backwards. This is also true of huchen from Czechoslovak streams and of the taimen from Mongolia (K. Hensel, unpublished). It should be added that the gas bladder is straight in the genera *Salmo, Salvelinus* and *Thymallus.*

COLORATION

The coloration of the huchen is rather variable, not only changing with advancing age, but varying according to the provenance of individual populations, as pointed out, e.g., by Ivaška (1951). Nevertheless, it is possible to distinguish essentially two colour variants, corresponding to the two subspecies of *H. hucho.* As mentioned above, this is the only external character by which the two subspecies can be separated. In the huchen, the characteristic reddish pigmentation is largely distributed over its flanks, exceptionally occurring on the ventral side of the body (but then only in the form of large reddish spots), whereas in the taimen the red invariably occurs on the belly as well. One may agree with Gritsenko (in a letter of 2 Sept. 1978) who states that "red-bellied taimen never occur west of the Ural range, nor red-flanked ones east of it". The third colour variant, confined to the range of the taimen, exhibits larger oval reddish blotches replacing the continuous red-coloured area (Fig. 13).

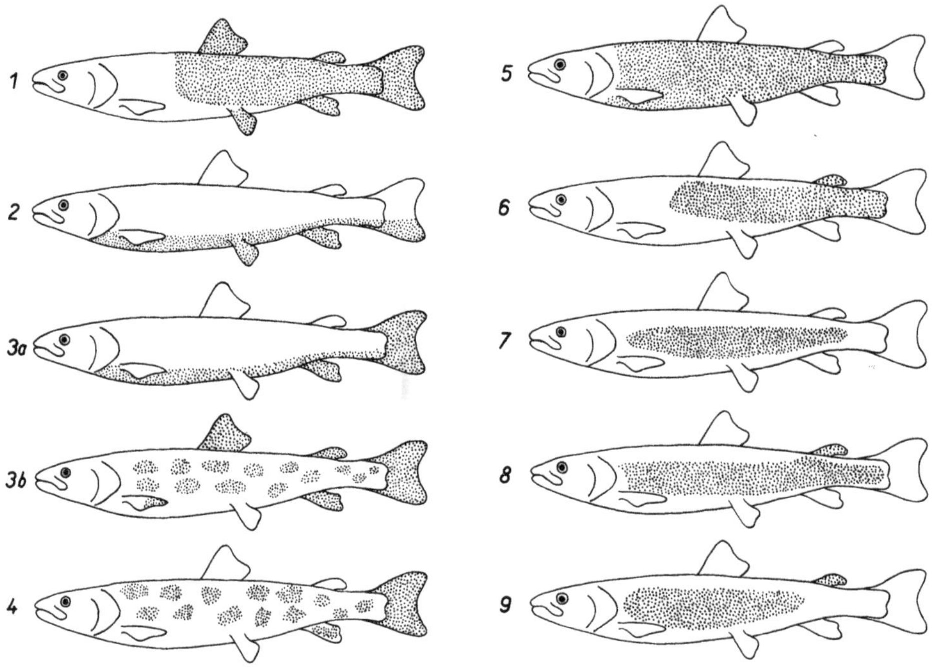

Fig. 13. Distribution of red pigmentation over the body of *Hucho perryi* (1), *H. h. taimen* (2 – Amur, 3 – Lena, 4 – Lake Baikal, Angara, 5 – streams of the Ural Mts.) and *H. h. hucho* (6 – Pielach, 7 – Váh, 8 – Orava, 9 – Turiec rivers). Compiled from various sources by Gritsenko (1–5) and Holčík.

In its basic, non-spawning coloration (Pl. 23.1) the nominate form varies from greyish-green to dark brown with a greenish tinge on the dorsal surface of the head and body. The flanks are paler, but the same colour as the dorsum, often with a purplish tinge. The fins are ashy yellow, with brownish to reddish ends; the adipose fin is golden-brown to reddish. Head, dorsum and flanks with minute, black or brownish spots of different shapes: round, oval or irregularly rounded on the head, often X-shaped on the flanks. Similar spots are also on the dorsal and the caudal fin. The adipose fin usually lacks such spots, but spotted adipose fins are rather frequent in certain huchen populations (e.g., in the Orava river; Pl. 4.1).

The coloration of sexually immature huchen is different. It was studied in detail by Bastl and Kirka (1959). According to their observations, the basic coloration of young about two months old is comparatively dark. A basic steel grey shade passes from the dorsum to the flanks in the form of dark bands. The abdomen and the spaces between the dark bands are silvery, with minute pigmented spots, the density of which decreases ventrally, the spots disappearing on the abdomen. The number of the dark lateral bands varies between 7 and 12, dropping with advancing age from an average of 9.7 at two months to 8.8 at one year of age. The dark bands persist rather a long time. According to Ivaška (1951), fifty per cent of young 15–18 months old still have them; they rarely occur in young fish about 2 1/2 years old. According to

Fig. 14. Coloration of young huchen. Above, young huchen from a pond culture (Kláštor pod Znievom), *Sl* 105 mm; below, young huchen from the Váh river (near the mouth of the Kvačianka stream), *Sl* 103 mm. Scale indicates 10 mm (adapted from Bastl and Kirka 1959).

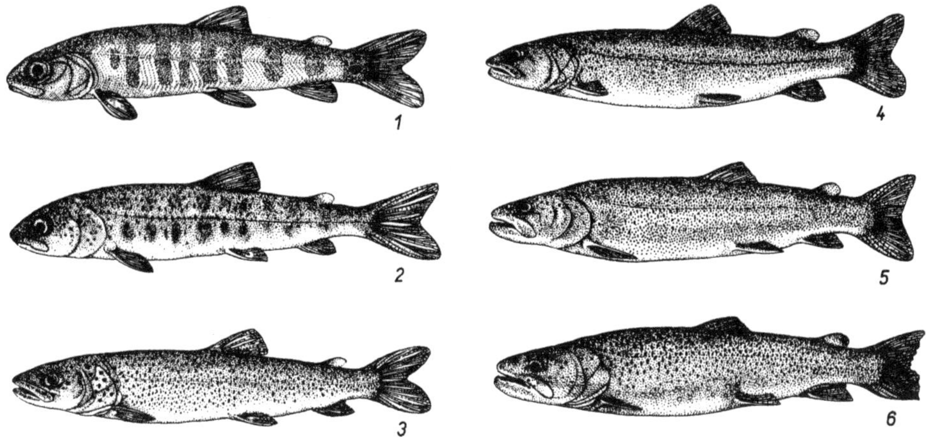

Fig. 15. Change of pigmentation and shape of head in the huchen in relation to age and growth. 1 – six weeks (0+), 40 mm (*Sl*); 2 – six months (0+), 90 mm; 3 – 1+, 210 mm; 4 – 2+, 402 mm; 5 – 5+, 602 mm; 6 – 11+, 1210 mm (drawing by M. Baradlaiová, based on original photographs by J. Nieslanik).

Bastl and Kirka (l.c.), minute spots, deeper in colour, occur on the body with increasing age. They are most densely aggregated on the dorsal surface, becoming thinner in the ventral direction and only rarely occurring below the lateral line. According to Dyk (in a letter of 17 May 1981), the dark pigment is first accumulated around the alimentary canal and on the upper surface of the head, thus forming a kind of a mask on the head of the young in lateral view. The black spots are round at first; on the flanks they are in the form of elongated juvenile spots situated transversely, and their size depends upon the nutritive condition of the fish. The coloration of young huchen is illustrated in Fig. 14 and Pl. 27, and its changes during growth are shown in Fig. 15.

The taimen differs from the nominate form first of all in that during the spawning period (but often also outside it; Kirillov 1972) the posterior and ventral side of its body is either bright orange-red or the whole area is raspberry red, or these parts of the body and the flanks have only large red blotches (Misharin and Shutilo 1971). It is worth noting that even the anal and caudal fins (sometimes only the lower part of the latter) are red, which has never been observed in the nominate form (Pl. 23.2 and 24.1). The dorsal fin is dark, the pectoral and ventral fins are yellowish. The remaining coloration resembles that of the nominate form (Nikol'skii 1956, Bukirev 1967, Kirillov 1972, Misharin and Shutilo l.c). Young taimen also have 7–10 dark bands on the flanks, which disappear with advancing age. Misharin and Shutilo (l.c.) state that taimen 2–4 years old no longer have these bands.

It is not without interest that the coloration of *Hucho perryi* resembles that of *H. hucho* and particularly that of *H. hucho taimen* (Gritsenko et al. 1974), which fact confirms the phylogenetical relation of the two species (Fig. 15).

SEXUAL DIMORPHISM

Much doubt and disagreement still exists as regards sexual dimorphism in *H. hucho*. It is generally maintained that the sexual dimorphism is incompletely developed, being apparent only just before and during the spawning period. At that time, the coloration of both sexes is much more intense, the males of the nominate form being darker, copper red or dark red on the flanks and bluish-green on the skull (Munda 1926a, Ivaška 1951, Nieslanik). In addition, the fins of the male are yellowish-grey, the dorsal and caudal fins showing undulated stripes and being darker in general coloration at that time. According to Nieslanik, the lower part of flanks of the female is slightly pinkish. According to Ivaška (l.c.) and Nieslanik, another character is the broad urogenital papilla of the female, which is tumescent, reddish and star-shaped. The papilla of the male is narrow, in the form of a longitudinal slit. Skácel denies the presence of such differences, however. In his experience the urogenital papilla is practically identical in both sexes and cannot be used for sexing. According to the same author, the most reliable character by which it is possible to separate males and females during four months is their coloration: the females are olive-green to greyish, with a distinctly white abdomen, whereas the males are copper red to dark purple, with a dark abdomen (Pl. 26. 2 and 29). However, a colour inversion occurs in some 10 per cent of cases so the female may be more intensively coloured than the male. In practice this may lead to mistaking one sex for the other visually. It may not even be true that the females are usually more robust and the males more slender in the spawning period.

No sexual dimorphism appears to exist in the taimen. Nikol'skii (1956) and Kirillov (1972) emphasize that there are no external differences between the two sexes. At spawning time both sexes are coloured more intensely. Their breeding colours (more pronounced red colour of the abdomen and fins, occasionally reddish spots occurring on the flanks) may occur long before the spawning period proper and also persist for a long time thereafter (Misharin and Shutilo 1971, Kirillov 1972).

Contrary to other salmonines males of the genus *Hucho* do not develop the kype which is known to occur in the genera *Salmo* or *Salvelinus,* nor a deformation of skull bones, such as that occurring in the genus *Oncorhynchus.* Only Schindler (1935), and after him Ladiges and Vogt (1965) mention the kype in the huchen, but this is undoubtedly an error. On the other hand, similarly to the genera mentioned above, there occurs a marked roughening and thickening of the skin of the males, while that of the females becomes more supple during the spawning period. While this character is so far known to occur in the nominate form (Ivaška 1951, Schindler 1935, Nieslanik, Skácel), it probably occurs in the taimen and other members of the genus *Hucho* as well. Apparently it is a phenomenon that occurs in all brood-hiding species of the family Salmonidae (Noakes 1980). This phenomenon was noted in greater detail by Stokłosowa (1966, 1970) in *Salmo trutta.* Pickering (1977) later extended its validity to the whole genus *Salvelinus,* thus confirming its regularity. During the spawning period the epidermis of the female is filled, due to hormonal influences, with a great amount of superficial goblet cells filled with mucus, which are raised above the surface. On the contrary, the male skin becomes markedly thicker at that time. Its epidermis consists almost exclusively of scale-like epithelial cells which become progressively flattened towards the surface of the skin. On the other hand, the number of mucous goblet cells is distinctly lower in the male epidermis. At spawning time, the female excavates a redd in a gravelly and rocky substrate and the mucous cells protect her skin from mechanical injury and subsequent infection. On the contrary, the thickened skin of the male (especially in the region of the anal fin in *H. hucho*) provides a better contact between the two sexes during the spawning act. More particulars are required on this phenomenon in *H. hucho* and in the genus *Hucho* in general.

So far, it is not exactly known whether *H. hucho* also shows such breeding tubercles as are known to occur, e.g., in lake charr *(Salvelinus namaycush),* Arctic grayling *(Thymallus thymallus arcticus),* but also in other species of the Salmonidae (Vladykov 1954, Wiley and Collette 1970, Collette 1977, Kratt and Smith 1978). In *H. hucho* it has only been briefly mentioned by Pappenheim (1909: 121: "In dieser Zeit die mit schwartenartigem Ausschlag"). Similarly as in lake charr, the tubercles should occur in the anal fin region of both sexes. Neither our own nor Polish breeders know such tubercles, however.

CYTOMORPHOLOGY

The karyotype of the huchen was studied by Sofradžija (1979), Ráb (1981) and Ráb and Liehman (1982); that of the taimen by Dorofeeva (1977) and Viktorovskii et al. (1985). As seen in Table 5, summarizing their results, the differences between *H. h. hucho* and *H. h. taimen* are minimal. Still smaller differences exist between populations of the nominate form from different river systems. The karyotype of the huchen from Slovakia differs from that of the huchen from Yugoslavia in having one submetacentric pair less and one acrocentric pair more (Pl. 4). This in turn causes a difference in the number of chromosome arms *(NF)*. Ráb end Liehman (l.c.) believe that it is not a Robertsonian fusion but a pericentric inversion which is responsible for this difference.

6. Protein Specificity

According to Nyman (1967), who analysed certain proteins of several salmonid species by electrophoresis, the protein characteristics of *Hucho hucho* and *Salvelinus namaycush* show no similarity (Fig. 16) except for serum esterase. In that character the two species are identical, but only under certain conditions (buffer system of Ashton and Braden, pH 8.7). Under altered conditions (acetate buffer system, pH 4.0) the esterase zone of the huchen differed in slower motion, while the esterase zones of all *Salvelinus* spp. and their hybrids compared were identical.

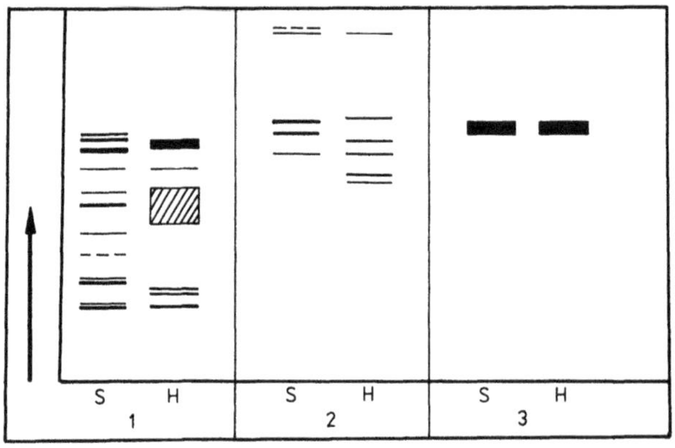

Fig. 16. Differences in the protein analysis of *H. hucho hucho* (H) and North American *Salvelinus (Cristivomer) namaycush* (S). 1 – serum proteins, 2 – liver esterase, 3 – serum esterase (adapted from Nyman 1967).

Part II

DISTRIBUTION

7. Geographical Distribution

The huchen, *Hucho hucho,* is a Eurasian species with a disjunct range (Fig. 17).

In Europe, it occurs in the Danube, Volga and Pechora river systems. It does not occur, however, in all streams of the above systems. In the Danube system it only inhabits submontane rivers draining the Alps, the Dinaric Alps, the Carpathians, and partly also the Bohemian-Moravian Highland, the Šumava (Bohemian Forest) Mountains and the Weinsberger Wald Hills. It does not occur in the waters of the Pannonian and Walachian lowlands. Thus the statement that the huchen inhabits exclusively the left tributaries of the Danube, or waters draining limestone mountain ranges, as endorsed, e.g., by Koller (1907) or Ivaška (1951), is erroneous. In the Volga and Pechora river systems, the huchen inhabits the submontane rivers of the western slopes of the Ural Mts. and those of the Severnye Uvaly Hills. Between the Danube river system on the one hand, and that of the Pechora river and the submontane rivers of the Volga river system on the other, there is the extensive region of the Eastern European Flatland. The water bodies of that region are not populated by the huchen. Thus they form a broad boundary zone between the ranges of two geographic races of the huchen: the endemic subspecies of the Danube river system, the huchen *(Hucho hucho hucho)* and the Ural-Siberia-Amur subspecies, the taimen *(Hucho hucho taimen).*

In Asia, the species is distributed from the rivers of the eastern slopes of the Ural Mts. in the west, in the rivers and rarely also lakes of the Central Siberian Plateau and the Verkhoyanskii Khrebet Mts. in the north, the Altai, Sayan, Khangai, Khentei, Yablonovoi Khrebet and Xingan mountain ranges in the south, up to the rivers rising in the Dzhugdzhur, Stanovoi Khrebet and Sikhote-Alin' mountain ranges in the east. However, it does not occur in the rivers of the Western Siberian Lowland, the Yamal and Gydansk Peninsulas, or the Taimyr Peninsula and the Northern Siberian Lowland.

The rivers populated by *Hucho hucho* belong to the systems of the Atlantic Ocean (the Black Sea), the Caspian Sea, the Arctic Ocean (the Barents, Kara and the

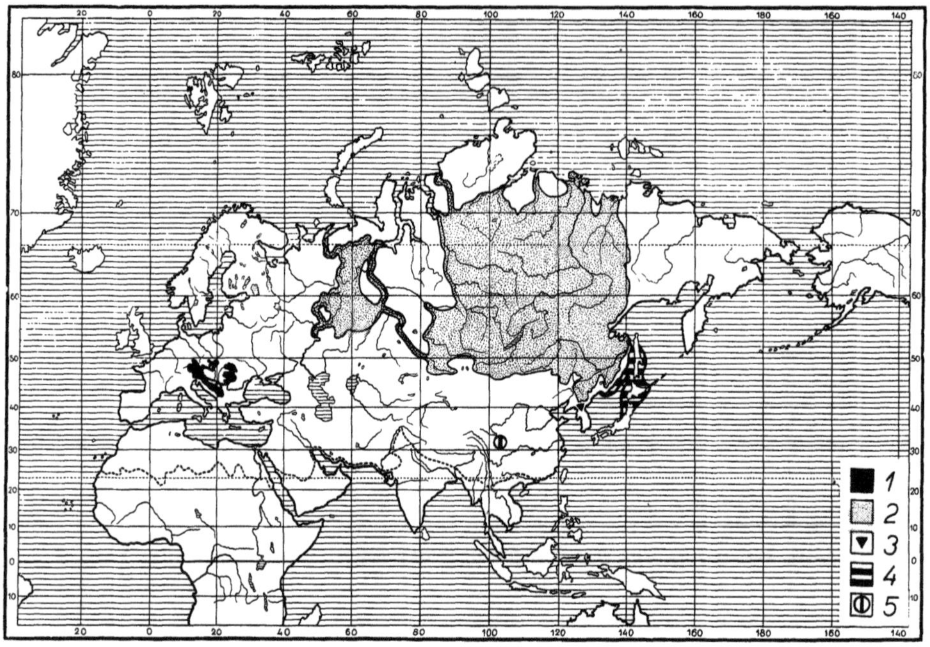

Fig. 17. Distribution of *Hucho* spp. 1 – *H. hucho hucho*, 2 – *H. h. taimen*, 3 – *H. ishikawai*, 4 – *H. perryi*, 5 – *H. bleekeri* (orig. K. Hensel).

Laptev seas), and the Pacific Ocean (Sea of Okhotsk). The river systems include those of the Danube (Black Sea), Volga (Caspian Sea), Pechora (Barents Sea), Ob', Enisei, Pyasina (Kara Sea), Khatanga, Anadyr, Olenek, Lena, Omoloi, Yana (Laptev Sea), Uda, Tugur and Amur (Sea of Okhotsk).

According to the zoogeographic classification of streams, the species occurs in the Holarctic (Palaearctic) and the Sino-Indian Regions. The huchen *(Hucho hucho hucho)* is limited exclusively to the northern Ponto-Caspian District of the Ponto-Caspian Province of the Euro-Mediterranean Subregion of the Holarctic. The taimen *(Hucho hucho taimen)* is distributed over the Circumpolar and Baikal Subregions of the Holarctic as well as in the Eastern Asian Subregion of the Sino-Indian Region (according to the classification of Bănărescu 1960).

In the section which follows we give all streams known to be, or have been, populated by the huchen (according to Hensel 1980, who also gives a detailed list of references).

DISTRIBUTION OF THE HUCHEN

The huchen is endemic to the Danube river system (Fig. 18). In the main stream of the Danube itself it occurs only in sections which in character correspond to the submontane and mountain zones; that is, in its upper reaches down to the

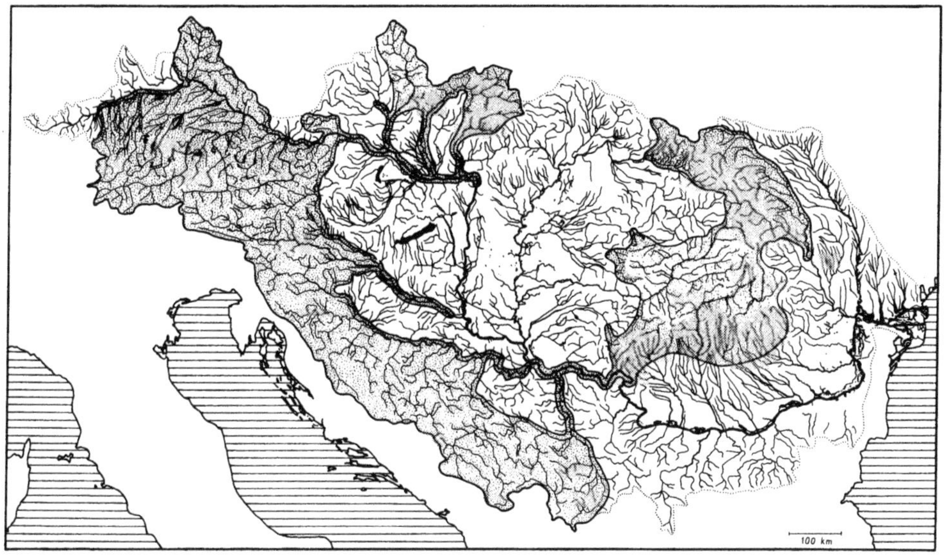

Fig. 18. Autochthonous distribution of huchen *(H. h. hucho)* (orig. K. Hensel).

Štúrovo–Budapest section, at the end of the middle reaches in the cataracts of the Iron Gate gorge. The huchen has never been recorded in the Hungarian section of the Danube below Budapest, as well as in its lower reaches in Rumania. There is only one record (1912) of a huchen 330 mm long caught in Lake Yalpukh. This lake lies on the left tributary to the Danube Delta of the same name (Jentsh 1912 from Berg 1948–1949).

At present, the huchen occurs in the main stream of the Danube only sporadically: it is now very rare in Swabia; in Upper Bavaria it practically does not occur in the Danube (except for few catches in the Ulm–Neustadt section). In the Austrian and Czechoslovak sections of the Danube only occasional individuals are caught (in the sections Linz–Vienna and Bratislava–Radvaň nad Dunajom). In the section of the Iron Gate, where it was abundant still at the turn of the century, it does not occur at present (Fig. 19).

In connection with the occurrence of the huchen, the following streams are mentioned:

1. Iller (R)[6]

Well populated by huchen even at present.

2. Lech (R)

The numbers of huchen in the Lech, particularly above Augsburg, are very high, similarly to the Upper Wertach (L Lech).

3. Altmühl (L)

While Seez (1939) states explicitly that huchen do not live in the river,

[6] The left (L) and right (R) tributaries are invariably denoted in the direction of the stream.

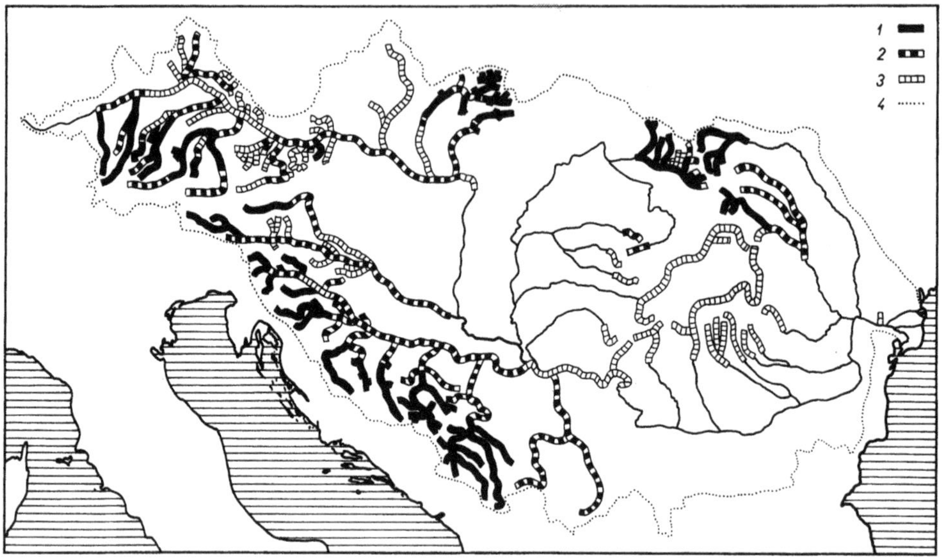

Fig. 19. Changes in the distribution of the huchen in its original range. 1 – present permanent occurrence, 2 – present sporadic occurrence, 3 – historically documented past occurrence, 4 – boundary of the Danube river system (orig. K. Hensel).

Prawocheński and Kolder (1968), Skácel (1976) and Harsányi (1982) include the Altmühl in their list of huchen streams. Huchen no longer live there.

4. Laaber (L)

No recent data are available on the occurrence of huchen in this stream. Formerly, however, they did occur there.

5. Naab (L)

Reports are available on the former abundant occurrence of huchen in the Naab and its tributary, the Vils (R).

6. Regen (L)

Huchen occur in the Schwarzer Regen and its tributary, the Teisnach, but they have disappeared from the Weisser Regen and the Schlossauer Ohe.

7. Isar (R)

Even today, this is one of the best huchen streams in Bavaria. Huchen are very numerous, especially in the Lenggries area, as in the Isarkanal near Unterfähring. It is worth noting that there is a good stock within the precincts of Munich. Only below Bad-Tölz and Freising is the occurrence of huchen threatened at least until a water treatment plant is built on the stream. There are even proposals to limit the numbers of huchen in the Isar to maintain the local trout and grayling populations.

Satisfactory numbers of huchen are also found in the Loisach (R Isar) below Kochelsee. Formerly huchen also occurred in the Amper (L Isar), but became extinct due to heavy pollution of the stream. Within the Amper river system, huchen occur sporadically in Lake Ammersee, but are more abundant in the lower reaches of the Ammer river flowing into the lake. The huchen has also been reported to occur in the Würm river (R Amper).

8. Vils (R)

Harsányi (1982) mentions previous occurrence of huchen in that stream. Huchen no longer occur in the Vils.

9. Inn (R)

The huchen population of the Inn river is abundant almost throughout the stream, but particularly in its lower reaches. (But Harsányi (1982) states that the occurrence of huchen is now only sporadic in the area of the mouth of this river into the Danube.) In the Inn river system, huchen occur in the lower reaches of the Mangfall (L Inn) and the Kalten (R Mangfall). Formerly they also occurred in the Alz river (R Inn) whence they even penetrated into Lake Chiemsee. However, abundant occurrence of huchen is reported all along the Grosse Ache (flowing into Lake Chiemsee), and in the Salzach river (R Inn) in the section from Salzburg down to the mouth into the Inn.

10. Ilz (L)

In the Ilz river, huchen used to occur throughout the stream, but at present they are only sporadic. They do, however, populate its tributary, the Mitternacher Ohe.

11. Erlau (L)

Huchen do not occur there any more.

12. Grosse Mühl (L)

No recent data are available on the occurrence of huchen in this river. They still occurred there at the beginning of the 20th century.

13. Traun (R)

Formerly there was a good stock of huchen in the Traun river. At present, they occur only sporadically in its lower reaches. Equally sporadically they occur in the lower reaches of the Krems (R Traun) and have also disappeared from the Ager (L Traun), the Vöckla (L Ager), the Aurach (R Ager) and apparently even from the Alm (R Traun).

14. Enns (R)

Huchen are abundant there. They are said to have disappeared completely from the Steyr (L Enns).

15. Ybbs (R)

Nowadays huchen are already very rare in the river.

16. Melk (R)

The sporadic occurrence of huchen has been recorded.

17. Pielach (R)

Huchen are common in the Pielach even today.

18. Krems (L)

The occurrence of huchen in the Krems has only been reported by Wendland (1965), who explicitly reports them as occurring in Lower Austria. Jungwirth (1978), on the other hand, reports on the occurrence of huchen in the Krems (R Traun), in Upper Austria.

19. Kamp (L)

Huchen occurred there until World War II.

20. Traisen (R)

Huchen were still recorded as occurring in the Traisen at the beginning of the 20th century.

21. Morava (March) (L)

Formerly huchen occurred in the upper reaches of the Morava river in the environs of Bludov, in a short section between Klášterec and the mouth of the Desná river. The occurrence of huchen has also been mentioned in Olomouc. Occasional huchen were caught even in the lower reaches where they apparently ascended from the Danube. Records are available of a huchen caught at Angern (opposite Záhorská Ves) in 1894 and at Kroměříž in 1900. Zbořil and Absolon (1916) mention the huchen as a rarity occurring in the Morava in the environs of Hodonín. From time to time, however, huchen occur in the lower reaches of the Morava even at present; obviously these fish ascended from the Danube. In 1975, a group of fishermen of the State Fisheries at Stupava caught a huchen 4 kg in weight in the Morava opposite Marchegg (Austria) (P. Kronauer, personal communication). The assumption of autochthonous occurrence of huchen in the Dyje river (R Morava) is based on a painting by Stilfried at the Biological Station at Lednice, depicting a huchen caught in the Dyje river near Ranšpurk on 2 Apr. 1893. The considerable weight (14 kg) of that fish suggests, however, that it probably did not come from an autochthonous population in the Dyje, but probably came there via the Morava river from the Danube.

In the present century, huchen were introduced into some of the streams of the Morava river system. Thus in 1909 and 1910, and also later in 1951–1953, 1960, 1961 and 1965, huchen young were stocked in the Bečva river (L Morava). Huchen live there to this day. In 1909 to 1910, huchen young were even introduced into the Bránecký potok Brook (L Bečva) where they survived till 1914. Since 1949, huchen young have been regularly released in the Dyje river in the section Znojmo–Vranov, as well as in the Znojmo and Vranov reservoirs. At present, huchen occur in the Dyje river above and below Znojmo and in the reservoir. In 1969, huchen were also introduced into the Mašovický potok Brook (L Dyje). The results of this introduction are unknown. Huchen were also introduced into the Svratka (L Dyje) and the Čtyřdvorský potok Brook, where they were released in 1936. For particulars on these introductions, see Chapter 21.

22. Leitha (Lajta) (R)

According to Koller (1907), huchen do not live in the right tributaries of the Danube, between the Traisen and the Drava rivers. Hartlieb (1948) and subsequent authors have reported huchen to occur in the Leitha in the territory of Austria.

23. Raab (Rába) (R)

In connection with the occurrence of huchen, this river has been mentioned only by Skácel (1976), apparently erroneously.

24. Váh (L)

At present, huchen occur in the Váh from the lower reaches of the Čierny Váh down to Trenčín. They are more abundant only in the section between Liptovský

Mikuláš and Ružomberok. Formerly huchen were abundant in the Váh, and they occurred even deep into the lower stretch of the river. For example, before World War I they were abundant near Piešťany. The last known catches in that stretch of the Váh date back to 1951. Even in 1924 a huchen was caught in the Váh near Hlohovec. Canalization of the Váh practically eliminated the huchen in the middle and lower reaches of the river; in those areas it may occur in the former river-bed: in April 1985, a 300 mm *(Tl)* huchen was caught near Madunice (J. Sedlár, personal communication).

As to the tributaries of the Váh river, huchen are reported to occur in the Belá (R Váh), Kvačianka (R Váh) and Ľupčianka (L Váh); those are the spawning streams of huchen. The occurrence of huchen is also mentioned in the mouth of the Ľubochnianka (L Váh); huchen are commonly known to occur in the Orava (R Váh) and Turiec (L Váh) rivers. They have already disappeared from the Podhradský potok Brook (L Váh). It is of interest, however, that at present they occur in the Revúca (L Váh), where they even spawn; this is indicated by observations made in early April 1981 in the Revúca above the Biely Potok, from which two pairs of parent fish and one surplus male were fished for breeding purposes, all 7–8 kg in individual body weight (J. Černý, personal communication). This is also where Pl. 9.2 comes from. Until the construction of the pumped storage hydropower plant, huchen migrated to spawn in the Čierny Váh up to Svarín (and occasionally up to the confluence with the Ipoltica (A. Kaliský, personal communication).

In the Orava river, huchen occurred all along the stream and also ascended its tributaries. The construction of the Orava reservoir exerted a very unfavourable influence on, and completely devaluated, the section of the Orava river below the dam down to the mouth of the Oravica river. Thus the huchen from the Orava are able to migrate to spawn only in the Oravica and the Studený potok streams and in the mouth of the Zázrivka stream (a weir in the stream prevents them from migrating further upstream). Huchen also occur in the Orava reservoir. From there they migrate to spawn mainly in the Biela Orava and Mutnianka (L Biela Orava) streams. Before the dam was constructed they had migrated from the Biela Orava as far as the Polhoranka stream (L Biela Orava). From the reservoir the huchen also migrate to spawn in the Čierna Orava river and in its tributaries, the Lipnica, Syhlec and Zubrzyca streams.

The Turiec river is one of the richest huchen streams. Now that the operation of the paper mill in Martin has been stopped, huchen have again occurred in the section Vrútky–Moškovec (formerly they occurred only upstream of Martin). Most of the stream of the Turiec is a protected territory with respect to huchen.

In the Váh river system, huchen were introduced into the Rajčianka river (L Váh). They became well established, but had to be removed from the river, as they exerted an unfavourable influence on the populations of trout and grayling. Huchen were furthermore introduced into the Kysuca river, where they also became established. Until recently, huchen spawning was observed every year in the Czadeczka stream (which is the Polish name for the upper reaches of the Čierňanka stream, a left tributary of the Kysuca river). Ivaška (1951) states that occasional huchen were formerly caught in the Kysuca river when they wandered there from the Váh. Sometime in the nineteen-twenties huchen were also introduced into the Nitra river (L Váh), in 1957 into the Nitrica river (R Nitra) and in 1961–1962 into the Nitrianske Rudno reservoir where spawning was also observed.

25. Hron (L)

Dudich (1958) does not consider the huchen to be autochthonous in the Hron river. The author states that huchen were introduced there at the beginning of this century. It is of interest, however, that the occurrence of huchen in the Hron river was described by Božena Němcová as early as 1858. Before World War II, huchen were more abundant in that river, in the nineteen-twenties especially in the section between Brezno and Banská Bystrica and even lower downstream, down to Zvolen. At present, however, only isolated short sections of the originally continuous population have been preserved and huchen occur only sporadically in the Hron river, and only in its upper reaches.

From the Hron river huchen used to migrate to spawn in some of its tributaries, including the Hiadeľský potok (Vážna), Ľubietovský potok (Hutná), Moštenický potok (Moštenica), Ľupčica and Drienka (Driekyňa) streams, as well as to the Neresnica, a tributary of the Slatina.

Within the Hron river system, huchen have been introduced into Lake Rossgrund in the Štiavnica lake system.

26. Drava (Drau) (R)

In Austria, the Drava river is well populated by huchen even at present. In its lower reaches in Yugoslavia, occasional huchen occur below Maribor down to Osjek. Of the tributaries, Möll (L) and Gail (R) are mentioned as containing huchen even at present, most of all in their lower reaches. Earlier data are available on the occurrence of huchen in such streams, as the Vellach, Gurk, Görtschach, Lavant, Feistritz, Hohenmaunten, Radel, Dravinja, Pesnica, Plitvica and Bednja. The present situation in these streams is unknown to us. The Mur (Mura) river (L Drava) is well populated in its upper (Austrian) reaches even at present, but huchen have disappeared from the Slovenian part due to pollution (according to Skalin 1983). Ivaška (1951) mentions the alleged occurrence of huchen even in the following tributaries of the Drava: Rog, Strug, Reiznerovac, Vir, and Neteča.

27. Tisza (Tisa) (L)

In the Tisza river itself, huchen occurred from the headwaters, i.e. from the Biela Tisa and Čierna Tisa, down to the confluence with the Rika river. At present, they are more abundantly found particularly in the boundary section between Rumania and the Ukraine.

In the Tisza river system, huchen occur in the Vişeul river (L Tisza) from its mouth into the Tisza, upstream to the town of Vişeul-de-Sus (individual fish up to Moisei), furthermore in its tributary, the Vaser (R) from Vişeul-de-Sus up to the village of Cozla. Huchen also populate a short stretch of the Novăţ stream (R Vaser). They also occur in the Ruscova (R Vişeul) from its mouth up to the village of Poienile-de-sub-Munte, in the Crasna and the Bistra (R Vişeul) where spawning grounds are found. Huchen also inhabit the Carpathian rivers Shopurka, Kosovskaya (= Kisva), Apshitsa, Teresva, Tereblya and Rika (= Veľký Ag) (all R Tisza). At present, huchen are reported to occur in these rivers very rarely.

According to a written communication by M. Papadopol (1977), huchen occur even in the Someşul Cald (L Tisza) and Crişul Repede rivers (the Körös (Crişul) river system, L Tisza); according to Frank (in litt. 1980), they apparently occurred even in the Crişul Alb and Crişul Negru rivers. At the beginning of this century, they even occurred in the Mureş river (L Tisza) down to the town of Alba Julia, as well as in its tributaries, the Streia, Ilvele and Lapuşna. They were re-introduced into the Ilvele and Lapuşna, but the results are problematical.

Huchen were also successfully introduced into the Hornád river (the Slaná river system, R Tisza). In the nineteen-sixties they were introduced into the area of Margecany, where they have remained up to present, as indicated by catches of individuals 7–9 kg and even 15 kg in body weight (R. Žitňan, personal communication), coming from natural reproduction. At the same time huchen were introduced into the Hornád river gorge above Spišská Nová Ves where, however, they allegedly exerted an adverse influence on trout and grayling populations, and were therefore removed from the river by systematic fishing. Since 1977, huchen have been released even in the Hornád river below the Ružín reservoir. It may be inferred from bag records that at present huchen occur in the Hornád river even in the stretches from Malá Lodina up to Krásna nad Hornádom or, as indicated by unsubstantiated reports, up to a distance of 8 km of the Hungarian section of the Hornád (J. Zubrický, personal communication).

28. Sava (R)

Of the river systems of the Danube drainage area, that of the Sava river is most abundantly populated by the huchen. Huchen are abundant in the main stream itself above Kranj, being more sporadic lower downstreams and absent from the environs of Trbovlje. Formerly they occurred down to Zagreb; in its lower reaches, sporadically down to Belgrade.

Of the left tributaries, the following are reported as huchen streams: the Kamniška Bistrica (an important spawning stream), Savinja (a permanent population, spawning between Celje and Ljubno), Krapina and Trebež; of the right tributaries: the Bohinja, Sora (one of the major spawning streams in the environs of Ljubljana, at present only above the cellulose mill at Goričan), Ljubljanica (particularly in the tributaries, the Malo Grabno, Hribski, Gradaščica, Bistra and Lubija where spawning grounds are also found), Krka, Kupa (with the tributaries Dobra, Kupčina, Trepča, Glina and Odra), Una (including its tributaries, the Sana, Sanica and Dabar), Vrbas (including the mouths of its tributaries, the Ungar and Vrbanja), Bosna (and, in its system, the Željeznica, Zujevina, Miljacka, Mokranska Miljacka, Dobrinja, Kasidolska Rijeka, Misoć, Ljubinja, Vogošće, Fojnica, Lepenica, Lašva, Usora, Gostović, Krivaja, Biosćica, Kalinj, Stupančica and Spreča) and, finally, the Drina, denoted by Ivaška (1951) as the richest huchen river of all. As to the Drina river system, huchen have been reported to occur in the Tara, Piva, Sutjeska, Bistrica, Čeotina, Prača, Lim, Uvac and Drinjača rivers. The Lim is the southernmost projection of the huchen's range.

29. Timiş (Tamiš) (L)

Huchen still occurred in the Timiş river at the turn of the century.

30. Velika Morava (R)

Huchen occur in the upper and middle reaches and in its left tributary, the Ibar river. They are said to be scarce there.

31. Cerna (L)

Huchen were still to be found in this river at the beginning of this century. They do not occur in it any more.

32. Jiul (L)

Huchen lived in its Transylvanian and Oltenian stretches. In 1900 to 1930 they were gradually removed, until they disappeared completely.

33. Oltul (L)

At the turn of the century huchen still lived in the Oltul in Transylvania, Oltenia and Multenia. At present they occur there no more. They used to occur even in the Lotru river (R Oltul), but gradually disappeared due to frequent timber-floating in 1890 to 1920.

34. Argeş (L)

Huchen still occurred in the Argeş river early in this century, similarly as in the Doamnei (L Argeş) and Tîrgului (L Doamnei) rivers. They were still reported from the Dîmboviţa river (L Argeş) in the nineteen-thirties, but died out even there due to frequent timber-floating and excessive pollution. They were probably re-introduced later, as they are found in the Lacul Vidraru reservoir on the Argeş river at present.

35. Ialomiţa (L)

Excessive pollution and deforestation caused huchen to disappear completely from the Ialomiţa river and its tributary, the Prahova stream (L), as early as the first quarter of the present century.

36. Siretul (L)

Huchen seem to have been common there at the beginning of the present century. Today they get there exceptionally during floods from a tributary, the Bistriţa river, also entering the right tributaries, the Suceava (Rădăuţi region) and Moldova (Cîmpulung and Făleiceni regions). Formerly, huchen probably occurred commonly in the two tributaries mentioned.

At present, huchen still occur in the right tributary of the Siretul, the Bistriţa river, from the town of Piatra Neamţ up to the Iacobeni dam in the Vatra Dornei region. They also occur in its right tributaries the Dorna (from Poiana Stampei down to the mouth into the Bistriţa), Neagra and Barnaul rivers.

Earlier papers mention the occurrence of huchen even in Lake Lacul Roşu, passed by the Bicaz river (R Bistriţa), as well as in the Buzăul river (R Siretul). More recently, they have also been reported from the Bicaz valley reservoir.

37. Prut

The occurrence of huchen in the Prut river is reported from its confluence with the Cheremosh river upstream to the villages of Delyatyn or Vorokhta. Formerly, the Cheremosh river (R Prut) was considered as a rich huchen stream, the same as its

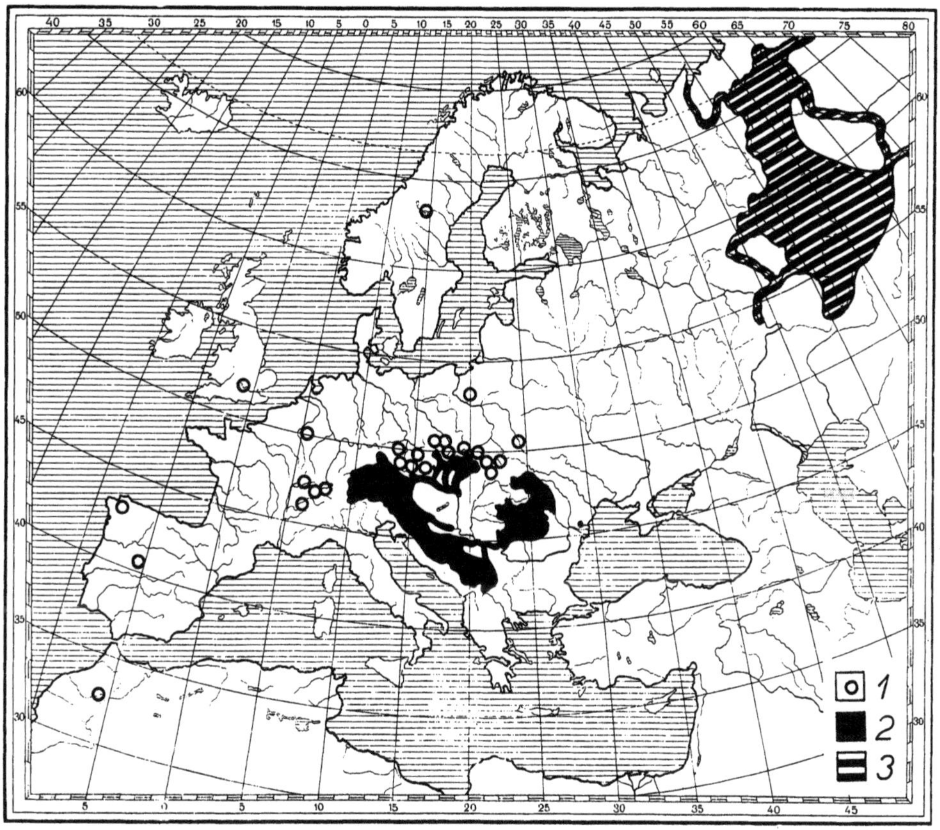

Fig. 20. Introductions of *Hucho hucho* into European and north African streams. 1 – introduction, 2 – autochthonous occurrence of huchen, 3 – ditto, of taimen (orig. K. Hensel).

tributaries, the Ludovy, Richka, Putila (all R), Shibeny, Dembronia, Bystrets, Iltsia and Bereznitsa (all L). The present distribution and numbers of huchen in the Prut river system are unknown.

38. Yalpukh (L)

A single huchen, 330 mm in length, was caught in Lake Yalpukh on the river of the same name, near the Danube Delta (Jentsh 1912, from Berg 1948–49).

Besides, huchen were introduced even outside the Danube drainage area into the following streams (Fig. 20):

Vistula (Wisła) river system: the Sola (R), Skawica (R), Dunajec (R), Poprad (R Dunajec), Rabski potok (L San, R Vistula), Bug (L Narew, L Vistula) and Drwęca (L Vistula);

Odra (Oder) river system: the Odra, Moravice (R Opava, L Odra), Olše (L Odra), Nysa Kłodzka (L Odra) and Stěnava (R Nysa Kłodzka);

Labe (Elbe) river system: the Vltava (L), Malše (R Vltava), Otava (L Vltava), Sázava (R Vltava), Berounka (L Vltava), Klíčava water supply reservoir on the Klíčava rivulet (R Berounka) and Ohře (L Labe) rivers;

Rhine (Rhein) river system: the upper reaches of the Rhine, Sarine (L Aare, L Rhine), Saane (R Sarine) rivers;

Indalsälven river system: the Indalsälven reservoir and the Rexforsån rivulet;

Meuse (Maas) river system: the Semois (R) and the Lesse (R) rivers;

Thames river system: the Thames;

Drinage areas of the rivers in the La Coruña region (the northwestern projection of Spain);

Duero (Douro) river system: the Tormes river (L);

Rhône river system: the Rhône, Usses (L Rhône), Doubs (L Saône, L Rhône) and Rodane rivers;

Umm-ar-Rbia river system: the Umm-ar-Rbia, Fellate, and Véd Gígú rivers (Morocco).

Finally it is necessary to mention the introduction of the huchen into the rivers and lakes of the Quebec Province, Canada.

For particulars on the introductions of the huchen into the water bodies mentioned above, see Chapter 21.

DISTRIBUTION OF THE TAIMEN

The taimen is distributed from the Volga and Pechora river basins east up to that of the Yana river in the north and that of the Amur river in the south (Fig. 21). Taimen do not occur in the Northern Dvina basin, although the Severnaya Kel'tma (a left tributary of the Vychegda river, Northern Dvina system) and the Yuzhnaya Kel'tma (a right tributary of the Kama river) are connected by swamps on their upper reaches. According to Kirillov (1955), taimen do not occur in the Indigirka river although, e.g., Berg (1948–1949) states the contrary on the basis of data provided by Dryagin (1935). Shaposhnikova (1964) doubts the information in Navozov (1912) on the occurrence of taimen in the Malyi Ik river (the Ural river system).

1. Volga

In the Volga river drainage area, taimen occur exclusively in the Kama river system. Formerly, individual taimen descended from the Kama even into the Volga (down to Tetyushi). At present, however, these places are covered by the Kuibyshev reservoir, and there are no reports of taimen occurring in it. In the Kama river, taimen are abundant, especially in its upper reaches, i.e. above the confluence with Chusovaya river. Lower downstream, between Perm and Sarapul', they are scarcer, and occur singly in the lower reaches. After the erection of the Kama dam, they even populate the reservoir which reaches roughly from the mouth of the Chusovaya river up to the town of Berezniki. Most of all, taimen inhabit the tributaries of the Kama river, namely the Veslyana, Lup'ya, Kosa, Yuzhnaya Kel'tma and its tributary, the Lopva, furthermore the Vishera (with its tributaries, the Veltsa, Uls, Yaz'va, Melmysa, Kolchina and Kol'va, with its tributary, the Berezovaya river), Yaiva, Kos'va, Chusovaya (and its tributaries, the Koiva, Kusa, Syl'va with its tributary, the Barda), Belaya (with its tributaries, the Ufa, Niaza, Kukozara and Yazyasha) and,

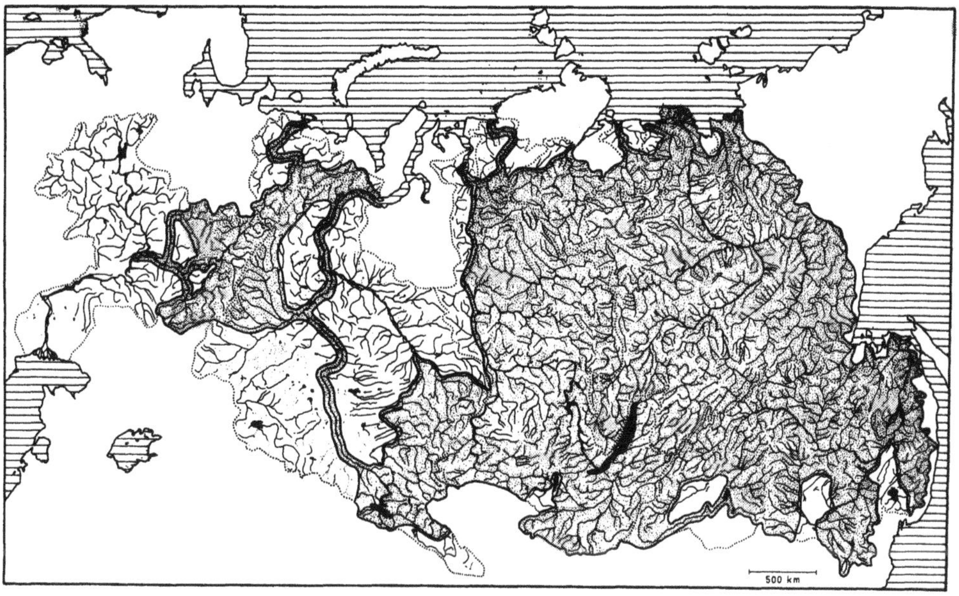

Fig. 21. Distribution of the taimen, *H. h. taimen* (orig. K. Hensel).

finally, the Vyatka river in the section above the town of Slobodskoe, with its tributary, the Kobra. In the Kobra river basin, taimen occur in the Pashnyak and Soz Yuzhnyi rivers.

2. Pechora

In the Pechora river, taimen occur all along the stream, but are more abundant in its middle and especially its upper reaches and, above all, in its tributaries the Ilych and Shchugor rivers. They also occur in the Usa river. They have abso been found in the Pechora Sea, on the shore of the Yugorskii Shar strait.

3. Ob'

The Western Siberian Lowland, passed by the Ob' river, does not offer suitable conditions for the life of the taimen. In the Ob' river itself, therefore, taimen occur either in the upper or in the lower reaches. They do not occur in the middle reaches from Kolpashevo (i.e. roughly from the confluence with the Chulym river) down to the mouth of the Irtysh river. They also scarcely occur in the Ob' Delta.

Of the tributaries, the following are mentioned as being populated by taimen: the Biya with Lake Teletskoe, the Tom', the Chulym (in its system: the Chernyi Iyus and Belyi Iyus rivers) and the Irtysh rivers. In the Irtysh, taimen occur from its upper reaches down to its mouth into the Ob' river. In the Irtysh river system, the occurrence of taimen has been reported in the upper reaches in the area of Lake Zaisan (the present Bukhtarma reservoir), in the Bukhtarma and the Uba rivers. In the lower reaches, taimen are known to occur in the tributaries of the Tobol river, viz. the Iset' (and its tributary, the Miass), Tura (with the tributaries, the Salda and Tagil, with the Barancha and Mara), the Tavda with the Loz'va and Sos'va (and its tributaries, the Kakva and Lyalya) rivers. Taimen also occur in the left tributaries of the lower reaches of the Ob' river.

4. Enisei

In the Enisei itself, taimen occur from the upper reaches down to the delta. They are abundant also in its tributaries; in the middle and lower reaches, only in the right ones. In the headwater of the Enisei, the rivers Bii-Khem (Great Enisei), Ka-Khem (Little Enisei), Kyzyl-Khem and Shishkhid, including the lakes of the Darkhat basin (Dood Cagaan, Targan) and Sharin river are mentioned in connection with the occurrence of taimen. Of the tributaries of the Enisei, taimen have been reported from the rivers Sistyg-Khem, Khemchik, Us, Abakan, Tuba (together with the Mocharsk Lakes), Sisim, Mana and Angara. In the Angara basin, Lake Baikal and its tributaries should be mentioned in particular. In Lake Baikal itself, taimen occur only scarcely in the littoral zone, being more abundant near the mouth of the Frolikha river only. Of the northern tributaries of Lake Baikal, the occurrence of taimen has been mentioned in the rivers Bertsia, Snezhnaya (and also in Lake Snezhinskoe), Kichera (with Lake Kulinda) and especially in the Frolikha (with Lake Frolikha). Furthermore, taimen occur in the rivers Tempa, Bol'shaya, Severnaya, Turka and Selenga. Of the tributaries of the Selenga, they have been reported from the Orkhon, including its tributaries, the Tuul, Dzhida and Uda, from Lake Ögii, as well as Lake Terkhiin Cagaan on the Khoit Terkhiin river. Taimen are absent from the Delger Mörön river and from Lake Khövsgöl (Khubsugul).

Of the tributaries of the Angara river, taimen are known to occur in the Irkut and Ilim rivers and in the tributaries of the latter, the Konchenga, Ireeka and Igirma rivers. They are abundant even in the tributaries of the middle reaches of the Enisei, e.g., in Podkamennaya Tunguska, Tynep (the Bakhta river system), Nizhnaya Tunguska (with its tributaries, the Kochechum and Severnaya rivers); in the lower reaches it is the Kureika and Khataika rivers.

Taimen have also been reported to occur singly in the following reservoirs: Krasnoyarsk (on the Enisei river), Irkutsk, Bratsk, and Ust'-Ilimsk (on the Angara river).

5. Pyasina

In the Pyasina river basin, taimen inhabit mainly the upper reaches, particularly the so-called Noril'sk lakes. They scarcely ever occur in the lower reaches of the Pyasina river.

6. Khatanga

Taimen are scarce in the river itself, but occur in its mouth (allegedly only sexually immature individuals). They are more abundant in the middle and upper reaches of the Kheta river (including Lake Ayan) and in the Kotui river (of its tributaries, the Voevorlikhan and Kotuikan rivers are mentioned).

7. Anabar

Taimen occur all along the stream, being more abundant in the upper and middle reaches, and single individuals occur in the lower ones.

8. Olenek

In this river, taimen occur from the upper reaches down to the delta and even in zones where fresh water contacts salt water. As regards the tributaries, the occurrence of taimen has been reported from the Arga-Sala, Tirekh-Yurakh, Ulakhan-Dyarkalaakh, Siligir, Senkyu, Ukukit, Nekekit and Beke rivers.

9. Lena

Taimen are common in the Lena river system. In the Lena itself they occur from the headwaters down to the delta, entering the sea in summer and staying there untill autumn when they migrate back to freshwater habitats. In the Lena river system, they inhabit the Kirenga river and its tributaries, the Okunaika and Vitim rivers. In the Vitim basin, they occur in the Bambuika river and in Lake Oron. Of the other tributaries of the Lena, the rivers Nyuya, Olekma (in its basin, the Charsk lakes and the Tokko river), Aldan (with its tributaries, the Maia and Uchur rivers and Lake Bol'shoe Toko), Vilyui (with its tributaries, the Chona, Akhtaranda, Tyung and Mukhtaryma rivers as well as the Vilyuisk reservoir), and finally the Lyampishka and Moloda rivers.

10. Omoloi

Taimen are known to occur even in this river system.

11. Yana

This is the easternmost Siberian river in which taimen occur. They inhabit the whole stream, being particularly abundant in its tributary, the Adych river.

12. Uda

Taimen occur in this stream.

13. Tugur

Taimen occur even in the basin of this river.

14. Amur

In the Amur river basin, taimen occur almost everywhere, but mainly in the mountain and submontane streams or in lakes. In the headwaters of the Amur river, this is the Onon river with its tributaries, the Baldzh and Adrag rivers, besides the Ingoda and Shilka (with its tributary, the Nercha river) and also the Argun' river. In the middle and lower reaches of the Amur, the occurrence of taimen has been mentioned in the rivers Zeya (and its tributary, the Selendzha), Sungari (and its tributary, the Shan-si), Ussuri (and its tributary, the Khor) and in the rivers flowing into the eastuary of the Amur. In the lower reaches of the Amur, taimen also occur in Lake Udyl'.

In the Amur basin, taimen occur even in Lake Dalai and Lake Buir, as well as in the Urshun, Khalkhin and Kherlen rivers. They do not live in the Ulz river, contrary to the statement of Dulmaa (1973).

8. Habitat

GENERAL CHARACTERISTICS

Hucho hucho is an exclusively freshwater salmonid species which predominantly populates running waters, but occurs even in lakes and reservoirs in places. In some waters it is distributed down to their estuaries, e.g., in the Lena Delta (Pirozhnikov 1955, Kirillov 1972), the Khatanga Delta (Luk'yanchikov 1967b), down "the places of contact of fresh and salt water", as stated by Kirillov (1972: 64). It may even invade salt water strongly mixed with fresh water. In fact Luk'yanchikov (l.c.) states that it sometimes occurs even in Khatanga Bay (the Laptev Sea). It prefers regions at altitudes of 200–600 m, but may occur even at lower or higher elevations where conditions permit. The highest occurrence of the nominate form has been recorded in the Lim river on the Yugoslavian-Albanian border (1007 m above sea level, Ivaška 1951, 1953), and in the Shishkhid river (upper reaches of the Enisei) in Mongolia (1560 m above sea level, Dulmaa 1979) for the taimen. Disregarding the obviously accidental finding of a young *H. hucho* in Lake Yalpukh in the Danube Delta (Berg 1948–1949), the lowest occurrence of the nominate form is in the stretch of the Danube in the Iron Gate (67.5 m above sea level, Ivaška 1951). As mentioned above, the taimen occurs down to sea level.

Except in the spawning period, streams in the foothill zone (formerly the so-called grayling and barbel zones) are the typical habitats of the huchen (Pl. 5 to 8). The species prefers cold streams with water rich in oxygen and with a sufficiently diverse bed. The water temperature of these streams does not usually exceed 15° C (Prawocheński and Kolder 1968), but the huchen can even tolerate short spells of water temperatures as high as 20–22° C (provided that the depth and velocity of the stream are sufficient) (Nieslanik).[7]

Compared to other salmonids, *H. hucho* appears to be less demanding as to dissolved oxygen content. Vlasova (1959) states that like the grayling *(Thymallus thymallus)*, it can tolerate a decrease in the oxygen content down to 5 mg l^{-1}, whereas the brown trout can tolerate such a decrease down to 7–8 mg l^{-1}. According to Prawocheński and Kolder (1968), the major factors governing the occurrence of *H. hucho* include well aerated water (dissolved oxygen content at least 8–9 mg l^{-1}), water temperature not exceeding 15° C, gravelly or gravel-sandy bottom, a rapid and strong stream alternating with pools in places, and sufficient food supply.

The environmental requirements of the taimen are similar to those of the

[7] In September, 1956, Nieslanik observed that huchen kept in storage ponds at Jablunkov at water temperatures of 18° C were breathing hard, standing in the inflowing water close below the water surface and ignoring food or objects moving near them. As soon as the water inflow was increased they descended into deep water and calmed down.

nominate form, as can be concluded from fragmentary items of information from Sabaneev (1911), Pravdin (1949), Nikol'skii (1956), Bukirev (1967) and Sigunov (1972). Contrary to the huchen, the taimen frequently inhabits even the lower reaches of northern rivers because they are cool. In many of them taimen occur almost continuously from their upper reaches down to their deltas, e.g., in the Lena, Enisei, Anabar, Yana and Olenek rivers (Pravdin 1949, Pirozhnikov 1955, Kirillov 1972, 1976). Ol'sḥanskaya (1965) considers the taimen a "lacustrine-riverine" species, but it actually prefers rivers.

OCCURRENCE IN RUNNING WATERS

Typically, adult *Hucho hucho* inhabit clean-cut habitat types, the character of the habitat depending on the size of the fish. In this respect there is no difference between the nominate form and the taimen. According to information gleaned from several papers (Sabaneev 1911, Nikol'skii 1956, Ivaška 1951, Svetina 1962, Bukirev 1967, Misharin and Shutilo 1971, Sigunov 1972, Olifer 1977 and many others), adult huchen prefer the following habitat types:

(1) Deep holes below rapids, cascades, weirs, sluices and waterfalls, often shaded by overhanging trees;

(2) Calm water behind rocky projections from banks and groynes, projecting into streams and forming whirlpools;

(3) Below and above lateral tributaries;

(4) In scours in and outside the stream line;

(5) Behind boulders, piers, ice-aprons and other obstacles in the stream bed;

(6) At the confluence of lateral branches of a stream bed and above their bifurcation;

(7) On the convex side of stream bends; and

(8) In abruptly narrowed places of the stream bed.

Some of these habitat types are illustrated in Fig. 22. Various circumstances influence habitat selection by *H. hucho*. During spells of high water levels or of turbid water, the fish keep closer to banks, swimming higher up in a calmer stream and when the water drops again. In the morning and in the evening they are usually found in stream bends, at the boundaries between rapids and pools and in shallow water, as a rule moving to the latter at night. During the day and in summer heat, huchen stay in deep holes below rapids or swim to the mouths of cool tributaries. In winter they seek deeper, in summer – shallower places in the stream.

Young, sexually immature huchen and taimen usually have no permanent habitats. Their young first stay in the vicinity of spawning sites and then gradually move downstream. According to Bastl and Kirka (1959), Misharin and Shutilo (1971) and Nieslanik, young fish prefer:

(1) Side branches of the main stream with flowing water, where the bottom is gravelly or even muddy in places and where shallows alternate with moderately deep pools;

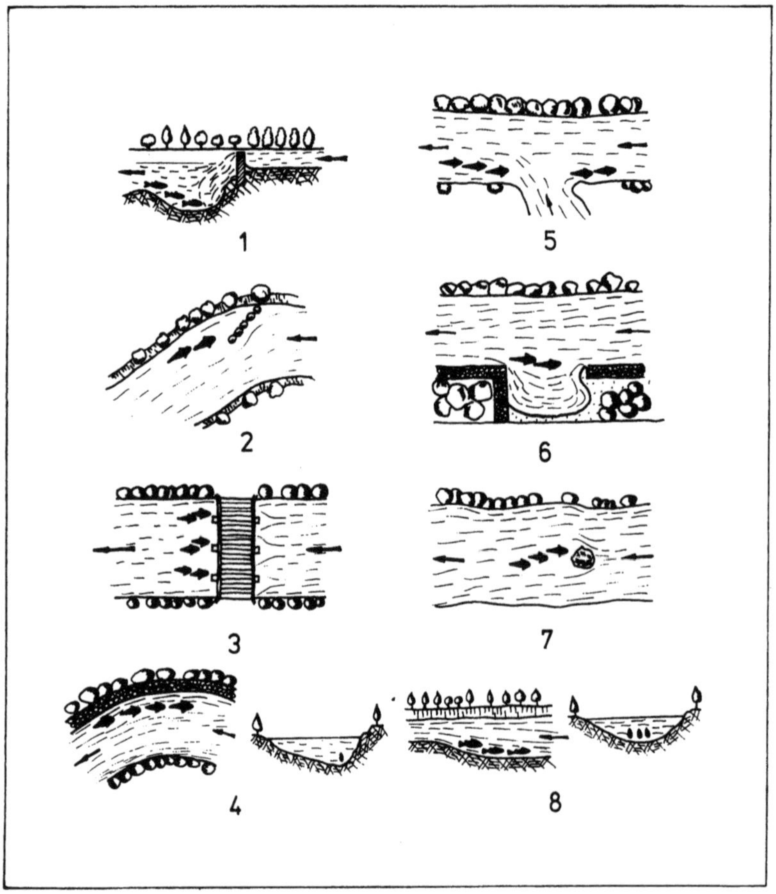

Fig. 22. The usual habitats of huchen. 1 – below a weir, 2 – behind a lateral rocky projection, 3 – below bridge piers, 4 – in a stream bend, 5 – above and below the mouth of a tributary, 6 – in a lateral excavation, 7 – behind an obstacle, 8 – in scours of the bottom (after Hartlieb 1961).

(2) Calm waters near banks, connected with the main stream on their lower end, often with a muddy bottom;

(3) Broad river beds divided by gravelly islets into several lateral branches with shallow, flowing water;

(4) Small tributaries.

Beyond doubt, their habitat selection is influenced by the presence of young of other fish species which are preyed upon by young *H. hucho.*

Bigger young huchen, 2–4 years old, inhabit the same habitats as the adult ones, a certain hierarchy being maintained. According to Sigunov (1972) and Nieslanik, the adults keep to the upper part of a deep pool or to a threshold and the younger and smaller ones are behind. More extensive pools and deep parts of the stream are often shared by up to several tens of big fish. This has been documented for the taimen by, e.g., Nikol'skii (1956) and Sigunov (1972), for the huchen by Nieslanik and also by

Rudolf (1964). The latter author, together with the skin-divers Bezděk and Macháček of Brno, have even recorded it on a film shot in Slovak rivers. Rudolf reports that in a single pool he observed around 16 huchen between 6 and 15 kg in estimated body weight, the longest around 140 cm in total length and approximately 25 kg in weight.

OCCURRENCE IN STAGNANT WATERS

The considerable adaptability of the Eurasian huchen is also indicated by numerous data on its occurrence in stagnant waters, such as lakes and reservoirs. The less frequent occurrence of huchen in lakes compared to that of taimen is due to the fact that the Danube basin is poor in natural lakes, the existing ones being either inaccessible to the huchen (Alpine and Carpathian lakes lying at high altitudes, their outlets often being fast-flowing and shallow) or not offering the necessary optimum living conditions together with the possibility of reproduction (e.g., the lakes Balaton and Neusiedler See). The capability of *H. hucho* to populate even stagnant waters is evidenced by their occurrence in reservoirs built on foothill rivers populated by the species (the Orava reservoir; Holčík 1966a, Holčík et al. 1965; the Bicaz and Vidraru reservoirs in Rumania; Iancu and Decei 1972; the Drava reservoir and Lake Zalatar in Yugoslavia; Vokač 1973, Kosorić et al. 1975). In the Orava reservoir, as well as in the Rudno reservoir on the middle reaches of the Nitrica rivulet (Kattoš 1966), huchen stay throughout their life except for the spawning period; even the young ones leave the native streams shortly after hatching and move down to the reservoir.

From observations made in Lake Baikal (Kozhov 1963, Tugarina and Kupchinskaya 1977), as well as in the Orava reservoir, one may conclude that in such water bodies both taimen and huchen predominantly populate the littoral parts and especially the mouths of tributaries. Only Luk'yanchikov (1967a) states that taimen occur throughout the Bratsk reservoir on the Angara. In the Ust'-Ilimsk reservoir, on the same river, young taimen stay near the mouths and in the nearby inshore zones of the rivers Ilim and Kezhma-Dubinskaya, whereas the adult fish stay in the tributaries, where they inhabit pools and holes below weirs and cascades (Olifer 1977). According to Ol'shanskaya et al. (1977), in the Krasnoyarsk reservoir on the Enisei, taimen occur only at low temperatures, inhabiting mainly bays, often even the pelagial parts, but moving into tributaries when the temperature rises. In the Irkutsk reservoir, taimen inhabit the upper and central parts (Tugarina et al. 1977).

In general, however, taimen occur at much lower densities in lakes and reservoirs than in rivers, and exceptionally rather than regularly (Nikol'skii 1956, Svetovidova 1960, Kozhov 1963, Pushkina 1977). No doubt the same is true of the huchen. This species appears to enter stagnant waters only when conditions in running waters are unsatisfactory for some reason or another. Here the water temperature may be the governing factor, or more precisely the difference in the water temperature between a lake and its tributary. This has already been suggested by Ol'shanskaya et al. (1977) for taimen in the Krasnoyarsk reservoir, and also by our own experience from the Orava reservoir (Holčík et al. 1966). As a matter of fact, its tributary the Biela Orava river has low water levels during summer (or may even cease to flow in places) and consequently its water becomes overheated.

Part III

BIONOMICS AND ECOLOGY

9. *Reproduction and Development*

The Eurasian huchen and beyond doubt the whole genus *Hucho* may be included in the ecological group of lithophilous, brood-hiding species (group A.2.1, Balon 1975a). In this respect there is practically no difference between the huchen and the taimen, nor between populations of these two forms in different river systems.

AGE AT SEXUAL MATURITY

Table 6 summarizes all available data on the age at which *H. hucho* attains sexual maturity and at which females spawn for the first time. It is evident that the taimen attains sexual maturity and spawns for the first time at a higher age than the huchen, despite the fact that their dimensions and particularly their body weights are approximately the same at that time. Thus in *H. hucho,* as in other fish species, it is the size and not the age that decides sexual maturation. In general, *H. hucho* may be characterized as a species which attains sexual maturity later than other salmonids. Males of *H. hucho* usually attain sexual maturity and spawn for the first time one year earlier than females and at smaller body dimensions. In general, sexual maturity is attained by the males of the nominate form at 3–4 years of age and 1–2 kg body weight, by the females at the age of 4–5 years and at 2–3 kg body weight. Male taimen attain sexual maturity at the age of 3–8 years and around 1 kg body weight, whereas females at the age of 4–8 years and around 2 kg body weight. As for the differences between the two geographic forms and between populations in different river systems in this point, they are undoubtedly due to the influence of different climatic conditions and geography (Prawocheński and Kolder 1968). Northern populations attain sexual maturity at a higher age than southern ones.

Observations of several authors (Köttl and Kafka 1927, Peňáz and Příhoda 1981, our data) indicate that in pond cultures huchen attain sexual maturity later than

TABLE 6

Attaining sexual maturity and first spawning of *Hucho hucho* in different localities

Locality	Age	Standard length* (mm)	Weight (kg)	Author
		Hucho hucho hucho		
Sava river system	—	627	3	Munda (1925, 1926a)
Ammer	—	—	1.5	Seez (1939)
Drava	4	627	2—3	Krauss (1933), Vokač (1959)
Danube river system	4	564	1.7	Siebold (1863), Svetina (1970)
Turiec, Orava	5	582	3	Ivaška (1951)
Fish farms	6	—	3	Köttl and Kafka (1927), Nieslanik
		Hucho hucho taimen		
Siberia, Far East, China	4—5	400—500	—	Pravdin (1949), Sigunov (1972), Wu (1979)
Amur	4	368	—	Nikol'skii (1956)
Lena	7	451	1.5	Karantonis et al. (1956)
Enisei	—	442	2.5	Podlesnyi (1958)
Vilyui	6	552	2	Kirillov (1962)
Kama	7	—	—	Bukirev (1967)
Khatanga	8	552	3	Luk'yanchikov (1967b)
Angara	6—7	552—644	2—2.1	Misharin and Shutilo (1971), Olifer (1977)
Yakutia	7	598	1.5	Kirillov (1972)
Anabar	7	568	1.9	Kirillov (1976)
Vitim	7	644	3.2	Kalashnikov (1978)

*For *H. h. hucho* the authors state the total length (*Tl*) while for *H. h. taimen* mostly the fork fength (*Fl*) is stated. For the sake of uniformity we hawe converted all data to standard length (*Sl*).

under natural conditions. This is as late as the 6th to 8th year of life. This is certainly connected with the quality and quantity of food.

If the Soviet authors are right in stating that taimen do not spawn each year, then this is a difference from the nominate form. Sabaneev (1911) writes that not all taimen do migrate to spawn, some of them remaining in their original habitats. According to Misharin and Shutilo (1971), taimen spawn only every third year, according to Sigunov (1972) every other year. The nominate form may be assumed to spawn every year, as the same fish are used every year for hand-stripping in huchen farms. The leaving out of some of the spawning periods by the taimen apparently occurs in northern populations and corresponds with the situation in other species of freshwater fishes occurring in northern regions, as pointed out, e.g., by Kennedy (1953). This probably pertains mostly to females, similarly as observed in coregonids (Reshetnikov 1980). Thus it may be assumed per analogiam that even in the taimen it is not a genetically fixed feature, but a mere adaptation of certain populations to specific environmental conditions. At any rate, attention should be paid to this problem in the future.

SEX RATIO

The problem of sex ratio in *H. hucho* is still considerably obscure in spite of its considerable practical importance, and it is among the least investigated items of the bionomics of the species and of the genus *Hucho* in general. Jungwirth (1977a) and Nikol'skii (1956) are the only authors to state explicitly that the sex ratio is 1:1. A 1:1 ratio can also be supposed from the statements of Vladykov (1926) and Kulmatycki (1931a) that *H. hucho* spawn in pairs. According to other authors on the other hand, males are more numerous than females. Thus, Sabaneev (1911) and Jagoditsch (1930) state that males predominate over females. According to Ivaška (1951), males usually predominate over females at a 2–3:1 ratio in spawning grounds. Shnarevich and Moshuk (1956) state that in the Cheremosh river males predominate over females at a 7:1 ratio. Vokač (1959) and Nieslanik (in litt.) also mention a predominance of males. Nieslanik writes that the predominance of males should be mainly ascribed to the fact that the females are more voracious and less cautious, and when searching for food, which they need to build up their voluminous ovaries, they fall an easier prey to anglers. Jagoditsch (1930) also believes that the predominance of males in the upper reaches of the Sava river is the result of man's activity; he explains it as resulting from stream regulation which changed the character of the bottom, impeding redd excavation and thus delaying the breeding season and causing overripening of eggs.

FECUNDITY

Fairly numerous data are available in the literature on the fecundity of *H. hucho*. In a female 6 years old and weighing 3.6 kg, Kulmatycki (1931b) found a total of 1600 eggs. According to Krauss (1933), a female huchen weighing 2 kg contained 2800 eggs and, since the fecundity increases by some 1200 eggs per each 1 kg of weight increment, the author estimates that a female weighing around 10 kg may contain approximately 12,400 eggs. Munda (1935) reports that by hand-stripping a female weighing 4–5 kg one may obtain 3–4 thousand eggs; a female weighing 6–8 kg may yield 5–6 thousand eggs; and a female over 10 kg in weight up to 10–15 thousand eggs. The largest number, 35 thousand eggs, was found by him in a female 16 kg in weight. According to Ivaška (1951), females from the Turiec river, weighing 3–5 kg, produce around 2400 eggs on average; those weighing 4–5 kg produce up to 4 thousand; those weighing 6–8 kg, up to 6 thousand; those weighing 10–12 kg produce 8–12 thousand; and at 18 kg up to 18,600 eggs. Of the other data it is necessary to mention that of Vokač (1959) according to whom a female 2 kg in weight will produce around 2800 eggs at first spawning. According to this author, the absolute fecundity increases by 1000–1200 eggs per each 1 kg increment in body weight. A similar statement is found in Svetina (1962), according to whom it is a rule with *H. hucho* that one egg is produced per approximately 1 g of female body weight. Skácel (1963) also states that the relative fecundity of female huchen from Slovak waters is approximately 1100 eggs per kilogram body weight. According to Jungwirth (1978), a female produces 1200–1500 eggs per kg body weight annually.

Naturally, the number of eggs is not constant throughout the life of a fish, but varies in dependence on age and size. Thus Shnarevich and Moshuk (1957) found, in the huchen from Soviet Bucovina, that the absolute fecundity of females 2.2–5.5 kg in weight was 4560–11,717 eggs, and similar values appear to be valid for the huchen in the Transcarpathian region of the Ukraine (Vlasova 1959) (Table 7). It is evident that the number of eggs increases with increasing age, length and weight of the female. In a female from the Orava reservoir, Holčík and Bastl (1968) found 11,113 eggs. A female huchen caught in the Tormes river (Spain), originating from eggs imported from Czechoslovakia, contained 16,900 eggs; the fish weighed 10.5 kg and was six years old (Lobón-Cerviá in a letter of November 1979). This review reveals considerable discrepancies between the data provided by different authors. They can be explained by the fact that Ivaška, Skácel, Vokač, Svetina and Munda consider the number of eggs which can be obtained from a female by hand-stripping (i.e. the so-called "obtained fecundity"), whereas the remaining authors report the absolute fecundity, including residual eggs which remain in the female's abdominal cavity after spawning or stripping.

TABLE 7

Total absolute and relative fecundity of huchen from streams in Soviet Bucovina and Carpatho-Ukraine (adapted from Shnarevich and Moshuk 1957 and Vlasova 1959)

Age (years)	*FL* (*Sl*) (mm)	*w* (kg)	Gonad weight (g)	Fecundity	
				Absolute	Relative
			Bucovina		
4	480 (442)	2.2	—	4560	2072
5	580 (534)	2.8	—	7225	2580
6	640 (589)	2.9	—	9200	3172
10	840 (773)	5.5	—	11717	2130
			Carpatho-Ukraine		
4	590 (543)	3.0	22	7600	2533
6	750 (690)	4.2	87	8540	2033
8	840 (773)	5.7	262	11384	1997

Very valuable data were obtained by Nieslanik during his long-term practical work on huchen farms. They are compiled in Table 8. It is evident from it that an increase in the size of the female is followed by an increase in both the absolute fecundity (in this case, the obtained fecundity) and egg diameter, while on the other hand the relative fecundity gradually decreases. Nieslanik's data were used to compute regression equations and correlation coefficients (Tables 9 and 10),

TABLE 8

Total absolute and relative fecundity of huchen, as obtained by hand-stripping in Slovak farms (adapted from Nieslanik). Numerators indicate average values, denominators indicate ranges

Length group (mm)	*Sl* (mm)	*w*[1] (kg) I	*w*[1] (kg) II	Absolute fecundity (No. of eggs)	Relative fecundity[2] (eggs kg^{-1}) I	Relative fecundity[2] (eggs kg^{-1}) II	Egg diameter (mm)	Volume of eggs (ml)	No. of fishes
501—600	576 573—582	3.2 3.0—3.5	2.5	8517 8100—9000	2680 2443—3000	3600	4.3 3.8—4.6	388 335—450	3
601—700	652 609—700	5.0 3.5—7.5	4.0 3.8—4.3	10535 7850—14000	2162 1160—2642	2823 2706—2943	4.3 3.8—4.7	558 350—940	12
701—800	741 709—799	6.3 4.5—8.8	4.8 3.6—7.2	10996 3580—15000	1849 430—2960	2244 675—3700	4.6 3.8—6.0	678 450—800	14
801—900	861 826—880	10.1 9.0—10.8	8.5 7.1—9.8	15229 8300—22900	1501 869—2130	1699 976—2617	4.9 4.8—5.1	1775 1600—1950	6
901—1000	961 907—989	14.1 11.8—16.4	11.9 9.5—14.3	15533 9490—22375	1162 772—1463	1415 825—1660	4.9 4.5—5.4	1400	8
1001—1100	1034 1016—1043	18.0 16.2—20.8	14.5	23558 1860—26700	1311 1088—1562	—	5.0	1750	3
Whole material	780 573—1043	8.4 3.0—20.8	7.7 2.5—14.3	12875 3580—26700	1772 430—3000	1974 675—3700	4.6 3.8—5.4	732 335—1950	46

[1]Weight before (I) and after (II) stripping. [2]Converted to weight before (I) and after (II) stripping.

TABLE 9

Equations of certain relationships characterizing huchen fecundity (same material as for Table 8)

	r	n
$F = -16602.8849 + 37.7031\ Sl$	0.655***	46
$F = 3311.0564 + 1.1431\ w_1$	0.693***	46
$F = 2057.8238 + 1.4200\ w_2$	0.585**	28
$F = 4104.2288 + 11.3193\ v$	0.911***	31
$FR_1 = 5622.5557 - 4.9346\ Sl$	−0.658***	46
$FR_1 = 3025.9508 - 0.1487\ w_1$	−0.666***	46
$FR_1 = 484.3070 + 0.1000\ F$	0.019	46
$FR_1 = 3087.0831 - 1.3435\ v$	−0.204	31
$FR_2 = 8209.9758 - 7.5947\ Sl$	−0.647***	28
$FR_2 = 3942.7071 - 0.2569\ w_2$	−618***	28
$FR_2 = -117.1898 + 0.1670\ F$	0.233	28
$FR_2 = 4514.6517 - 2.4992\ v$	−0.156	14
$D = 2.2056 + 0.0031\ Sl$	0.614***	45
$D = 3.8277 + 0.0001\ w_1$	0.480***	45
$D = 4.3601 + 0.00001\ w_2$	0.426*	28
$D = 3.2755 + 0.00001\ F$	0.271	45
$D = 3.6921 + 0.0011\ v$	0.563***	31
$v = -1923.3156 + 3.7198\ Sl$	0.863***	31
$v = 45.9764 + 0.1067\ w_1$	0.787***	31
$v = 162.4105 + 0.1170\ w_2$	0.810***	14

* $p < 0.05$, ** $p < 0.01$, *** $p < 0.001$.

F — absolute number of eggs, Sl — standard length in mm, w_1 — body weight in g before stripping, w_2 — ditto, after stripping, FR_1 — number of eggs per kg of w_1, FR_2 — ditto, of w_2, v — volume of eggs obtained by stripping in ml^3, D — egg diameter in mm before swelling, r — correlation coefficient, n — number of cases.

TABLE 10

Multiple regression equations of certain relationships characterizing huchen fecundity (same material as for Table 8; same symbols as for Table 9; R = multiple regression coefficient)

	R	n
$F = 5786.8769 + 1.0959\ Sl + 0.7514\ w_1$	0.694	46
$F = 8854.3862 - 3.9694\ Sl + 0.9579\ w_2$	0.587	28
$FR_1 = 3400.8069 - 1.4778\ Sl - 0.0567\ w_1$	0.666	46
$FR_2\ \ 5668.9331 - 4.3116\ Sl - 0.0204\ w_2$	0.639	28
$D = 1.2199 + 0.0056\ Sl - 0.0001\ w_1$	0.716	45
$D = 3.7978 + 0.0015\ Sl - 0.00002\ w_2$	0.483	28
$v = -1651.3680 + 3.3890\ Sl - 0.0056\ w_1$	0.955	31
$v = -1186.6401 + 2.5463\ Sl - 0.0144\ w_2$	0.887	14

TABLE 11
Equations characterizing fecundity of taimen from the Angara river and huchen from Soviet Bucovina and Carpatho-Ukraine (calculated from data by Olifer 1977, Shnarevich and Moshuk 1956 and Vlasov 1959)

Relationship	$r(R)^1$	n
Angara (718—966 mm *Sl*, 4.6—12.7 kg)		
$F = -35123.9682 + 61.6234\ Sl$	0.877*	6
$F = 631.1481 + 1.8877\ w_1$	0.009	6
$F = -9357.1396 + 2201.5735\ A$	0.955**	6
$FR_1 = 5414.7612 - 4.2714\ Sl$	−0.009	6
$FR_1 = 2936.5968 - 0.1309\ w_1$	−0.017	6
$FR_1 = 322.8813 + 152.5653\ A$	0.153	6
$F = -1567969.4537 + 1950.2648\ Sl + 1.6579\ w_1$	0.889	6
$F = 17672.4189 + 4205.3838\ A - 60.5310\ Sl$	0.982	6
$F = 22081.6551 + 5222.3816\ A - 2.7237\ w_1$	0.992	6
$FR_1 = 32.1892 + 0.9861\ Sl - 18574.3732\ w_1$	0.048	6
$FR_1 = 7106.3429 + 459.5537\ A - 12.5554\ Sl$	0.698	6
$FR_1 = -1475.7666 + 746.5422\ A - 0.6313\ w_1$	0.911	6
Bucovina + Carpatho-Ukraine (442—773 mm *Sl*, 2.2—5.7 kg)		
$F = -3480.6726 + 19.4730\ Sl$	0.943**	7
$F = 1906.2314 + 1.7826\ w_1$	0.898**	7
$F = 1643.9186 + 1132.9821\ A$	0.902**	7
$FR_1 = 4453.8137 - 3.3747\ Sl$	−0.363	7
$FR_1 = 3520.1528 - 0.3089\ w_1$	−0.511	7
$FR_1 = 3565.5584 - 196.3221\ A$	−0.289	7
$F = -4814.1615 + 25.9171\ Sl - 70.9464\ w_1$	0.824	7
$F = 2415.8016 + 1016.1343\ A - 0.0414\ Sl$	0.904	7
$F = -1713.2362 + 311.8052\ A + 13.5385\ w_1$	0.854	7
$FR_1 = 439.1255 + 8.8004\ Sl - 0.9424\ w_1$	0.611	7
$FR_1 = 3261.5576 + 40.6569\ A - 1.8559\ Sl$	0.140	7
$FR_1 = 2809.7113 + 169.4715\ A - 0.3969\ w_1$	0.407	7

[1] *R* valid for multiple regressions.
Explanations as for Tables 9 and 10. *A* — age (years).

quantifying the so-called "obtained fecundity" of the huchen. In all cases, huchen from Slovak waters, mainly from the Orava river system, are involved. We treated in a similar way the data available on the total fecundity of the huchen from Soviet Bucovina and the Transcarpathian region of the Ukraine (Table 11), which, however, provide only an approximative picture of the character of correlation between the length, weight and age of the fish and the absolute as well as the relative fecundity.

All data available on the fecundity of the taimen pertain to absolute fecundity. Pravdin (1949) states that a female taimen produces 10–34 thousand eggs, but does

TABLE 12

Comparison of total theoretical (I) absolute and (II) relative fecundity of taimen from the Angara river and of huchen from the streams of Bucovina and Carpatho-Ukraine, based on equations given in Table 11

		Taimen		Huchen	
		I	II	I	II
Sl (mm)	700	8012	2425	10150	2092
	750	11094	2211	11124	1923
	800	14175	1998	12098	1754
w (kg)	3	6294	2544	7254	2593
	4	8182	2413	9036	2285
	5	10070	2282	10819	1976
	6	11957	2151	12601	1667
Age (years)	4	?	933	6176	2780
	6	3852	1238	8442	2388
	8	8255	1544	10708	1995

not mention the size of the female. According to Soldatov (1915, from Nikol'skii 1956 and Bukirev 1967), the fecundity of the taimen in the Amur river is 10,316–33,343 (avg. 21,763) eggs, and in the Enisei river 8–22 thousand eggs; however, this author does not mention the respective sizes of the females either. Wu (1979) also states the fecundity of female taimen from the Chinese tributaries of the Amur river to be 10–30 thousand eggs. In five female taimen from the Angara river (of unstated size), Misharin and Shutilo (1971) found 15,900–35,000 eggs, the relative fecundity being 1230–1700 eggs per kilogram body weight. In four female taimen from the Vilyui river (of unstated size), Kirillov (1962) found 7440–19,700 eggs, and Bukirev (1967) found, in two females from the Kama river, weighing 7.5 and 9.5 kg, 14,000 and 19,700 eggs, respectively. Finally, Olifer (1977) reports, for female taimen from the Angara river, 4.6–12.7 kg in body weight and 78–105 cm in fork length, 6372–23,600 eggs. We used his data, though not numerous, for regression and correlation analyses to obtain at least an approximate idea of the nature of correlations characterizing the fecundity of the subspecies (Table 11). A direct comparison with the nominate form is not quite possible, since the sizes of the fish examined were different in the two cases. However, both the absolute and the relative fecundity of the two forms appears to be practically identical for the overlapping values (i.e. the length group of 700–800 mm *Sl* and the weight group of 4–6 kg, respectively), which fact, too, indicates their specific identity (Table 12).

According to observations made by Skácel, the "obtained" (and thus certainly the absolute) fecundity is considerably limited by age. The highest relative fecundity is usually found in females 6–10 kg in weight, being up to 20% higher than in other weight groups. The data summarized in Table 8 show, however, that the relative fecundity decreases distinctly with increasing size (and age) of the fish. Female huchen do not lose fecundity at even quite a high age and large size. Therefore females 20 kg in weight are still used for artificial reproduction. Thus the belief of Jagoditsch (1930) that huchen females over 12 kg in weight are sterile is erroneous.

EGG CHARACTERISTICS AND GONAD WEIGHT

Several rather different data are available on the size of the eggs of *H. hucho.* Vlasova (1959) states the diameter of huchen eggs as 2–5 mm; Svatoň (1966) as 4.8–5.7 mm; Bastl and Holčík (1968) as 3.9 mm; Peňáz and Příhoda (1981) as 3.95 mm. Disregarding differences caused by preservation (as is the case of the data by Vlasova, Bastl and Holčík), the egg size is positively influenced by the size of the female. The dimensions measured by Nieslanik (Table 8) show that the diameter of huchen eggs varies between 3.6 and 6.0 mm (averaging 4.6 mm); smaller eggs come from smaller fish and vice versa, the positive correlation between body length or body weight and egg diameter being statistically highly significant (Table 9). At the same time, the egg size increases progressively with increasing volume of the ovary. From this point of view one must also consider the egg diameter in the taimen which Kifa (1974) and Korovina (1978) state to be 5.3–5.9 mm and which apparently relates to eggs of bigger fish.

The size of the eggs increases after activation and contact with water. According to observations made by Nieslanik, the eggs increase in size within one or two hours after activation and addition of water; their diameter increases by 10–15% and their volume by 30–50%. This process was studied in detail by Peňáz and Příhoda (1981). According to them, the process of swelling begins about 30 minutes after activation and proceeds intensively for about two hours. Thereafter the rate of increase in the egg diameter slows down, but continues for another three days, whereupon the egg size becomes stabilized. Peňáz and Příhoda (l.c.) found that the egg diameter first increased by 9.4% and then by 15% as compared with that of unfertilized eggs; the volume of the eggs increased by 53%. According to these authors, a one-litre vessel will hold 31,000 unswollen and 12,500 swollen huchen eggs.

Vlasova (l.c.) states the weight of huchen eggs as 10–20 mg. Bastl and Holčík (l.c.) weighed eggs of a huchen (from the Orava reservoir) that had been preserved for several years and found that their weight averaged 40 mg. According to Kifa (1974), the weight of taimen eggs varies between 83 and 93 Mg. The considerable differences between the values given above are due both to different sizes of the fish and to preservation.

Live, unfertilized eggs of the huchen and the taimen are orange-red, amber, or sometimes even red (Sabeneev 1911, Peňáz and Příhoda l.c., Nieslanik) and, like those of *Oncorhynchus, Salmo, Salvelinus and Brachymystax* spp., which have the same type of embryogeny as the salmon-lenok group of fishes (Soin 1980), they are slightly sticky, oligoplasmatic, with a large amount of yolk and a low protoplasm

TABLE 13

Gonad weight and maturity coefficient of huchen from Slovak hatcheries in relation to size and weight

	Sl (mm)	w_l (g)	wg (g)	% wg
$\bar{x}$	819.93	8997.11	1409.61	15.53
s	124.32	4155.78	1006.01	7.38
$s_{\bar{x}}$	23.49	785.37	190.12	1.39
$c.v.$	15.16	46.19	71.37	47.52
min. — max.	573 — 1016	3000 — 16245	253 — 4010	2.15 — 35.06

$wg = -5225.1477 + 8.0919\ Sl$ $r = -0.640^{**}$
$wg = -768.3754 + 0.2421\ w_l$ $r = -0.641^{**}$
$\%\ wg = -33.1461 + 0.0594\ Sl$ $r = 0.022$
$\%\ wg = -0.4496 + 0.0018\ w_l$ $r = 0.008$
$n = 28$

w_l — body weight before stripping, wg — gonad weight, % wg — maturity coefficient.

concentration (Pl. 26.1). The yolk consists of liquid, homogeneous and transparent ichthulin lacking yolk granules but containing a quite large number of fat droplets 10–180 μm in diameter, which are uniformly distributed over the yolk surface close beneath the yolk membrane. The egg shells are very tough and quite transparent in live eggs. The stickiness of the egg shell is very slight and disappears with rinsing in water (Peňáz and Příhoda 1981). The perivitelline space is small, which increases the specific gravity of the eggs so that they rapidly sink to the bottom when discharged from the female's body. A high content of carotenoids (beta-carotene, tunaxanthin, lutein, taraxanthin, astaxanthin; Czeczuga 1975) and the large amount of yolk are adaptations to development under poor oxygen conditions.

Compared to the eggs of *Salmo trutta* m. *fario* and the remaining species of the subfamily Salmoninae, the eggs of *Hucho hucho* are somewhat smaller. On the other hand, the intensity of swelling is greater in the eggs of *H. hucho* (Peňáz and Příhoda l.c.).

Very few data are available on the weight of ovaries. According to Vlasova (1959), the weight of ovaries of three huchen from the Transcarpathian region of the Ukraine, varied from 22 to 262 g, i.e. from 0.73 to 4.61% of total body weight. It follows from the values obtained by Nieslanik that in sexually mature female huchen 573–1016 mm in body length and 3–16.3 kg in prespawning weight, the gonad weight amounts to around 250–4010 g and the coefficient of maturity (gonadosomic ratio), i.e. $100 \times \frac{\text{gonad wt}}{\text{body wt}}$, varies within the range 2.15–35.06. While the gonad weight increases significantly with increasing body length and weight, the maturity coefficient shows no changes in this respect (Table 13).

According to Nieslanik, the weight of male huchen testes (including sperm-ducts) in full ripeness varies from 2.5 to 3.5% of their live weight. According to Harsányi (1982), a pond-reared male 9 kg in body weight yielded at stripping time 6 ml milt repeatedly (but the author does not mention how many times the male was used within one season). According to the same author, 1 mm^3 of milt contains around 12 million spermatozoa. The dimensions of the spermatozoa (in μm) are the following: head 3.84 × 3.10; tail 36.7.

SPAWNING

Spawning Migrations

When spawning time comes, both the huchen and taimen leave their usual habitats and migrate into smaller and shallower tributaries where the water is 0.3–1.2 m deep (Ivaška 1951, Pravdin 1949, Bukirev 1967, Vokač 1959). On migrations to their spawning grounds, they overcome quite high obstacles – up to 150 cm according to the observations of Ivaška (1951). Naturally, *H. hucho* cannot overcome large waterfalls and weirs and its ability in this respect is probably inferior to that of other salmonids. Referring to observations of a Buryat hunter, Kozák (1972) writes that in the Snezhnaya river (a southern tributary of Lake Baikal), taimen do not occur because they are unable to overcome a waterfall several tens of metres high, above the mouth of that river although grayling *(Thymallus baicalensis)* can easily do so (!) on their annual spawning migrations. The erection of a dam on the Orava river in 1956 actually interrupted the spawning migrations of huchen from the stretch below the Orava dam and caused the migrating fish to stop and accumulate in the stilling basin below the dam where over 100 adult huchen, 3–16 kg in weight, and an additional 20–30 smaller ones were caught (Nieslanik). The decrease in the numbers of huchen in the Turiec river is due, among other things, to the erection of two high weirs on its lower reaches, blocking the upstream migration of huchen from the Váh river (Köttl and Kafka 1927, Novák 1971). Ivaška (1951), too, complains that weirs impede the progress of huchen migrating from the Váh into the Orava river.

The spawning grounds are usually situated on the boundary between the foothill river and foothill brook zones (formerly the grayling and trout zones), but spawning may also take place directly in a bigger river. Before the canalization of the Váh river, for example, spawning grounds were found in the whole stretch of the river between Žilina and Kráľova Lehota, especially in the environs of Strečno, Sielnička, Teplá, near the Ľupča bridge, around Bešeňová, below Kráľova Lehota (Kuna in litt., Nieslanik). In the Turiec itself, the major spawning grounds are situated in the main stream: near Košťany, in the stretch between Košťany and Príbovce, and on the confluence of the Vríca with the Turiec river (Svatoň 1966). Similar situations are also found elsewhere, as reported by Vokač (1959) and Ivaška (1951). Some authors state, however, that in minor streams the huchen may even migrate to spawn downstream (Vladykov 1926, Shnarevich and Moshuk 1956), which can obviously be ascribed to the absence of suitable tributaries or to a lack of satisfactory substrate and other conditions necessary for optimal reproduction.

As a rule, however, both huchen and taimen spawn in small streams (Pl. 9.1). This was the case in the Orava river where the spawning grounds were known to exist in the Zázrivka, Studený potok, Oravica, Polhoranka, Hruštínka and Mutnianka streams. In the Váh river basin, the spawning grounds were known to exist in the lower section of the Kvačianka, Ľupčianka and Belá streams. The relatively small dimensions of these streams are indicated by the following data: in 1961 we observed huchen spawning in one of the side branches of the Studený potok stream above Podbiel; it was about 2 m wide and 40–60 cm deep. One of the spawning grounds of the Váh huchen is in the Jánošíkovský potok stream, which is 2–3 m wide and around 40 cm deep. Spawning huchen are often only half-submerged, as can be seen in the photograph taken in the Pielach, Austria, and published by Jungwirth (1977a). A similar situation is found in the taimen which spawn in mountain streams or in tributaries where the hydrological regime resembles that of mountain streams (Kirillov 1972). According to Olifer (1977), taimen migrate to spawn in tributaries (of the Angara river) of the first and second order.

Very little is known of the length of the spawning migrations. Sabaneev (1911) thought that taimen migrated to spawn at distances of whole hundreds of versts. Ivaška (1946) also assumed long migrations to take place, as he wrote that huchen migrated from the Danube into the Váh river up to Púchov where they were stopped by the local weir – a distance of about 190 km. In fact, however, the migrations are considerably shorter, as ascertained by Nieslanik, who himself followed the migration of some pairs. According to his observations, the length of a spawning migration does not exceed 10–25 km, depending on the age and size of the migrating fish. Big huchen, around 15–20 kg in weight, undertake only short migrations and mostly spawn in the main stream. According to Vokač (1959), the spawning migration in the Drava river is often very short – from one river bank to the other. The length of the migration may be greatly affected by the water level. Thus in the Kysuca river, exceptionally high water levels occurred in 1957, enabling huchen 5–7 kg in body weight to migrate up the Čierňanka stream to the Polish territory, thus covering a distance of 30–40 km (Nieslanik).

From the fact that the spawning migrations of the huchen are regularly repeated every year and that the fish occur in identical places, one may conclude that, like other species, the huchen have a homing instinct, compelling them to return to the places where they were born. This is indicated both by observations made by Vladykov (1926) in the streams of the Transcarpathian region of the Ukraine, and by our own in the Orava river, but exact data based on tagging are still lacking. It should, however, be noted in this connection that in the spring of 1980, immediately after the cessation of the efflux of waste water from the paper mill at Martin, which had prevented fish from inhabiting the lower reaches of the Turiec river for several decades, huchen occurred in it, having migrated there from the Váh river. We therefore assume that either the homing instinct is genetically fixed and persists throughout many generations, or part of the population lacks this instinct and selects spawning tributaries at random.

Spawning Site

The selection of a spawning site is apparently governed by the suitability of the spawning substrate on the one hand (according to observations of various authors, preferably a clean, i.e. not muddy but gravelly, sandy or rocky and gravelly bottom) and, by hydrological conditions on the other. The size of the fish certainly also plays a part. The spawning sites are usually situated at the boundary of water flows, especially at the lower end of a longish pool, in creeks between islets, in the mouths of tributaries, and behind bridge piers (Nieslanik; Bukirev 1967). According to observations of the latter author, the spawning sites of taimen in the Kama river were in water 0.5–1.2 m deep, with stream velocity 0.6–1.0 m s^{-1}. In the Amur river, too, the spawning sites are at depths of 0.5–1.5 m (Kifa 1974). In the Frolikha river, taimen spawn in cascading stretches (Misharin and Shutilo 1971).

The opinion that huchen instinctively select sites that will subsequently be used for spawning by *Chondrostoma nasus,* their chief prey (Haempel 1910, Stölzle 1910, Munda 1935, Prawocheński and Kolder 1968) must be rejected on the basis of the present state of our knowledge, even if it is beyond any doubt that when the spawning sites of the two species are identical it is of considerable importance for a successful development of *H. hucho* young. Pölzl (1905) was the first to notice the conspicuous coincidence of the spawning sites of *H. hucho* and *C. nasus* and to draw the conclusion of the intentional selection of the spawning site by *H. hucho.* According to Pravdin (1949), the spawning sites of taimen are situated in those sections which are also used by *Oncorhynchus keta.* The spawning sites are not invariably situated in identical places, but change according to the water level, as observed by Nieslanik: at low water levels they are situated lower than those of *C. nasus,* whereas at high levels they are above, which is most probably connected with the height of the water column required for undisturbed spawning.

Spawning Redd

According to observations of some authors (Jagoditsch 1930, Hollý 1934, Ivaška 1951, Vokač 1959), a redd prepared by the female in a suitable substrate is an elliptical depression with the longer axis along the stream line. According to the above authors, a redd may be 120–200 cm long, around 60 cm broad and 10–40 cm deep; according to Robida (1902), up to 100 cm deep. However, Nieslanik states that the dimensions mentioned refer to fish of smaller sizes, less than 7 kg in weight. Big fish excavate considerably bigger redds, the length of which varies between 2–3 m and breadth between 1.5 and 2 m. When spawning has ended the lower part of the redd is usually 5–10 cm shallower than its upper part.

The most exact data so far on the redd of the huchen have recently been presented by Schulz and Piery (1982) from the Drava river. They found the following dimensions: maximum length 6.5 m, maximum breadth 2.4 m, area 13 m^2, maximum depth 23 cm, mean depth 8 cm; the substrate was gravelly, the predominant diameter of particles being 16–70 mm, the stream velocity in the spawning site varying between 5 and 70 cm s^{-1}, but in the redd proper less than 10 cm s^{-1}, with a back flow of 0–5 cm s^{-1} in its middle and near its bottom. According to the authors, the female excavating the redd must have removed about 1.2 tonnes (!) of material. Unfortunately, the authors do not mention the size of the female.

Ivaška (l.c.) reported that nine huchen spawned in the Turiec river in a stretch around 30 m long, where they excavated 4–5 redds. He mentioned an observation made by Köttl, who found 30 redds in a section 6 km long in the Orava river. According to Nieslanik, the redds of separate pairs are 10 m or more apart. This prevents mutual disturbance or attacks. This author observed huchen pairs spawning closest to one another in the Oravica river above the bridge at Tvrdošín (seven pairs in a section about 150 m long), in the Mutnianka river, a left tributary of the Orava (five pairs in a section 500–600 m long), above the mouth of the Ľupčianka into the Váh river (four pairs in a section 150 m long) and in the Kvačianka river (five pairs spawning in a section some 300 m long). It may be assumed that the rather long distance between the redds (the above data averaging 70–75 m) is due to a low population density of the species, caused by anthropic factors, and that under normal circumstances a larger number of pairs would spawn in the same space.

Judging from the data by Kifa (1974), the redds of taimen practically do not differ from those of huchen: their dimensions are stated as 1.0–1.5 m, the depth of the layer of eggs in the substrate as 20–30 cm.

Spawning Season, Thermal Regime and Duration of Spawning

Hucho hucho is among the salmonid species which spawn in spring. Various authors correlate the arrival of *H. hucho* in spawning sites with the departure of ice or waters after the thaw (Pravdin 1949, Sigunov 1972, Wu 1973, Olifer 1977, Jungwirth 1977a) although there are records of the fish having appeared in their spawning sites at the time when the stream was still covered with ice (Misharin and Shutilo 1971). Kulmatycki (1931c) mentions observations made by local inhabitants and verified by himself, according to which the spawning time of the huchen coincides with the occurrence of the first alder *(Alnus)* leaves. In central Europe, the spawning time usually falls in April and May; in more southerly regions in February (e.g., in the streams of the Transcarpathian region of the Ukraine spawning takes place in March to May; Komárek 1955, Vlasova 1959); in the Krka, Drava and Ljubljanica rivers, Yugoslavia, it takes place in February and March (Munda 1935). The spawning of the taimen, a form living in the far north, is shifted to later spring and summer months. Thus in the Lena river taimen spawn in May–July (Kirillov 1972), in the Vilyui river usually in early June (Kirillov 1962); in the Aldan river in May and June (Pirozhnikov 1955); in the Angara and Frolikha rivers in May and June (Misharin and Shutilo 1971, Olifer 1977). In the same months taimen also spawn in the Kama river (Bukirev 1967). Only in the Amur river and in the streams of the Altai and northern Ural Mts. does spawning take place in April and May (Sabaneev 1911, Nikol'skii 1956).

The time at which spawning takes place is mainly determined by temperature, but also by the weather in general. Jungwirth (1977a) notes that spawning is short and spontaneous in warm and sunny weather and at water temperatures around 10° C. Waters from thawing snows can delay the spawning by as much as several weeks. If the winter is mild and the temperatures are relatively high in February and March, spawning in Austrian streams begins as early as mid-March. After a long and cold winter, or if the early spring is cold and rainy, spawning is shifted to late April, or even to early May. Observations made by Nieslanik show that in Slovak waters huchen begin to spawn fairly regularly 5–6 weeks after the departure of ice. In 1953, huchen spawned in the main stream of the Váh river from 28 March to 16 April; in 1954, from 15 April to 9 May; and in 1955, from 27 March to 10 May. Nieslanik also reports that a long winter may delay spawning till the first half of May, whereas a warm and mild winter may accelerate it, so that it takes place in late March.

Both huchen and taimen arrive at their spawning sites at the time when the water temperature has risen to 6–10° C and sometimes even before that. Butkovský (1930) and Novák (1932) recorded a spawning migration of huchen from the Orava into the Mutnianka rivers at water temperatures of 4.6 and 5.0° C, respectively. In the Frolikha river in Siberia, the first, still unripe taimen appeared in the spawning sites at only 3° C of water temperature (Misharin and Shutilo 1971). During the spawning time itself, the water temperature in the spawning sites varies within a comparatively wide range. This is clearly seen in Table 14, compiled from the data gathered by Kulmatycki (1931c) from the Cheremosh river. Similarly, the spawning of huchen in the streams of the Transcarpathian region of the Ukraine takes place at water temperatures ranging from 6 to 8° C (Vlasova 1959), in Austria at 10° C (Jungwirth 1977a). The spawning of taimen takes place at similar water temperatures: in the Chinese tributaries of the Amur river, at 5–10° C (Wu 1979), in Transbaikalia at 6–8° C (Misharin and Shutilo 1971), and in the Angara river at 6–12° C (Olifer 1977). According to Nieslanik, the thermal regime during the spawning time of huchen is characterized by considerable variation – it often happens that at night the air temperature drops below 5° C, rising above 15° C during the day.

The duration of the spawning time is different, depending primarily on the water temperature. If the water temperature is relatively constant, spawning is spontaneous and short (Jungwirth 1977a). According to Nieslanik, the total duration of spawning in a stream section lasts 10–14 days. Kulmatycki (1931c) and Svatoň (1966) mention only 2–3 days, but their data are probably erroneous and pertain to the spawning of only one pair, as will be seen below. In unfavourable weather, spawning may be prolonged up to three weeks.

Spawning is intermittent, since a female may lay eggs for 2–5 days, as observed by Nieslanik. Thus the eggs are released in small portions. Portional spawning is also mentioned by Vlasova (1959). This, however, is not the true portional spawning, caused by asynchronous ripening of ova, as it is known to occur in other (especially

TABLE 14

Water temperatures (°C) at spawning time of huchen in the Cheremosh river system in 1930 (from Kulmatycki 1931c)

Stream	Date	°C				
		$\bar{x}$	s	$s_{\bar{x}}$	c.v.	min.—max.
Probina	15.4.—5.5.	5.44	1.08	0.24	19.94	3.6—7.4
Bereznica	19.—26.4.	8.38	1.68	0.60	20.08	6.6—11.3
Bystrzec	24.—30.4.	6.58	0.66	0.27	10.06	5.9—7.3
Iltsia	24.4.—4.5.	10.40	3.54	1.44	33.99	6.4—14.4
Cheremosh river system	15.4.—5.5.	6.94	2.47	0.39	35.54	3.6—14.4

cyprinid) fishes, but a successive discharging of ripe gametes. In fact, the females ovulate during a rather long period of time (Peňáz and Příhoda 1981).

The spawning act itself takes place mostly at night (Ivaška 1951, Bukirev 1967); only Nieslanik observed the most intensive spawning in afternoon hours, when the water was warmest, particularly on sunny days.

Prespawning and Spawning Behaviour

Having shown territorial behaviour for most of the year, before spawning time *H. hucho* begin to aggregate into bigger shoals and successively migrate to their spawning sites. Smaller fish are the first to depart, followed by bigger ones. Thus at the spawning sites in the Frolikha river, first the young, still unripe taimen weighing 0.4–3.5 kg appeared in early May, followed by females 6–28 kg in weight later in the month (Misharin and Shutilo 1971). Males apparently are the first to arrive in the spawning sites, females coming later. The arrival in the spawning sites takes some time, 3–7 days according to Nieslanik. Bigger fish are the first to spawn, followed by those which are spawning for the first time (Nieslanik, Bukirev 1967). The aggregation of huchen in the prespawning period is also observed in ponds.

So far, a detailed description of the course of natural spawning is lacking, which is natural, since the spawning sites are usually situated well away from river banks (Kirillov 1972). Thus most observations of spawning behaviour refer to captive fish; it does not seem to differ much from that occurring in the wild. On this point the reports of huchen breeders, particularry Ivaška (1951) and Nieslanik, are available.

According to Ivaška, sexually mature huchen begin to form groups of 2–3 individuals. Several days later, when the fish show increasing restlessness, a male and a female separate themselves in each group (Pl. 9.2), whereas the surplus males start fighting one another. The sign or signs deciding the selection of a mate are unknown.

Režný (1951) states that about two weeks before spawning time a male selects a female and then "guards" her incessantly; on the other hand, according to Ivaška it is the female that selects a particular male and then drives away other suitors. According to Nieslanik, a male which succeeds in overcoming a weaker rival will then stay with the female. The duels of males are very intensive and last for several days; the rivals often inflict serious injuries upon each other by biting; taimen exhibit similar behaviour, as observed by Sigunov (1972). The injured parts include the dorsal and the caudal fins but also the flanks and even the head (Pl. 10.2 and 11.3). The victorious male then joins the female, which has been staying at some distance, and they migrate to the spawning site together. Typically, the female at this time gapes widely from time to time as if yawning. There is no correlation between the sizes of the female and her mate – it often happens that a female weighing 15–20 kg will spawn with a male weighing 3.5–5 kg, and vice versa (Studený potok, our observation in April 1961).

The pair stops in a selected site and prepares the spawning redd. The female periodically descends to the bottom, using the rear part of her body to examine by touching, and to select a place suitable for the purpose, which takes her 2–4 days as a rule. It often happens that the female does not stay in one place but changes sites to select a more suitable one. Big fish often select a site near a spacious cover or at a greater depth. In that period the water level in the spawning sites is lower during morning hours and rises in the course of the day, the water sometimes becoming somewhat turbid from thawing snow. All this compels the female to change sites, and it sometimes happens that she will release eggs in several places according to the water level.

In the selected site, the pair then starts excavating the redd. The female is the more agile, but is assisted by her mate from time to time. They excavate the redd by violent strokes of their tails and caudal fins. Ivaška (1951) states that the female also uses her mouth. The excavation is so intensive that the female's ventral part is often covered with surface lesions; scratches are frequent especially in the regions of the anal and the caudal fin. In excavating the redd, the fish remove not only the surface impurities, sand and coarse gravel, but even bigger stones up to the size a human head (Vokač 1959). The amount of material removed from the redd is considerable (pp. 73–74).

At that time, the female shows increasing aggressiveness, actively attacking and driving away, together with her mate, other males which approach the site. If another pair prepares to spawn nearby, fighting may even occur. Most frequently this happens in ponds where several pairs attempt to spawn naturally in a limited space. The intensity of the mutual attacks may be so great as to cause serious injuries and even the death of the adversaries. This happened in a farm of the Slovak Anglers' Union at Dubná Skala where a male (15 kg) and a female (18 kg) of two different pairs attacked each other and shortly thereafter succumbed to wounds caused by their bites.

After several days of preparations, spawning proper takes place. It proceeds at short intervals. The female arches her body over the redd, thrusting her abdomen downwards. She opens her mouth in a wide and counvulsive gape and releases eggs, vibrating her caudal fin intermittently. At the same moment the male, pressing the rear part of his body to the female, releases the sperm, which is visible in the form of a whitish cloudlet. The male sometimes covers the rear part of the female from above or from one side. During the spawning act the pair slowly moves forward, the female moving her tail so as to cover the fertilized eggs with sand and pebbles.

The intervals between successive spawning bouts are longer during morning hours; during afternoon hours they become shorter, apparently even due to the effect of rising water temperature, so that they last only several minutes. During the intervals the fish quieten down; they stand side by side, sometimes moving downstream and returning shortly thereafter. At that time the male often leaves his position next to the female, circling round her and energetically chasing away any intruding males. According to Skácel's observations of huchen spawning in the Orava river in 1981, the female withdrew during the nuptial play about 2 m away from the redd at most, but the male withdrew to a distance up to 10 m, most frequently when driving away grayling that spawned nearby. Attacking them, the male sallied forth so violently that it sometimes got up to the offshore shallows and one third of his body emerged from the water. The female pays no heed to the approach of other males or other fishes. Both sexes become less shy, so that their spawning may be easily observed by a person carefully approaching the water, provided that the stream is not too broad. The fish often are as if stunned (Kulmatycki 1931b) and sometimes allow one to approach almost to touching distance (our observations in the Studený potok Brook in April 1961). Ivaška (1951) and Režný (1951) even describe a case in which a robust male weighing 20 kg seized in his mouth a pole with which a keeper tried to drive apart the fish aggregated in a pond, and shook it so violently that the keeper almost lost hold of it. Skácel found that in nature the spawning pair invariably returned to the redd after being driven away several times. The intervals at which they did so were of varying duration, the longest lasting 30 minutes. It is not suprising that at that time huchen fall an easy prey to poachers, a fact which several earlier authors complained (e.g., Köttl and Kafka 1927, Kulmatycki 1931).

According to Nieslanik, the fish usually calm down with approaching evening and do not continue spawning until about 9 a.m. the next morning as a rule. Spawning at night was observed only exceptionally, when the weather was very warm and the water temperature rather high.

Vokač (1959) asserts that a female will spawn with several males, but neither Ivaška nor Nieslanik nor any other observers made any such observation.

After having spawned, the two fish do not part immediately but stay together for up to two weeks. First they stay near the redd, later the partners start to return slowly to their usual summer habitats. As a rule the fish drift downstream, their tails first, side by side, often touching each other, thus moving from one pool to another. They may stay up to several hours in each pool, the male carefully examining the surroundings, in particular the lower, shallower part of the pool and looking for a place offering a quick retreat. This is invariably made head first. Nieslanik, on whose records this section is mostly based, states that he observed different pairs of huchen behaving in this way for two or three weeks, within which period they descended 5–10 km downstream. In the main stream or near their permanent habitats, the pairs then part definitely and occupy their original habitats or select new ones.

At spawning time, food intake is decreased, especially in females, but does not cease completely (Ivaška 1951, Bukirev 1967).

Insemination

As in most fishes, insemination is external and occurs at the time of synchronous release of eggs and sperm in water. Apparently the process of insemination is short; Haempel (from Wunder 1936) and Pavlík (1957) observed that the spermatozoa retain their motility in water for only 30–45 seconds, but according to Harsányi (1982), individual spermatozoa retain motility for as long as 120 seconds at 10° C.

DEVELOPMENT

The incubation period from activation to hatching depends on water temperature, which is why the values reported by different authors are considerably different. Ivaška states 167–190 temperature units (*TU*; day-degrees) from activation to the so-called "eyed egg" stage at temperatures of 4–11° C, or 203–232 *TU* from activation to hatching. Prawocheński and Kolder (1968) state 228.4 *TU* at 6–7° C up to the eyed egg stage, or 294.1 *TU* up to hatching. According to Nieslanik's observations at Jablunkov, the number of *TU* from activation to the eyed egg stage was 176.7 at water temperatures of 6–7° C; hatching began after 278.9 *TU* and ended at 300.1 *TU*. The yolk sacs were absorbed at 485.6 *TU*. According to observations made by Skácel, the development from activation to the eyed egg stage takes 270 *TU*, up to the end of hatching at water temperatures ranging from 4 to 14° C 290 *TU*. According to Peňáz and Příhoda (1981), the duration of the incubation period for rearing at water temperatures ranging from 4.8 to 5.5° C is as follows (in *TU*): onset of hatching, 266; mass hatching, 301; end of hatching, 331. Hatching lasted 4 days, beginning on the 31st and ending on the 34th day of development. According to the above authors, the whole embryonic period lasts about 45 days, i.e. 457 *TU*. The most exact data on the length of incubation, however, have recently been published by Jungwirth and Winkler (1984). On the basis of their experiments, they concluded that the concept of Q_{10} or day-degrees is at least questionable and that the dependence of the length of incubation on temperature is described by the equation $y = 2647.4/(x + 3.222)^{1.7865}$. At temperatures 4, 6, 8, 10, 12 and 14° C the respective lengths of incubation are 77.4, 50.0, 35.2, 26.3, 20.4 and 16.4 days.

Similar relations apparently exist in the taimen, as Misharin and Shutilo (1971) state the duration of the incubation period as 30–35 days (not mentioning temperature) and according to Kifa (1974), incubation of eggs of the taimen from the Amur river lasts 28–38 days at temperatures 5–7° C, i.e. approximately 190–196 *TU.*

Considering the embryonic period (i.e. from the moment of activation to the beginning of mixed nutrition), there are marked differences between the genera *Brachymystax, Hucho* and *Salvelinus,* among other things, in its duration. The data provided by Smolyanov (1961), Balon (1980b) and Peňáz and Příhoda (1981) and summarized in Table 15 show that *Hucho hucho* is closer to *Brachymystax* than to *Salvelinus* (except *S. malma*) where the duration of the embryonic period is markedly longer.

TABLE 15

Comparison of *TU* values at the end of successive developmental steps of the embryonic period (beginning with activation) in the lenok, huchen and several *Salvelinus* spp. (compiled from Smol'yanov 1961, Balon 1980b, Peňáz and Příhoda 1981)

Step	*B. lenok*	*H. hucho*	*Salvelinus*				
			namaycush	*fontinalis*	*alpinus alpinus*	*alpinus oquassa*	*malma*
1	2	3	5	6	?	8	8
2	7	15	23	33	?	22	16
3	19	33	92	100	66	60	90
4	54	53	134	161	?	146	?
5	?	99	165	?	?	193	?
6	?	135	232	?	?	313	?
7	?	189	294	?	?	409	?
8	150	229	326	444	406	451	390
9	195	288	493	510	?	563	?
10	280	339	650	696	576	653	550

The early development of this species is known in the nominate form only. In preserved material and therefore in a rough outline only, it was studied by Schindler (1935) and Schnakenbeck (1940). Balon (1956) studied the development of more advanced stages of the embryonic period, and only Witkowski and Kokurewicz (1981) and Peňáz and Příhoda (1981) succeeded in studying the development of the huchen starting from activation. The former authors, however, give only a very brief description of development, not respecting its stepwise character, nor the terminology used in recent papers of this kind. For this reason, with respect to the possibility and necessity to compare the development of *H. hucho* with that of the related genera *Brachymystax* and *Salvelinus,* we have adopted here the results obtained by Peňáz and Příhoda (1981).

In the development of the huchen, Peňáz and Příhoda distinguish 11 embryonic and two larval steps, or stanzas; after the end of the 2nd larval step, the young pass into a juvenile period of life (Table 16).[8] Some of the developmental phases are illustrated in Fig. 23.

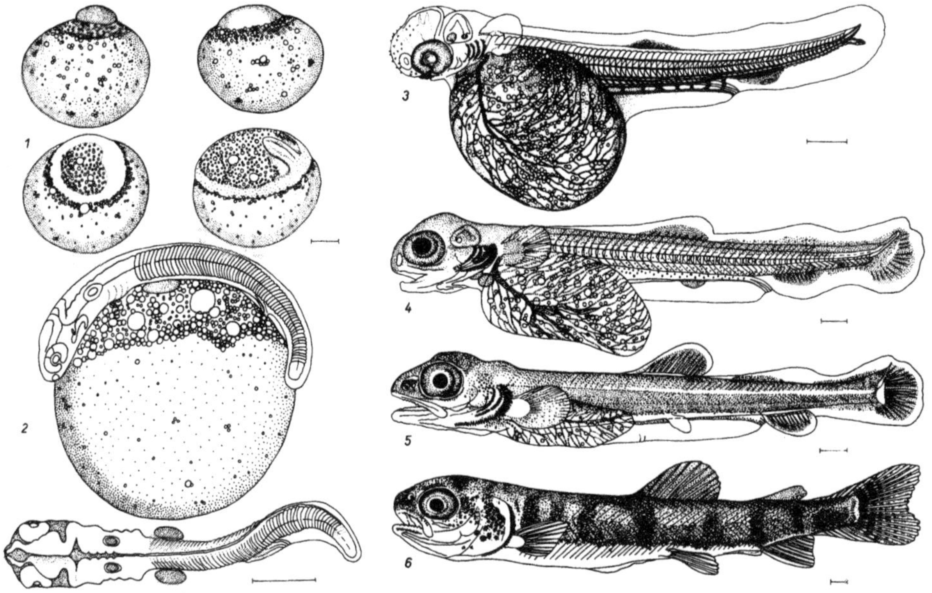

Fig. 23. Some early developmental steps of the huchen. 1 (from left to right and from top to bottom) – blastula (age 2 days and 17 hours); epithelial blastula (4 days and 8 h); early gastrula (5 days and 5 h); neurulation (7 days and 5 h); 2 – embryo at the age of 13 days and 19 h, total length 6.1 mm; 3 – embryo at the age of 33 days, immediately after hatching (10th embryonic step, total length 16.1 mm); 4 – embryo at the age of 35 days (11th embryonic step or F^1g eleutheroembryonic step, total length 18.3 mm); 5 – alevin at the age of 46 days (1st larval or 1st alevin step, total length 21.9 mm); 6 – fingerling at the age of 96 days (juvenile period, total length 33.0 mm) (after Peňáz and Příhoda 1981).

Compared to the developmental steps of other genera of the subfamily Salmoninae (Pavlov 1979, 1980, Balon 1980b), the embryos of *H. hucho* show some marked differences. Thus the periblastic sinus develops much later (as late as the 5th step) in *H. hucho,* in which point this species resembles *Brachymystax lenok* and differs from other salmonids, as pointed out by Witkowski and Kokurewicz (1981). The number of somites is significantly higher in *H. hucho* (72, compared to 54–59 in the genus *Salmo* and 60–62 in *Oncorhynchus*) and pigmentation begins much later than in other salmonid genera. For example, newly hatched embryos of *H. hucho* are not pigmented, the first melanophores occurring as late as the 10th embryonic step, approximately on the 34th–36th day after activation, when the young have attained the length of 15–16 mm. Melanin is not deposited at the margins of eyes until the end of the 8th embryonic step, when the embryos are around 10 mm long, and the

[8] Balon (1975b) denies the presence of a larval period in salmonids (replacing it by the term "alevin period"), since he does not find here the presence of typical larval structures. Moreover, at the end of the embryonic period, he recognizes an eleutheroembryonic phase which includes the interval between the moment of hatching and change to mixed nutrition. Similarly, at the beginning of development he discerns a cleavage phase which lasts from the moment of activation to the beginning of organogenesis.

TABLE 16

Periods and steps of early ontogeny in the huchen (adapted from Peňáz and Příhoda 1981). 1 — steps indicated by authors; 2 — comparable steps according to Balon (1985), C — cleavage phase; E — embryonic phase; F — eleutheroembryonic (free embryo) phase

Step 1	Step 2	Age from activation (h — hours, d — days)	*TU* (degree-days)	*Tl* (mm)	Characteristics
					Embryonic period
1	$C^{1}1$	0—10 h	0—2.5		Formation of periviteline space; bipolar differentiation of cytoplasm; early development of blastodisc
2	$C^{2}2$	10—63 h	2.5—15.3		Cleavage; formation of morula
3	$C^{3}3$	63—123 h	15.3—33.3		Blastula; onset of protoplasmic motion
4		123—190 h	33.3—52.8		Gastrula; onset of epiboly; formation of term ring and embryonic shield
5	$E^{1}4$	9—13 d	52.8—99.3		End of epiboly; brain, optic and auditory vesicles and tubular heart formation; separation of tail end; formation of first somites
6	$E^{2}5$	13—17 d	99.3—134.5	5.4— 7.5	Separation of tail; heart pulsation; onset of pectoral fins and finfold formation; tail and trunk movements
7	$E^{3}6$	17—23 d	134.5—188.6	7.5— 9.3	Onset of subintestinal-vitelline blood circulation; separation of head
8	$E^{4}7$	23—27 d	188.6—299.1	9.3—10.4	End of segmentation; hepatic-vitelline yolk respiratory system; first movements of pectoral fins; onset of eyes formation
9	$E^{5}8$	27—32 d	229.1—287.6	10.4—13.4	Differentiation of embryonic finfold; perforation of mouth orifice; complete vascularization of yolk-sac surface; formation of hatching glands; onset of hatching
10	$F^{1}9$	32—36 d	287.6—339.1	13.4—16.2	Free embryo; onset of branchial respiration; lower jaw movements; notochord curved up; beginning of embryonic finfold reduction and separation of dorsal fin; pigmentation of upper part of head and dorsum
11	$F^{2}10$	36—45 d	339.1—457.1	16.2—21.4	Reduction of embryonic finfold; start of calcification of skeletal structures and jaw teeth points; rapid yolk depletion

			Alevin period		
1	$A^1$11	45—53 d	457—545	21.0—24.0	Mixed nutrition; separation of unpaired fins; reduced yolk-sac covered by somites; calcification of oral cavity bones, opercle, pectoral girdle, branchiostegal rays and lepidotrichiae of caudal and anal fins
2	$A^2$12	54—90 d	545—	23.2—30.0	Complete resorption of yolk-sac; efficient exogenous feeding; gas-bladder filled; calcification of axial skeleton; increase in pigmentation
			Juvenile period		
1		90— d		30.0—	Parr marks; complete reduction of finfold vestiges; formation of first scales

eyes are not fully pigmented (the so-called eye-spot stage) until the embryos are 13 mm long, during the 9th embryonic step. The transverse juvenile marks, called the "parr marks", typical of all young Salmoninae, do not occur in young *H. hucho* until the beginning of the juvenile period, whereas young trout, salmon and charr already exhibit these marks at the end of the embryonic or at the beginning of the larval period. The development of scales in the huchen has certain differences, too. According to Witkowski et al. (1984), it also belongs to the oral-caudal type but the first scales do not occur until a length of 42.9 mm *Tl* and in the region below the end of the adipose fin, whereas in other Salmoninae, origins of scales occur at smaller body sizes and in the region between the dorsal and the adipose fin. On the whole the young of *H. hucho* appear to be less advanced at the end of the embryonic period than the young of *Salmo, Oncorhynchus* and *Salvelinus* spp. of equal age and are more similar to those of *Brachymystax* in the successive stages.

Further data and some particulars important from the breeder's point of view are given in Chapter 26.

10. Age and Growth

AGE AND SIZE

The greatest age so far ascertained in *H. hucho* is 20 years in the nominate form (Krauss 1933, Schulz 1985) and 55 years in the taimen (Podlesnyi 1958). Considering that the fish were 115 and 144 cm in *Tl* and 19 and 34.8 kg in weight in the cases of the huchen, and approximately 172 cm in *Fl* and 56 kg in the case of the taimen, and that the maximum size this species can attain is much greater, it is obvious that the data mentioned above do not indicate the maximum longevity of the species. On the basis of parameters in a von Bertlanffy equation, calculated from the mean values of linear growth (Table 24), one may estimate that the nominate form may live as long as 60 years (or even 100 years in the past), while the potential longevity of the taimen may be estimated at 100–170 years (Table 27). In both cases the higher longevity value was obtained by introducing in the von Bertalanffy equation the actual maximum body length instead of the calculated average value of asymptotic length (L_∞).

The capability of *H. hucho* to attain gigantic body dimensions has long been known. Mongols even tell a tale of a huge taimen found icebound in a river, from which nomadic tent dwellers cut off pieces of flesh throughout the winter, which saved them from starving to death. Next spring, after the thaw, the taimen swam away without any detriment, as it was so big that the people did no harm to it (Nikol'skii 1956, Sigunov 1972). Table 17 shows that catches of huchen weighing over 20 kg are not unusual. It is worth noting that 73% of all recorded and published catches of huchen weighing 20–40 kg come from the period after World War II.

TABLE 17

Review of catches of record huchen and taimen (20 kg and over in weight). Query (?) indicates derived length

Tl (cm)	*Sl* (cm)	*w* (kg)	Locality and date	Author
			Hucho hucho hucho	
127	114	20.0	Turiec, 1926	Novák (in litt.)
129 ?	116 ?	20.0	Danube near Melk, 9 Feb. 1920	Krall-Kralsberg (1930)
129 ?	116 ?	20.0	Tributary to the Danube near Grein, 1930	Anonymus (1937b)
129 ?	116 ?	20.0	Lech, 1961	Anonymus (1976a)
125 ?	112 ?	20.0	Danube near Weitenegg, 1930	Anonymus (1976a)
129 ?	116 ?	20.0	Isar, 1961	Anonymus (1976a)
129 ?	116 ?	20.0	Lech, 1962	Harsányi (1982)
129 ?	116 ?	20.0	Drava, 1980	Kainz (in litt.)
129 ?	116 ?	20.0	Lech, 1980	Kainz (in litt.)
120	108?	20.1	Danube near Linz, 1932	Anonymus (1932)
132	119	20.1	Drava near Rosegg, Feb. 1959	Adlassnig-Jaser (1959)
129 ?	112 ?	20.2	Inn, 1979	Kainz (in litt.)
129 ?	112 ?	20.3	Inn near Mühldorf, 1978	Kainz (in litt.)
127	114 ?	20.4	Orava near Oravský Podzámok, 1975	Szabo (1976)
120	108 ?	20.5	Inn, 1910	Anonymus (1910)
125	112 ?	20.5	Váh near Bešeňová, 23 Dec. 1981	Lacko (in litt.)
130 ?	113 ?	20.5	Mur near Teufenbach, 1978	Kainz (in litt).
130	117 ?	21.0	Lech, 1964	Anonymus (1976a)
131	118 ?	21.0	Lech near Schongau, 1961	Anonymus (1976a)
150	135 ?	21.0	Danube, 1908	Haempel (1910)
130	117 ?	21.0	Turiec near Martin, 1930	Novák (in litt.)
131 ?	114 ?	21.1	Drava near Greifenstein, 1975	Kainz (in litt).
138	118 ?	21.2	Orava near Krivá, 1955	J. E. (1975)
130	117 ?	21.2	Orava, 1974	Grman (1980)
120	108 ?	21.2	Dunajec, 10 Dec. 1981	Lacko (in litt.)
131	118	21.5	Isar near Lenggries, 1970	tt. (1970)
134 ?	121 ?	22.0	Drina, (?)	Krauss (1933)
134	121 ?	22.0	Váh near Ružomberok, 1946	Grman (1980)
134	121 ?	22.2	Váh, 1961	Grman (1980)
134 ?	116 ?	22.5	Váh, 21 Nov. 1920	Novák (1969)
135	121 ?	22.5	Lech, 1964	Anonymus (1976a)
129	116 ?	23.0	Inn near Wasserburg, Nov. 1932	Anonymus (1933)
135	121 ?	23.5	Lech, 1965	Anonymus (1976a)
128	115 ?	23.8	Turiec, 1977	Novák (in litt.)
128	115 ?	24.0	Drava, 1 Jan. 1973	Anonymus (1973)
136 ?	123	24.0	Confluence of Orava and Váh rivers, 1946	Bako (1946)
137 ?	119 ?	24.0	Turiec near Košťany, 1937	Ivaška (1951)
128 ?	115 ?	24.0	Lech, 1962	Anonymus (1976a)
138 ?	124 ?	24.6	Tisza, 1947	Vlasova (1959)
128	111	25.0	Loisach, 7 Dec. 1969	W. (1970)
138 ?	124 ?	25.0	Lech, 1962	Anonymus (1976a)
130	117	25.0	Drava, Dec. 1973	Anonymus (1973)
138 ?	120 ?	25.0	Mur, 1976	Kainz (in litt.)

TABLE 17

(Continued)

Tl (cm)	*Sl* (cm)	*w* (kg)	Locality and date	Author
139 ?	125 ?	25.5	Enns, 1970	Anonymus (1976a)
136	122 ?	26.0	Drava, 28 Feb. 1965	Hanny (1965)
141 ?	123 ?	26.5	Drava, 1976	Kainz (in litt.)
142 ?	128 ?	27.0	Drava, 1960	Anonymus (1976a)
130	117 ?	27.0	Drava, winter 1973	Anonymus (1973)
143 ?	129 ?	27.5	Drina, (?)	Krauss (1933)
143 ?	129 ?	27.5	Drava, 1974	Anonymus (1976a)
144 ?	130 ?	28.0	Turiec near Košťany, 1916	Novák (1969)
142	128	28.0	Drava, 1969	Anonymus (1976a)
144 ?	130 ?	28.2	Drava, 1969	Anonymus (1976a)
144 ?	125 ?	28.3	Drava, 1977	Kainz (in litt.)
145 ?	126 ?	29.0	Drava, 1976	Kainz (in litt.)
146 ?	127 ?	29.5	Drava, 1976	Kainz (in litt.)
136	122 ?	30.0	Drava near Spittal, 1975	Anonymus (1976b)
136	122 ?	30.1	Drava, 1975	Anonymus (1976a)
150	131 ?	32.0	Váh near Skalka, December 1930	Novák (in litt.)
132	119 ?	32.0	Drava, 1976	Löbl (1976)
153 ?	133 ?	34.0	Iller near Waltenhofen, 1928	Anonymus (1928)
144	130 ?	34.8	Drava near Spittal, 19 Feb. 1985	Schulz (1985)
130	117 ?	35.0	Drava, 1927	Vokač (1959)
158 ?	142 ?	38.0	Inn, 1956	Anonymus (1976a)
164 ?	147 ?	43.0	Drava near Zamlaka, 1937	Dyk (1937)
169 ?	153 ?	46.0	Hron, 1949	Ivaška (1951)
173 ?	156 ?	50.0	Inn near Rosenheim, prior to 1900	Anonymus (1976a)
179 ?	161 ?	52.0	Danube at Tulln, end of 19th century	Haempel (1910)
183 ?	165 ?	60.0	Danube at Tulln, 9 Jan. 1873	Anonymus (1897)
183 ?	165 ?	60.0	Danube at Tulln, 9 Jan. 1983	Anonymus (1897)
			Hucho hucho taimen	
140	129 ?	34.0	Amur, December 1936	Nikol'skii (1956)
145	134 ?	36.0	Amur, December 1936	Nikol'skii (1956)
167 ?	154 ?	40.0	Kama, (?)	Bukirev (1967)
187 ?	173 ?	56.0	Enisei near Krasnoyarsk, 1944	Podlesnyi (1958)
193 ?	178 ?	61.0	Mouth of Kochechum r. into Enisei, 1952	Sigunov (1972)
211 ?	195 ?	80.0	Yakutia, (?)	Kirillov (1976)
213 ?	196 ?	80.0	Amur near Innokentievka, winter 1945	Nikol'skii (1956)
216 ?	200 ?	86.0	Khatanga, 1952	Sigunov (1972)
224 ?	207 ?	95.0	Uda, (?)	Misharin and Shutilo (1971)
210	193 ?	105.0	Kotui, 1943	Sigunov (1972)

Except for fish weighing over 50 kg, all the rest are catches made by anglers. The above data justify the assumption that in the past, when the human influences on fish communities in running waters were much less intense and there were incomparably more professional fishermen and catches of big fish were only rarely recorded, big huchen were more frequent in the catches than they are today and atained bigger dimensions. Two huchen weighing 60 kg each, caught in the Danube at Tulln on 9 Feb. 1873, should be considered the biggest catches documented in the literature. According to the length-weight equation 1 in Table 20, their standard length may be estimated at 165 cm and their total length at 183 cm. In that same locality, another big specimen, weighing 52 kg, was caught several years later (Haempel 1910, Neresheimer 1937); its standard length may have been around 159 cm and its total length approximately 176 cm. In the territory of Slovakia, the biggest huchen was found after the fish community of the Hron river was killed by poisoning; the alleged weight of that fish was 46 kg (Ivaška 1951) and the assumed standard length and total length were 153 and 169 cm, respectively. The second biggest huchen from our waters is one 32 kg in weight, caught in the Váh river at Skalka in December, 1930 (Table 17). According to an unverified report, a huchen weighing 36 kg was shot in the Turiec river at Valentová in October, 1979 (Novák in a letter of 25 Oct. 1980).

V. Dyk (1937a) reports that a huchen 43 kg in weight and 248 cm in length was caught in the Drava river at Zamlaka in 1937. This is an obvious error, since a fish of such size would be expected to be around 157 kg. The correct length *(Sl)* of this fish was probably 148 cm, which is in good accord with the stated weight.

However, reports are available suggesting that the nominate form may attain greater length. There are two items of information, unfortunately unverified, by coincidence coming from Slovakia. According to one of them (Nieslanik), a flood washed ashore a huchen about 70 kg in weight from the Váh river near Ružomberok in 1870; according to another version, the huchen under consideration was pushed ashore by a landing raft and one of the crew killed the fish with an axe (Bél in a letter of 18 May 1978). A record of this event is said to be contained in the chronicle of a former parsonage at Partizánska Ľupča. According to the other report (I. Novák in a letter of April 1981), O. Mizera, of the staff of the State Fisheries at Kláštor pod Znievom, found a dead huchen (probably wounded by lead shot) in the Turiec river near Socovce in the summer of 1955 or 1956. It was in a fairly advanced stage of decay. The length of its head was 38–40 cm (measured with the aid of a matchbox), which means that its standard length may have been 1446–1800 mm and the corresponding average weight 40–79 kg or, taking the upper limit according to the appropriate length-weight equation, 58–116 kg. Unfortunately, neither of the two reports can be considered reliable. It remains a fact, however, that big huchen may still have occurred in the Váh river not long ago. Besides the catch in 1930 of a huchen weighing 32 kg, mentioned above, this is also suggested by the information obtained from J. Kuna by Nieslanik (in a letter of 30 March 1957), according to which several huchen 30–40 kg in weight (one of them estimated at 50 kg) were seen at spawning sites in the Váh river-bed in 1927–1928. Thus, Rothschein's (1980) assumption that in historical times the huchen in Slovak waters may have attained 160–170 cm in total length (i.e. 55–66 kg in weight) is by no means exaggerated.

Naturally, the sizes of huchen currently caught are smaller. According to the records of the Slovak Anglers' Union in Žilina, the average weight of the catches in Slovak waters varied between 3 and 8 kg. Munda (1925) summarized Jagoditsch's

TABLE 18

Percentage size structure of a catch of 527 huchen in the Sava and Ljubljanica rivers. Calculated from data by Munda (1925)

Length group (*Tl* in cm)	%	Weight group (kg)	%
31— 40	0.2	0.5— 1.9	40.8
41— 50	8.9	2.0— 3.9	34.7
51— 60	32.8	4.0— 5.9	11.2
61— 70	24.8	6.0— 7.9	6.8
71— 80	14.6	8.0— 9.9	3.6
81—90	9.5	10.0—11.9	1.7
91—100	5.9	12.0—13.9	0.2
101—110	2.8	14.0—15.9	0.6
111—120	0.5	16.0—17.9	0.4

and his own records on catches in the waters of Slovenia, collected over more than 20 years, and found that 37% of them included fish 1–2 kg in weight, 24% included fish 2–3 kg in weight, and the rest was other fish. We have re-calculated Munda's data in view of length groups and give the results in Table 18. It is evident that two thirds of the catch included huchen up to 70 cm in total length, or up to 4 kg in weight. From a report by Mednyansky (1844), stating that "notable huchen, often 30–35 pounds in weight are caught" in the Váh near Kraľovany (i.e. weighing 16.8–19.6 kg, since the above statement was obviously referring to the Vienna pound, equivalent to 560.012 g), it follows that huchen over 10 kg in weight were not very frequent even at the beginning of the 19th century, to which period the above report relates.

Data on record catches of taimen are much sparser, particularly as in the USSR and Mongolia, where this form occurs, the fish are predominantly caught on a commercial basis. It follows from the available data, however, that taimen attain greater sizes than huchen. The biggest fish of this form and hence also of the species *Hucho hucho* and the genus *Hucho* caught so far is a taimen 210 cm in length (probably fork length) and 105 kg in weight, caught by a Yakutian fisherman in the Kotui river (a right tributary of the Khatanga river) in 1943 (Sigunov 1972). Even other data, summarized in Table 17, indicate the gigantic sizes attained by this form.

The fact that taimen attain bigger size than the nominate form is due to two causes. The first is the space factor: fishes inhabiting larger streams attain greater sizes, and taimen, as mentioned in Chapter 7, inhabit large Siberian rivers, often down to their mouths. The second cause is the effect of the Bergmann's rule according to which the size of the members of a genus or a species (including fish and other cold-blooded vertebrates, as demonstrated by Lindsey 1966) is greater at higher latitudes than at lower ones. But there is practically no difference between huchen and taimen in the most frequented size groups. Nikol'skii (1956) states that the usual size for the taimen in the Amur river is 5–10 kg; it is 2.9–7.4 kg in the Khatanga river (Luk'yanchikov 1967b); according to Sigunov (1972), who specially searched for large-sized taimen, fish heavier than 30–40 kg have never been caught with hook

TABLE 19

Maximum weights of selected salmonid species (rounded to nearest kg). Compiled from Schäferna (1928), Berg (1948—1949), Berg et al. (1949), Dyk (1956), Clemens and Wilby (1961), Carlander (1970), Sigunov (1972), Kirillov (1972), Khatkevich (1973), Scott and Crossman (1973), Behnke (1980), Reshetnikov (1980), Martin and Olver (1980)

Species	Weight (kg)
Hucho hucho	105
Hucho perryi	60
Oncorhynchus tschawytscha	57
Salvelinus namaycush	55*
Salmo trutta caspius	51
Stenodus leucichthys	49
Salmo salar	47
Salmo trutta m. *lacustris*	36
Salmo trutta labrax	24
Salmo gairdneri	24
Salmo trutta trutta	22
Salmo trutta m. *fario*	20
Salvelinus confluentus	18
Salmo ischchan	17
Salvelinus alpinus	16
Salmo mykiss	11
Brachymystax lenok	8
Salvelinus fontinalis	7

*According to Martin and Olver (1980), this value is dubious; a fish 46 kg in weight, caught in Lake Athabasca in 1961, is recognized as the officially valid record.

and line. According to Pirozhnikov (1955), the seine net catches obtained in summer in the Lena river in the area of Zhigansk, Govorevo, Tit-Ary and Kysam, are dominated by fish 81 – 120 cm *(Fl)*, 20% of which is taimen 31 – 50 cm in length; in the area of Zhigansk, the average weight is 3.9 kg, in that of Govorevo 6.7 kg. Misharin and Shutilo (1971) also report that seine net catches are dominated by taimen 50 – 75 cm in length *(Fl)*. Teplov (1951) states that the usual weight of taimen caught in the Ilych river (the Pechora river basin) is 1.5 – 4 kg, rarely 6 – 8 kg, fish of greater weights (16, 20 and 30 kg) occurring very rarely. In the rivers of Yakutia, 15 – 20 kg is the most frequent maximum weight of taimen in the catches (Kirillov 1976). It is remarkable that fish of greater average sizes were hardly ever caught even in the distant past. Tsepkin (1980a) found that the sizes of subfossil taimen from the settlements of Neolithic man on the Angara river were 50 – 80 cm *Fl,* which agrees with the average size of taimen caught in the Angara even at present.

Table 19 contains a synopsis of maximum weights of some salmonid species recorded so far. It is evident that the species of the genus *Hucho* considerably surpass other species and genera even in this respect; they may rightly be included among the biggest freshwater fishes at all, such as *Silurus glanis* (300 kg), *Pangasianodon gigas*

TABLE 20

Parameters of regression equations of the length-weight relationships in various huchen and taimen populations. Unless stated otherwise, the information pertains to material with sexes not distinguished

Locality	Length range (mm)	log $w = a + b$ log Sl log $w = a + b$ log Tl log $w = a + b$ log Fl a	 b	n
	Hucho hucho hucho			
Sava and Ljubljanica[1]	390—1180 *Tl*	−5.47791	3.14420	527
	347—1060 *Sl*	−5.25958	2.11973	
Váh, Turiec, Orava[2]	55— 140 *Sl*	−5.313	3.216	50
Tisza river system[3]	145— 850 *Fl*	−5.73588	3.19889	—
	133— 782 *Sl*	−5.64350	3.20661	—
Dunajec[4]	492—1000 *Tl* ♂♂	−5.167	3.139	41
	495—1095 *Tl* ♀♀	−4.869	3.066	37
Drava[5]	325—1440 *Tl*	−5.04163	3.03004	45
Czechoslovak streams[6]	54—1215 *Sl*	−4.880305	2.99662	243
	54—240 *Sl* juv.	−4.85148	2.98598	63
	289—1018 *Sl* ♂♂	−5.463433	3.21075	52
	253— 955 *Sl* ♀♀	−5.698177	3.28874	26
Kláštor pod Znievom — ponds[2]	25— 120 *Sl*	−4.756	2.900	606
Podsuchá — ponds[7]	19.9—57.7 *Sl*	−5.06274	3.10455	?
	21.7—66.8 *Tl*	−5.09151	3.00618	?
Hatcheries and farms of the	573—1016 *Sl* ♀♀	−4.40946	2.86707	46 I
Slovak Anglers' Union[8]	573—1016 *Sl* ♀♀	−5.43456	3.18580	28 II

I — pre-spawning, II — post-spawning relatioship.

(250 cm), *Arapaima gigas* (200 kg) or *Catlocarpio siamensis* (250 cm) (Smith 1945, Nikol'skii 1974).

At present, catches of bigger individuals can hardly be expected, as the continuously increasing interference of man with river ecosystems limits the living space of *Hucho hucho*. The species is being increasingly confined to small submontane streams where it cannot attain bigger sizes.

LENGTH–WEIGHT RELATIONSHIP

Table 20 summarizes equations characterizing the length-weight relationships in various huchen and taimen populations. Except for Bastl and Kirka (1959), Witkowski et al. (1985) and Schulz (1985), who calculated and published this relationship themselves, we calculated it from the data contained in the papers of the remaining authors. At the same time, we wish to point out that in the case of the huchen only the length and weight data sub 1, 4, 5 and 6 from Table 20 come from live specimens, the rest coming from fish preserved for various periods of time. As for the taimen, the data most probably come from live fish.

With regard to the above facts it is evident that the various populations do not differ much from one another. Taimen populations appear to have lower mean

TABLE 20

(Continued)

Locality	Length range (mm)	log w = a + b log Sl log w = a + b log Tl log w = a + b log Fl a	b	n
		Hucho hucho taimen		
Enisei[9]	510—1170 *Fl*	−4.78150	2.88954	—
	469—1077 *Sl*	−4.46392	2.81729	—
	510—1060 *Fl* ♂♂	−3.94267	2.58317	—
	469— 975 *Sl* ♂♂	−3.84653	2.58238	—
	630—1170 *Fl* ♀♀	−3.86695	2.59652	—
	580—1077 *Sl* ♀♀	−3.77766	2.59811	—
Amur[10]	767—1200 *Fl*	−5.68910	3.23867	—
	706—1103 *Sl*	−5.58242	3.24216	—
Kama[11]	397— 996 *Fl*	−5.67671	3.23686	42
	365— 915 *Sl*	−5.56166	3.24774	42
Anabar[12]	405— 730 *Fl*	−4.79209	2.88371	26
	373— 672 *Sl*	−4.67521	2.87909	26
Angara[13]	280—1050 *Fl*	−5.25396	3.08716	123
	258— 966 *Sl*	−5.37673	3.17688	123
Vitim[14]	220—1005 *Fl*	−5.63608	3.22231	138
	212— 925 *Sl*	−5.51721	3.22111	138

[1]Munda (1925); [2]Bastl and Kirka (1959); [3]Vlasova (1959); [4]Witkowski et al. (1985); [5]Schulz (1985); [6]Holčík et al. (1977); [7]Peňáz and Příhoda (1981); [8]Own data; [9]Pravdin (1949); [10]Nikol'skii (1956); [11]Bukirev (1967); [12]Kirillov (1976); [13]Olifer (1977); [14]Kalashnikov (1978). Only the data sub 2 and 6 have been published, the remaining ones have been calculated from those provided in the papers of the respective authors.

weights than do huchen. A comparison of the weight increments of huchen with that of taimen is given in Table 21. The weight of the huchen was derived from an equation calculated from the data of Munda (1925) and given in Table 20, where the 95% confidence limits of coefficient b are 3.06786 for the lower and 3.17160 for the upper limit. We calculated a separate equation for the taimen by averaging the respective parameters of the length-weight equations *(Sl)* of the various populations of this form. The calculation of confidence limits is based on the assumption that they bear a similar percentage relation to the mean value of b to the huchen. The calculated equation is $\log w = -5.19619 + 3.09738 \log Sl$; the 95% lower limit is $b = 3.04588$ and the upper one is $b = 3.14888$. The table clearly shows the small differences between the huchen and the taimen, which may be considered non-significant and the weights corresponding to any particular body length are practically identical.

It follows from the equations summarizing the data from Czechoslovak waters that (1) the weight growth of young in the late embryonic, larval, and early juvenile

TABLE 21

Comparison of weights for some size classes in *Hucho hucho hucho* and *H. hucho taimen*. Explanation in text

Sl (mm)	*Hucho hucho hucho* Weight (kg) Average	*Hucho hucho hucho* Weight (kg) Range	*Hucho hucho taimen* Weight (kg) Average	*Hucho hucho taimen* Weight (kg) Range
200	0.08	0.06— 0.11	0.09	0.06— 0.11
300	0.30	0.22— 0.40	0.30	0.22— 0.40
400	0.72	0.53— 0.98	0.73	0.54— 0.99
500	1.45	1.05— 2.00	1.44	1.06— 2.01
600	2.56	1.83— 3.56	2.56	1.84— 3.56
700	4.13	2.94— 5.81	4.13	2.95— 5.79
800	6.27	4.43— 8.87	6.25	4.43— 8.82
900	9.05	6.36—12.89	9.00	6.34—12.78
1000	12.58	8.79—18.00	12.47	8.74—17.80
1100	16.93	11.78—24.35	16.76	11.68—24.03
1200	22.21	15.38—32.09	21.94	15.23—31.61
1300	28.52	19.66—41.36	28.11	19.43—40.67
1400	35.93	24.68—52.32	35.36	24.35—51.36
1500	44.56	30.49—65.12	43.79	30.05—63.82
1600	54.50	37.17—79.91	53.48	36.57—78.20
1700	65.86	44.77—96.85	64.53	43.99—94.65
1800	78.70	53.35—116.10	77.02	52.36—113.31
1900	93.16	62.98—137.82	91.07	61.73—134.34
2000	109.33	73.71—162.17	106.75	72.17—157.89

periods (standard length 20–58 mm) is faster than that of juvenile huchen (standard length 54–240 mm); (2) compared to males, females have higher initial weights but a slower growth rate; and (3) females attain higher weights than males at identical body lengths. The equations also indicate the fact known to farmers in adult huchen and observed in juveniles by Bastl and Kirka (1959), that the growth rate of huchen reared in fishponds lags behind that of huchen living free indicating an insufficient quality or quantity of food supplied to the captive fish. It is interesting to note that young huchen reared in ponds at Podsuchá (Peňáz and Příhoda 1981) increased in weight more rapidly than young reared in ponds at Kláštor pod Znievom (Bastl and Kirka 1959) although the linear growth was slower in the former (Fig. 27). The differences, however, may have been due to the fact that the young from Kláštor pod Znievom were weighed after preservation in ethylalcohol, whereas those from Podsuchá were weighed in a fresh condition.

RELATIONSHIPS BETWEEN PARTICULAR LENGTHS

In the literature, the lengths of huchen and taimen caught or examined as to growth are stated in several ways. Intimations on legal regulations of fishing, valid in most European countries, state the legal size of the fish as the total length *(Tl)*, i.e. from the tip of its snout to the most distant edge of the longest ray of its caudal fin. Catches and data pertaining to the taimen are given in the Soviet literature in values of the Smitt length (fork length; *Fl*), i.e. from the tip of the snout to the end of the medial rays of the caudal fin. Finally, in the Czechoslovak ichthyological literature the length of the huchen is expressed as the standard length *(Sl)*, i.e. from the tip of the snout to the end of hypuralia. To enable comparison, we give here equations by means of which any of the above lengths can be converted to any other:

$Sl = -4.749 + 0.903\ Tl,$

$Sl = 0.098 + 0.920\ Fl,$

$Fl = 0.331 + 1.0803\ Sl,$

$Tl = 8.073 + 1.102\ Sl.$

The equations were derived from 174 specimens of huchen of Slovak origin, 42–940 mm in *Sl*, and were also applied to determine the *Sl* of the taimen. The following equations are valid for huchen young 59–171 days old, 19.9–57.7 mm in *Sl:*

$Sl = 0.7742 + 0.8491\ Tl,$

$Tl = -0.8281 + 1.1754\ Sl.$

The equations were derived from the values given by Peňáz and Příhoda (1981) and pertaining to huchen reared in a pool after artificial spawning.

LENGTH AND WEIGHT GROWTH, PONDERAL INDEX

Fairly numerous data are available on the length growth of huchen and taimen. However, not all of them provide reliable information, as already pointed out by Kirka (1963). For example, Haempel (1910), but also Kulmatycki (1931a) and Shnarevich and Moshuk (1957) estimate the smallest fish to be younger (by a year on an average) than they are in reality. Another shortcoming of the growth data is that only a minor part of them relate to individual annuli; mostly they are empirical lengths.

Data on size *(Sl)* attained by the huchen in successive years of life are given in Table 22, those for the taimen in Table 23.

It is evident that the variability is considerable even if the shortcomings stated above are taken into account. To obtain at least an approximate idea of the length growth of the two forms, we pooled the data in the two tables and calculated the values of Student's *t*-test and the coefficients of difference *(C.D.)* for successive years of life. It is evident from Table 24 and Fig. 24 that in successive years of life the huchen actually attain greater lengths than the taimen, the differences between the mean values for age groups II–VII being statistically significant. On the other hand,

TABLE 22

Growth rate (*Sl* in mm) of the huchen in different European rivers

	1	2	3	4	5	6	7	8	9	10	11	12	13	14	15	20
Danube, Aurach, Traisen, Traun, Pielach[1]	185	284	402	537	682	790	682	907	943	980	1016	1052	1079	1142	1260	
Isar, Alz, Inn, Lech, Enns, Iller, Ammer, Vertach, Loisach, Amper, Drau[2]	160	343	508	664	799	935										
Sava, Ljubljanica[3]	—	—	—	507	606	631	727	751	793	817	907	911	979			
Drava[4]	154	311	447	560	650	718	786	822	844	876	944	962	971	1007	1102	1147
Dobra[5]	266	370	462	549	592	666	737	849	958	973						
Kupa[5]	293	343	428	503	592	657	731	859	925	1070	1079					
Una[5]	302	320	414	477	582	606	657	802	883	952						
Danube[6]	236	352	512	556												
Váh[7]	128	313	406	479	546											
Orava[7]	135	313	445	528	602	675	686									
Turiec[8]	138	281	404	550	600	670	754	832	890							
Hron[9]	183	351	469	543	600											
Poprad[10]	130	243	348													
Dunajec[11]	193	342	464	562	647	720	780	832	889	927	961					
Klíčava reservoir[12]	136	282	354	406	444	486										
Vltava[13]	91	267	423	620	734											
Cheremosh[14]	311	402	474	537	600	663	709									
Tisza r. system[15]	269	414	542	626	718	745	754	782								

[1]Haempel (1910); [2]Seez (1939); [3]Munda (1925); [4]Krauss (1933); [5]Pažur et al. (1982); [6]Balon (1968); [7]Kirka (1963); [8]Kirka (pers. comm.); [9]Combined after Sedlár and Stráňai (1975) and Holčík et al. (1977); [10]Combined after Kirka (1975) and Holčík et al. (1977); [11]Witkowski et al. (1985); [12]Holčík (1970); [13]Havelka and Vostradovský (1974); [14]Kulmatycki (1931a, 1935); [15]Vlasova (1959). Values 3 calculated by Andraščík (1980), 4 and 13 interpolated from growth curves. Values 6 to 12 back-calculated to particular annuli, the rest are empirical data converted to *Sl* from *Tl* or *Fl* stated by respective authors.

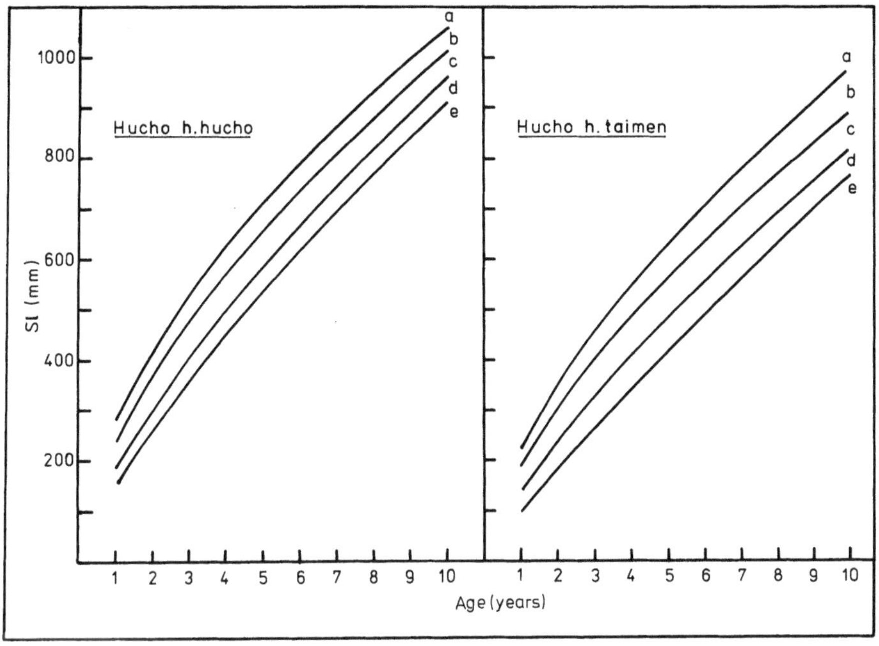

Fig. 25. Growth zones of huchen and taimen. Curves indicate limits of successive growth zones: a – very fast; b – fast; c – average; d – slow; e – very slow growth.

slow growth in the Váh and Turiec rivers, average growth in the Orava river and the Czechoslovak section of the Danube, and very slow growth in the Sava, Ljubljanica, Hron and Poprad rivers. As for taimen, the population in the Olenek river exhibits very fast growth; that in the Angara river (according to the data of Misharin and Shutilo) and the Shishkhid river, average growth, and that in Lake Frolikha very slow growth. At the same time, the opinion of Kirillov (1972) that the taimen population in the lower reaches of the Vilyui river exhibits inferior growth to that in the upper reaches is confirmed.

Furthermore, it is necessary to mention data not included in the above tables. Iancu and Decei (1972) report extremely fast growth of huchen in the valley reservoirs Bicaz (on the Bistriţa river) and Vidraru (on the Argeş river), Rumania. A fish 5 years old was 81 cm long *(Sl)* and weighed 5.7 kg; another fish, 3 years old, was 50 cm long and weighed 1.5 kg. As mentioned by these authors, in the Bistriţa river huchen 5 years old are only 67 cm long and 2.9 kg in weight; at the age of 3 years they are 46 cm long and 0.9 kg in weight. Very fast growth was also recorded in Morocco where 2-year-old fish attained 46–57 cm (probably *Tl;* Preudhomme 1955). Similarly fast-growing huchen have also been reported from the Tormes river, Spain (Anonymus 1973, Mondejar-Reyna 1981, Lobón-Cerviá in a letter of 2 Nov. 1979): at the end of the 4th year of life, the huchen attained weights of 5.8–6.3 kg (i.e. approx. 87–90 cm *Sl*), at the age of 6 years, 930 mm *Fl* and

10.5 kg, at the age of 8 years, 1026 mm *Sl* and 14.5 kg; two-year-old huchen averaged 1.8 kg, i.e. around 540 mm in *Sl.*

Svetina (1968, 1970) states that during the juvenile period (up to the 3rd year of life), the length growth and weight growth of huchen is very fast, slowing down between the 3rd and 4th year of life (in the period of gonad formation), and becoming slightly faster again after sexual maturity has been attained. According to this author, the period between the 5th and 10th year of life of the huchen is one of greatest activity, characterized by dynamic growth and intense increase in weight. Huchen older than 10 years begin to grow old, the reproduction capacity of males decreasing earlier than that of females. However, these statements should be considered as unsubstantiated. The growth of the huchen is almost equally intense until old age, and older fish often grow faster than younger ones. This is best evidenced by the fact that for most huchen and taimen populations analysed by ourselves, we failed to calculate the von Bertalanffy growth equation simply because the growth of the older age groups was equally intense to that of the younger ones. This uniform absolute length increment is characteristic of long-lived freshwater and marine fishes inhabiting cold waters, as pointed out by Ricker (1975).

The information on growth intensity in certain huchen and taimen populations provided by the von Bertalanffy growth equations (Table 26), should not be taken without reserve. In calculating it was necessary to exclude some of the values, as they were equal or even higher for higher age groups than for the preceding ones.

TABLE 26

Parameters of the von Bertalanffy equation* characterizing linear growth of huchen and taimen from selected localities in Europe and Asia (calculated from data in Tables 22 and 23)

Hucho hucho hucho				*Hucho hucho taimen*			
Locality	L_∞	K	t_0	Locality	L_∞	K	t_0
Isar, Alz, etc.	2431	0.08	0.17	Amur	3018	0.04	0.45
Vltava	1688	0.14	0.67	Lena	2061	0.06	0.32
Danube, Aurach, etc.	1341	0.14	0.21	Vilyui (upper reaches)	1500	0.11	1.33
Turiec	1118	0.18	0.41	Vilyui (lower reaches)	1093	0.14	1.17
Dunajec	1106	0.18	0.07	Shishkhid	999	0.14	0.50
Váh	874	0.20	0.02	Angara	1004	0.16	0.50
Tisza	916	0.23	0.87	Loz'va	911	0.21	0.58
Drava	1024	0.24	0.58				
Hron	832	0.28	0.37				
Orava	771	0.35	0.54				

*$l_t = L_\infty (1 - e^{-K(t - t_0)})$.

The localities were ordered according to increasing growth intensity *(K)*. Table 27 contains values of length groups of an "average" huchen and an "average" taimen according to data summarized in Table 22, the second column for each form containing values obtained by introducing maximum actual standard length (in the

TABLE 27

Comparison of growth intensity and maximum hypothetical *Sl* (mm) attained at different ages (years) by huchen and taimen, as indicated by the von Bertalanffy equations

	Hucho hucho hucho			*Hucho hucho taimen*	
Age	L_∞ = 1252 K = 0.13 t_0 = 0.52	L_∞ = 1650 K = 0.08 t_0 = 0.84	Age	L_∞ = 1447 K = 0.08 t_0 = 0.27	L_∞ = 1930 K = 0.05 t_0 = 0.85
1	153	127	1	111	94
2	286	244	2	214	184
3	404	352	3	309	269
4	507	452	4	396	350
5	598	544	5	477	427
6	678	629	6	552	500
7	747	708	7	620	570
8	809	780	8	684	636
9	863	847	9	743	699
10	910	909	10	797	759
20	1158	1317	20	1155	1220
30	1226	1500	30	1315	1499
50	1249	1620	50	1420	1772
60	1251	1635	60	1435	1834
80	—	1647	80	1445	1895
100	—	1649	100	1446	1917
102	—	1650	101	1447	1917
			167	—	1930

case of the huchen calculated from body weight). Plotting the respective values in a diagram produces growth curves, that for the taimen being less steep than that for the huchen. Consequently, having attained 20 or 30 years of age in subsequent years taimen overtake huchen in length; this is a distinct manifestation of growth compensation, known to occur in many other species of fishes. As expected, the weight increments decrease slightly but constantly after a certain age has been attained, but the decrease is so small that the growth curve in that part almost resembles a straight line.

Kirillov (1972) explains the rapid growth of the taimen in the Olenek river by a rich food supply and by lack of competitors, since the taimen is the only predatory species there. The slower growth in the lower reaches of that river is due to competition for food from lenok, inconnu, pike, perch, and burbot.

The generally slower growth of taimen during the first 10–15 years of life (Fig. 26) can be explained by an unfavourable thermal regime in the Siberian rivers. For example, according to Kirillov (1962), the Vilyui river (a left tributary of the Lena) is frozen for 213–231 days a year on the long-term average. Temperatures above zero occur there as late as the second half of May, when the temperature rises to 0.1° C. In late September, the temperature drops to 3–4° C again, and down to zero during the second decade of October. The ice is 20–25 cm deep within 10 days of the onset of frosts. On the other hand, maintaining a high growth rate at higher age is made

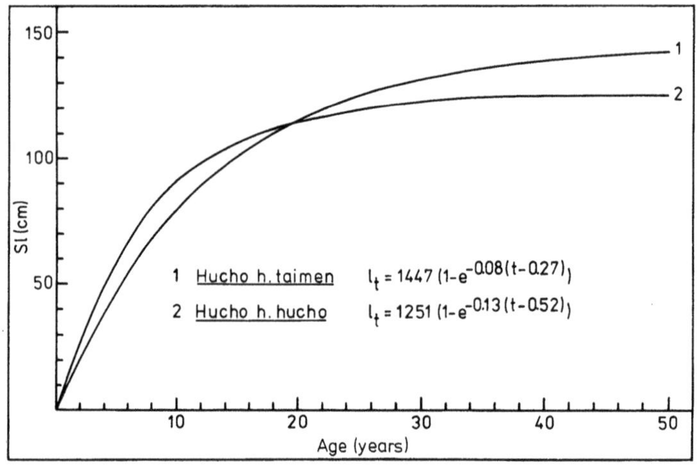

Fig. 26. Growth curves (based on von Bertalanffy equations) of (1) taimen and (2) huchen.

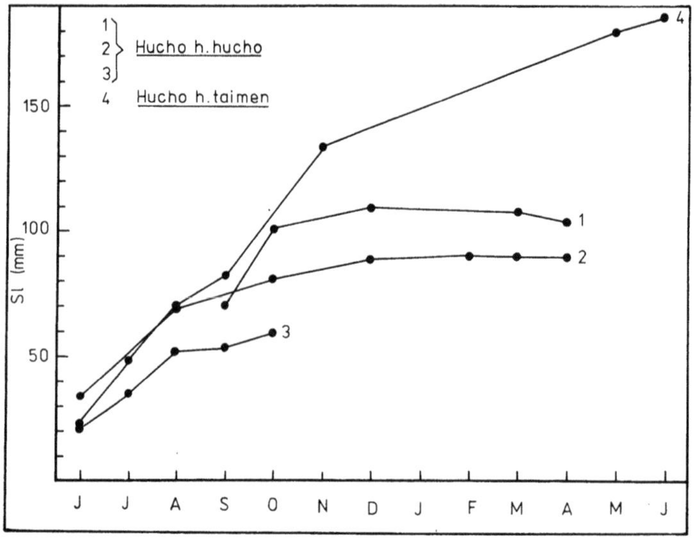

Fig. 27. Linear growth of young (1–3) huchen and (4) taimen during the first year of life. 1 – Váh river; 2 – ponds at Kláštor pod Znievom; 3 – ponds at Podsuchá; 4 – Khor river (Amur river system) (based on data by Levanidov 1951, Bastl and Kirka 1959 and Peňáz and Příhoda 1981).

possible by better trophic conditions compared to those of the huchen which, being confined to short sections of submontane streams, have at their disposal a smaller amount of available food than the taimen, which populate extensive plains of Siberian rivers.

The growth of young huchen was studied in detail by Bastl and Kirka (1959). They found that in the first year of life the young first grow very rapidly, their growth slowing down with the onset of winter, until it stops completely. The most intense growth lasts until mid-August. In natural environments, young huchen grow faster than in ponds (Fig. 27), which fact the authors ascribe to a rich food supply in the form of young of other fishes, especially minnow. Peňáz and Příhoda (1981), who studied the growth of young huchen during the first 5 months of life (in a pond culture), found a sigmoidal curve of length growth with depressions, or with regressions on the 66th, 145th and 164th day after activation. They ascribe the growth retardation to insufficient sampling or to ichthyophthiriosis, and the rapid increase on the 171st day to increased cannibalism. The growth retardation that occurred between the 59th and 66th days after activation, however, is apparently connected with a complete change to exogenous nutrition (Fig. 28).

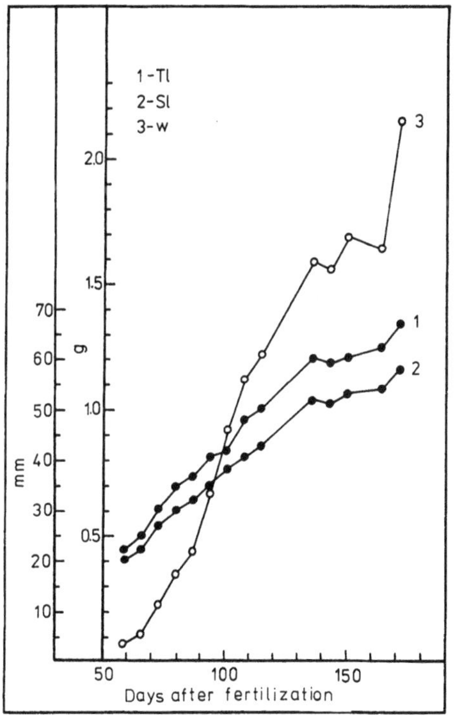

Fig. 28. Length (1, 2) and weight (3) growth of huchen juveniles during the first 5 months of life.

Levanidov (1951) informs on the growth of taimen in the first year of life. A comparison of his data (converted to *Sl*) with those of Bastl and Kirka (1959) and of Peňáz and Příhoda (1981) reveals (Fig. 27) that young taimen in the Khor river (a tributary of the Sungari river, Amur river basin) grow faster than young huchen in Slovak waters, their growth during the first 4–5 months being faster than subsequently. From the comparison, one may also conclude that the slower growth of the taimen during the first few years of life is not specific to this form, but results from the fact that it mostly involves northern populations which affected by severe environmental conditions, and biasing the overall average.

TABLE 28

Body weights (rounded to nearest kg) of huchen (1, 2) and taimen (3—8) in successive years of life. Data for (2) are back-calculated, the rest are empirical ones

	1	2	3	4	5	6	7	8	9	10	11	12	13	14	15	16
Tisza[1]	0.09	0.37	0.83	1.59	2.8	4.3	5.9	8.1	—	—	—	—	—	—	—	—
Váh[2]	0.03	0.40	0.86	1.42	2.1	—	—	—	—	—	—	—	—	—	—	—
Orava[2]	0.03	0.40	1.14	1.90	2.8	4.0	4.2	—	—	—	—	—	—	—	—	—
Turiec[2]	0.03	0.29	0.85	2.15	2.8	3.9	5.5	7.4	9.1	—	—	—	—	—	—	—
Hron[2]	0.08	0.56	1.31	2.07	2.8	—	—	—	—	—	—	—	—	—	—	—
Poprad[2]	0.02	0.17	0.54	—	—	—	—	—	—	—	—	—	—	—	—	—
Amur[3]	—	—	—	—	5.0	—	—	7.2	9.4	10.1	—	9.0	—	—	14.0	23.2
Enisei[3]	—	—	1.00	2.07	1.8	3.0	3.8	—	8.0	9.7	11.0	—	—	—	—	—
Lena[4]	—	0.07	0.34	0.72	1.2	2.3	3.8	4.8	6.6	9.5	11.6	—	—	—	—	—
Vitim[5]	0.09	0.26	0.47	0.78	1.2	2.1	3.4	4.6	5.7	—	—	11.9	11.0	—	—	—
Bratsk reservoir[8]	—	—	—	0.59	0.9	3.0	—	—	—	—	—	—	—	—	—	—
Angara[7]	0.18	0.38	0.54	0.80	1.5	2.5	3.4	4.7	5.8	6.8	8.0	9.4	9.8	10.5	12.6	—
Kama[8]	—	0.56	0.75	1.70	2.0	3.9	3.8	10.2	8.0	—	—	—	—	—	—	—

[1]Vlasova (1959); [2]Own data; [3]Pravdin (1949, averaged); [4]Pirozhnikov (1955); [5]Kalashnikov (1978); [6]Luk'yanchikov (1967a); [7]Olifer (1977, mean for 3 years); [8]Bukirev (1967).

No difference in growth intensity has been observed between males and females (Misharin and Shutilo 1971). In the Vis river, however, the linear growth was higher, but the weight growth was lower in males than in females (Bukirev 1967), whereas in the Enisei the linear growth was slower in males (Pravdin 1949).

Far fewer data are available on the weight growth of the huchen, those on the taimen being more numerous. Some of the available data are reviewed in Table 28, while Table 21 contains data on the weight growth of an "average" huchen and an "average" taimen.

The ponderal index ($K = w \, . \, 10^5/Sl^3$), calculated in all cases from the standard length and weight at the time of capture, shows (Fig. 29) a steady increase with increasing size, except for young 50–150 mm in length, where the regression of this relationship appears to be rather negative. According to the slope of the regression lines, the ponderal index of females increases faster than that of males, which may be ascribed to a higher weight of female gonads. The ponderal index in this material, including huchen from various (predominantly Slovak) rivers, averages 1.35 for females and only 1.27 for males. In bigger fish, however, the ponderal index appears to decrease rather than increase with increasing length. This may be concluded from the results obtained by re-calculating data in Table 17 (we included only those fish where the respective authors stated their length also). The ponderal index of 31 huchen 1060–1360 mm in *Sl* and 20–35 kg in weight varied from 0.98 to 2.19, averaging 1.43 at 1195 mm, but the regression of *Sl* to *K* is negative ($y = 2.36155 - 7.81597\,x$) although statistically non-significant ($r = -0.207$). It is certainly not without interest that (judging from such big huchen as above) there are certain differences between populations in different rivers; the highest ponderal indices were found for the huchen from the Orava, Inn and Turiec rivers, the lowest for those from the Danube and Váh rivers, as indicated by the following values (the first one indicating *K*, the second the standard length in mm, and the third one the number of cases examined): Inn – 1.60, 1100, 2; Drava – 1.60, 1300, 10; Turiec – 1.43, 1175, 4; Orava – 1.32, 1180, 4; Lech – 1.30, 1192, 4; Váh – 1.26, 1217, 3. It can indirectly be concluded from the above data that the trophic conditions in the Drava and Turiec rivers are better than in other river systems. Better weight growth of huchen in the Drava has also been pointed out by Schulz (1985) who states, for 45 huchen from that river with lengths between 325 and 1440 mm *Tl*, an average $K = 1.12$ (converting the length to *Sl*, the *K* would be considerably higher).

The most representative view of this index in the huchen can be obtained from the numerous material published by Munda (1925). It follows from his material that *K* for fish of 384–1055 mm increases fluently from 0.88 to 1.58, averaging 1.22.

As is evident from the papers by Nikol'skii (1956) and Bukirev (1967), the ponderal index increases with increasing size and age in the taimen, too. In the Amur river, the *K* of taimen 767–1199 mm in *Sl* and 4.4 kg in weight averages 1.37 (ranging from 1.25 to 1.50), and in the Kama river, the *K* of taimen 365–916 mm in *Sl* and 0.6–10.2 kg in weight averages 1.34 (1.14–1.74). These values pertain to fish whose size has been converted to *Sl*. In addition, we may state the values of *K* calculated by ourselves for the taimen from the following localities: Bratsk

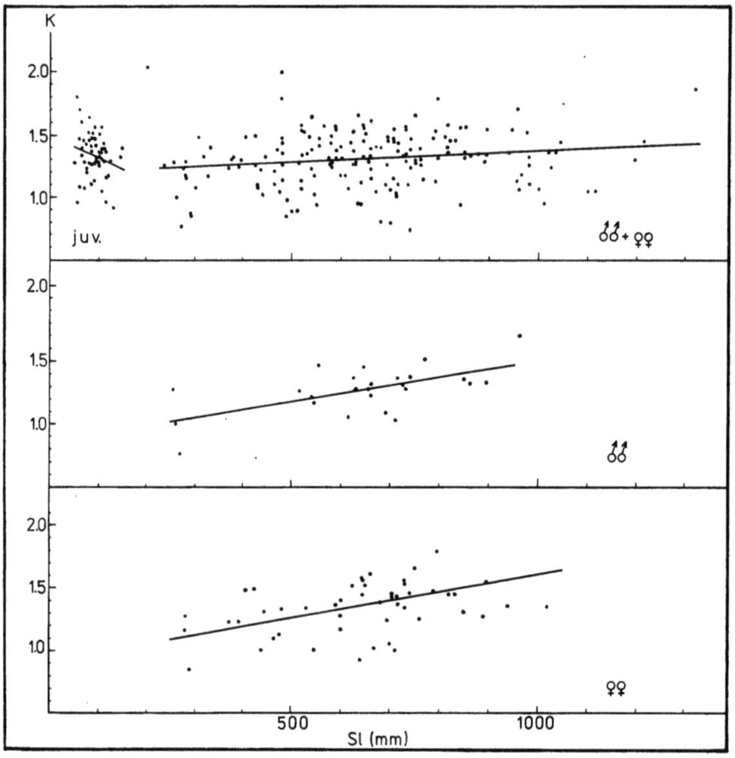

Fig. 29. Correlation between length (*Sl.* in mm, abscissa) and condition coefficient (*K*, ordinate) in the huchen. Equations of the regression lines shown: juv., $y = 1.4890 - 0.0018\ x$, $r = -0.23$, $n = 61$; both sexes pooled, $y = 1.1804 + 0.0002\ x$, $r = 0{,}17^{*}$, $n = 181$; males, $y = 0.8637 + 0.0007\ x$, $r = 0.61^{**}$, $n = 26$; females, $y = 0.9149 + 0.0007\ x$, $r = 0.74^{**}$, $n = 53$.

reservoir on the Angara (data from Luk'yanchikov 1967a), 1.36–1.87 for fish 4+–6+, 344–543 mm avg. *Sl,* the Enisei (Pravdin 1949), males 0.76–1.11 for fish 3+–9+, 469–975 mm avg. *Sl,* females 0.88–1.34 for fish 4+–11+, 580–1076 mm avg. *Sl,* the Vitim (Kalashnikov 1978), 0.92–1.50 for fish 0+–13+, 212–1005 mm in avg. *Sl;* the Angara in the area of the future Ust'-Ilimsk reservoir (Olifer 1977), where *K* varied from 1.02 to 1.46 (mean, 1.31) in taimen 258–952 mm *Sl* and 175–12,613 g, its value increasing with increasing size of the fish even there. In young taimen during the first year of life, the values of *K* vary from 0.69 to 1.28, averaging 1.15 (for fish 22.6–176 mm in *Sl* and 0.08–63.2 g in weight), as can be found by re-calculating data for the population in the Khor river, published by Levanidov (1951).

It is evident that the ponderal index is identical or very similar in the two forms of *H. hucho.*

In conclusion, this chapter can briefly be summarized as follows. *Hucho hucho* is the biggest member not only of the genus but of the family Salmonidae, and it is among the biggest species of freshwater fishes. At the same time, it is among the species with the greater longevity. There are practically no differences between the nominate form and the subsp. taimen, and the few differences that do occur (slower growth in the first third part of life but later faster in *H. h. taimen* than in *H. h. hucho*) should be ascribed to the fact that *H. h. taimen* lives in different environmental conditions. The growth of males and females differs only in the ponderal index, which is slightly higher for the females of both forms. *H. hucho* grows very fast and retains this high growth intensity for very long.

11. Trophic Ecology

Few data are available on the food of the Eurasian huchen, most of which pertain to the taimen. A similar situation is found as regards problems of the trophic relations of the species, despite the fact that this aspect is of great importance in view not only of management, but also of acclimatization experiments. Earlier reports are mostly based on empirical observations (Hollý 1934a, Sabaneev 1911, Tejčka 1934, Ivaška 1951). Exact data based on analyses of contents of stomachs, pertaining to the huchen, have been provided by Vladykov (1925), Kulmatycki (1931a, b, 1935), Shnarevich and Moshuk (1957), Bastl and Kirka (1959), Bastl and Holčík (1968), Nagy (1976), Witkowski and Kowalewski (1982) and Schulz (1985). Their common feature is scarcity of underlying material which, moreover, often came from diverse localities. Better information is available on the food of the taimen which was studied in detail by Levanidov (1951, 1959) and Teplov (1951); valuable information can also be found in other papers (Pirozhnikov 1955, Bukirev 1967, Misharin and Shutilo 1971, Kirillov 1972 and other Soviet authors).

QUALITATIVE COMPOSITION OF THE DIET

Diet of Young

Having changed to exogenous feeding (i.e. about 10–15 days after hatching), young huchen feed predominantly on invertebrates. According to Nagy (1976), young huchen from Slovak waters (the Váh, Mutnianka, Biela Orava, Orava, Turiec) and from the Orava reservoir, 49–325 mm in standard length and 1.5–319 g in weight (age 0+, 1+), preyed especially on larvae and pupae of certain insect groups (Chironomidae, Orthocladiinae, Dixinae), less so on mayfly larvae (Ephemeroptera), freshwater shrimps *(Gammarus)* and caddis-fly larvae (Trichoptera). In terms of weight, however, fish young predominated in their diet (Table 29).

According to detailed investigations, carried out by Levanidov (1951), the diet of young taimen in the Amur river system (the Khor river) consists, shortly after the change to exogenous feeding, of algae *(Spirogyra),* diatoms and terrestrial insects

TABLE 29

Qualitative and quantitative composition of the diet of young huchen from Slovak streams (after Nagy 1976)

Prey organism	Percentage frequency of occurrence	Percent-age weight	No. of prey organisms ingested	Prey size (mm)	Prey size in per cent of predator size
Araneidea	7.69	0.006	1	3	—
Gammarus fossarum	15.38	0.263	3—5	3—9	—
Ephemeroptera larvae	23.08	0.057	3—7	2—5	—
Trichoptera larvae	7.69	0.051	4	—	—
Orthocladiinae larvae	23.08	0.031	3—5	4—7	—
Dixinae larvae	15.38	0.034	2—3	5—7	—
Chironomidae larvae (indet.)	7.69	0.014	1	—	—
Chironomidae pupae	30.77	0.051	1—7	3—8	—
Nematocera pupae (indet.)	38.46	0.064	2—7	5—6	—
Nematocera imagines	23.08	0.014	2	6	—
Diptera (indet.)	7.69	0.039	2—5	5	—
Insecta (indet.)	23.08	0.027	6	—	—
Phoxinus phoxinus	15.38	30.841	2	39—80	27—34
Pisces (indet.)	15.38	68.507	1—3	20—110	22—34

but mainly of nymphae of mayflies (*Ecdyonurus arsenjevi, Ephemerella taeniata, E. basalis, Rhitrogena unicolor,* and *Baëtis* sp.). The fish were 23–26 mm in average size and 80 mg avg. weight. At that time, the young mostly stay inshore and in calm sections with a slow current. In late June, however, the young begin to withdraw from the shores and their diet begins to include, besides larvae of mayflies, *Baëtis* sp., *Procloeon* sp., *Ephemerella taeniata, E. basalis, E. lenoki, E. levanidovae* and *Ecdyonurus arsenjevi,* also larvae of the caddis-fly *(Apatelia zonella).* In July, at 51.8 mm avg. size and 1.41 g avg. weight, the young taimen begin to change to pisciphagy. Forty per cent of specimens examined contained fish in their stomachs *(Phoxinus lagowskii, Acanthorhodeus macropterus)* and 60 per cent had stomachs filled with benthic invertebrates (nymphae of Ephemeroptera). In August, having attained an average length of 75.5 mm and a weight of 3.98 g, young taimen fed exclusively on fish young; in addition to the species mentioned above, there also occurred *Gobio gobio cynocephalus, Orthrias barbatulus toni* and *Mesocottus haitej.* In September and in subsequent months, the diet also contained *Pungitius pungitius sinensis* and the only difference from the summer diet was that the size of the prey fishes was relatively bigger.

Levanidov emphasizes the fact that among the young taimen examined he did not find a single case of a mixed diet, i.e. one composed of invertebrates and fishes at the same time. On the contrary, Nagy (1976) found in his material 10 per cent of young huchen containing a mixed diet, 20 per cent consumed fish, and 70 per cent had invertebrates in their digestive tracts.

The change to pisciphagy takes place at differents sizes and ages. Levanidov

(1951) states that in the Khor river, young taimen begin to prey on fish after having attained around 50 mm in length *(Fl);* according to Nagy (1976), this occurs at about 90 mm *(Sl)* and at the age of 0+–2+. Olifer (1977) also found the change to pisciphagy to occur very early, at the age of 1–2 years; similarly, Tugarina and Kupchinskaya 1977, Pirozhnikov (1955) and Bukirev (1967). According to other authors (Nikol'skii 1956, Pravdin 1949, Misharin and Shutilo 1971, Kirillov 1972, Sigunov 1972), the change to a diet consisting predominantly of fish does not take place until the fish are 3–5 years old, i.e. at a considerably higher age. Apparently, the relative amount and quality of available food are the decisive factors, besides the size attained by the huchen and the taimen at a certain age.

Besides the invertebrates mentioned above, the diet of young taimen and huchen contains practically all aquatic and terrestrial organisms of suitable sizes. Thus the reports of the individual authors mentioned even crustaceans (Cladocera, Copepoda, Amphipoda, Decapoda), worms, molluscs, and aquatic and terrestrial insects.

It continues to be true of the huchen that immediately after having changed to exogenous feeding, the young of cyprinid fishes are its main prey. Among them are often the young of *Chondrostoma nasus,* which fact has given rise to the belief that the presence of young of this species is an inevitable condition for successful growth and development of the huchen, as mentioned earlier. In fact Krauss (1933) maintains that young huchen are directly dependent on young *C. nasus* and that they consume crustaceans and insect larvae only in extreme scarcity of more suitable food. That the young of *C. nasus* or other species are not inevitably necessary for the nutrition of young huchen is evidenced not only by the results of food analyses carried out in young huchen and taimen, but also by the successful acclimatization of the huchen in those river systems where *C. nasus* does not occur (e.g., the Labe (Elbe) system). The predominance of young *C. nasus* in the digestive tracts of young huchen is a local phenomenon depending on the concentration of *C. nasus* in stream sections populated by huchen; at the same time, one should bear in mind that *C. nasus* is, or used to be, the most abundant species of submontane rivers of the Danube system.

The influence of the relative amount of available food can be inferred on from the results of observations published by Levanidov (1951): in the Khor river, the abundance of taimen is low, but that of other species is high; the diet of young taimen is dominated, in terms of frequency of occurrence and weight, by *Phoxinus lagowskii, Gobio gobio cynocephalus,* and *Rhodeus sericeus.* On the contrary, in the Bir river, also belonging to the Amur river system, taimen are relatively abundant, but prey fish are scarce. The diet of young taimen is dominated by the young of *Phoxinus phoxinus,* the most numerous species. Moreover, in the Khor river, the author found no young of *Oncorhynchus keta* in the stomachs of young taimen nor a single case of cannibalism, whereas in the Bir river, with a relatively low amount of available food, the young of *O. keta* were of considerable importance and cases of cannibalism were not rare; for example, seven out of 25 stomachs of young taimen, collected on 22 June, contained from one to three individuals of young of the same species.

Diet of Older Individuals

The species composition of the diet of older individuals is very varied and it is not surprising that many authors, particularly the earlier ones, describe *Hucho hucho* as a voracious and insatiable predator (Herman 1887, Sabaneev 1911, Munda 1926, Sigunov 1972). Besides invertebrates which, however, are only a supplementary and rather exceptional component of the diet of older individuals, *H. hucho* predominantly feeds on fish. The list of fish species found so far in the stomachs of *H. hucho* contains 64 items (Table 30); the diet of the nominate form containing 24 species and that of the taimen 51 species of cyclostomes and fishes. The larger number of fish species preyed upon by the taimen is not a specific trait of this form but results from the fact that the species composition of ichthyofauna in the range inhabited by this form is much more varied. Besides, very few papers have been published on the diet of the huchen.

Again, the preference for certain species of fishes by the huchen is a function of the quantitative and qualitative structure of the trophic base in a particular stream inhabited. Thus the huchen is not a selective predator, which fact has also been pointed out by Schulz (1985) and Witkowski and Kowalewski (1984). Those species which are most frequent, most numerous or most readily available in a given stream predominate in the diet. That is why the reports of different authors differ considerably on this point. Ivaška (1951), V. Dyk (1956) and many others state *Chondrostoma nasus* to be the most important prey species; Kulmatycki (1931a, c, 1935) states that it is the minnow *(Phoxinus phoxinus)*, the barbel *(Barbus barbus)* and the sculpins *(Cottus gobio* and *C. poecilopus)*; Shnarevich and Moshuk (1957) state the barbel *(Barbus barbus)*, the spotted barbel *(B. meridionalis petenyi)* and the spirlin *(Alburnoides bipunctatus)*. From the paper by Nagy (1976) it follows that the diet is dominated, in terms of frequency and weight, by the roach *(Rutilus rutilus)* and the minnow, followed by the dace *(Leuciscus leuciscus)*, the gudgeon (*Gobio* sp.) and the vimba *(Vimba vimba)* (Table 31). In the Dunajec river, the species most frequently preyed upon include *Barbus meridionalis petenyi* and other cyprinid fishes *(Gobio gobio, Leuciscus cephalus, Alburnus alburnus)*, which occurred in the stomachs in largest numbers, whereas the trout and the grayling were found in only 9.8 and 7.8%, respectively of the huchen examined and only singly at that (Witkowski and Kowalewski 1982, 1984). In the Drava river, 44% of the diet (in terms of frequency of occurrence) consists of Cyprinidae, 30% of burbot *(Lota lota)*, 9.7% of pike *(Esox lucius)* and 17.3% of Salmonidae (grayling and brown trout) (Schulz 1985). In the Tormes river, Spain, where the huchen is an introduced and acclimatized species, the major and in fact the only prey is the local species, *Chondrostoma polylepis*, which is allegedly the only fish species that occurs in that river system (Lobón-Cerviá in a letter of Nov. 1979). Almost all authors point out a small percentage of salmonids in the diet of *H. hucho;* however, there are exceptions which then frequently lead to an erroneous evaluation of the economic importance of *H. hucho*. For example, an anonymous author (r. 1970) states that a huchen 15 kg in weight, caught in the Ach river, in the Tyrol, had six specimens of

TABLE 30

Review of qualitative composition (prey fish species) of the diet of huchen and taimen. Compiled from data by various authors. Only species (not geographical forms) are listed

	Hucho h. hucho	*Hucho h. taimen*
Petromyzontidae		
Lethenteron japonicum		+
Acipenseridae		
Acipenser ruthenus		+
Acipenser baeri		+
Esocidae		
Esox lucius	+	+
Esox reicherti		+
Salmonidae		
Oncorhynchus keta		+
Oncorhynchus gorbuscha		+
Salmo salar		+
Salmo trutta m. *fario*	+	
Salvelinus sp.		+
Hucho hucho	+	+
Brachymystax lenok		+
Stenodus leucichthys		+
Coregonus sardinella		+
Coregonus tugun		+
Coregonus autumnalis		+
Coregonus ussuriensis		+
Coregonus lavaretus		+
Coregonus muksun		+
Thymallus baicalensis		+
Thymallus thymallus	+	+
Cyprinidae		
Rutilus rutilus	+	+
Leuciscus leuciscus	+	+
Leuciscus cephalus	+	
Leuciscus idus		+
Leuciscus souffia	+	
Leuciscus waleckii		+
Phoxinus lagowskii		+
Phoxinus phoxinus	+	+
Leucaspius delineatus	+	
Chondrostoma nasus	+	
Chondrostoma polylepis	+	
Xenocypris macrolepis		+
Pseudorasbora parva		+
Gobio gobio	+	+
Saurogobio sp.		+
Barbus barbus	+	
Barbus meridionalis	+	
Hemibarbus labeo		+
Alburnus alburnus	+	
Alburnoides bipunctatus	+	

TABLE 30

(Continued)

	Hucho h. hucho	*Hucho h. taimen*
Vimba vimba	+	
Erythroculter mongolicus		+
Hemiculter leucisculus		+
Rhodeus sericeus		+
Acanthorhodeus macropterus		+
Carassius carassius		+
Carassius auratus		+
Cyprinus carpio	+	
Cobitidae		
Orthrias barbatulus	+	+
Cobitis taenia		+
Bagridae		
Pseudobagrus fulvidraco		+
Anguillidae		
Anguilla anguilla	+	
Gadidae		
Lota lota	+	+
Percidae		
Perca fluviatilis		+
Gymnocephalus cernuus		+
Gasterosteidae		
Pungitius pungitius		+
Gasterosteus aculeatus		+
Eleotridae		
Perccottus glehni		+
Cottidae		
Mesocottus haitej		+
Cottus poecilopus	+	+
Cottus gobio	+	+
Cottus sibiricus		+
Paracottus kessleri		+

brown trout in its stomach. Another author (R. 1970) reports that in the course of 40 years of fishing in the Inn river, he caught 27 huchen, stomachs of most of which contained grayling, followed by brown trout and only then *C. nasus*. The author concludes from that observation that it is the grayling and not *C. nasus* which is the "favourite" food of the huchen. Jungwirth (1980), too, considers the grayling the major item in the diet of huchen in the upper reaches of the Mura river. The alleged preying on trout by huchen introduced into the Hornád river was the reason for suppressing the occurrence of huchen in that stream in subsequent years.

Similar relationships also hold for the taimen, the diet of which consists predominantly of small accompanying species of fishes. However, valuable salmonid species become prey of taimen more frequently, since both the Amur river system and the rivers emptying into the Arctic Ocean contain a large number of salmonids in the wider sense of that term (i.e. members of the subfamilies

TABLE 31

Qualitative and quantitative composition of the diet of huchen (age 2+ to 7+) from Slovak rivers (after Nagy 1976)

Prey organism	Percentage frequency of occurrence	Percentage weight	No. of prey organisms ingested	Prey size (mm)	Prey size in per cent of predator size
Ephemeroptera larvae	10.5	> 0.1	1—3	10—12	—
Plecoptera larvae	10.5	> 0.1	3	21—23	—
Plecoptera imagines	5.3	> 0.1	1	20	—
Coleoptera larvae	5.3	> 0.1	1	6	—
Salmo trutta	5.3	5.4	1	160	28
Rutilus rutilus	10.5	20.8	1—2	150—170	28—31
Leuciscus leuciscus	5.3	19.4	1	240	41
Chondrostoma nasus	5.3	20.7	1	260	34
Vimba vimba	5.3	12.3	1	230	30
Gobio sp.	10.5	4.2	1—2	100—110	17—24
Orthrias barbatulus	5.3	1.5	2	105—122	25—29
Cottus sp.	5.3	0.4	1	60	13
Pisces (indet.)	10.5	15.3	1	140—215	24—45

Salmoninae, Thymallinae and Coregoninae), which often occur in masses. Typically, however, the occurrence of such species in the stomachs of taimen is more or less seasonal (as these species are mostly diadromous or their numbers are seasonally high). Thus Teplov (1951) has found that in the Ilych river (the Pechora river system) the diet of taimen contains young of the salmon *(Salmo salar)* and of the grayling *(Thymallus thymallus)*, which is apparently due to the fact that their numbers are considerable: the author states that in seine net catches there are 150 grayling and 50 young salmon per each taimen caught. On the other hand, in the Amur river no chum salmon *(Oncorhynchus keta)* has been found in the stomachs of taimen even during mass migration times, as the concentration of this species is low even at that time, compared to other species that inhabit the river (Nikol'skii 1956). In Yakutia, young of the coregonids, *Coregonus tugun* and *C. lavaretus pidschian*, often rank foremost in the diet of taimen; in the lower reaches of the Khor river (the Amur river system), it is the young of chum salmon *(Oncorhynchus keta)* and pink salmon *(O. gorbuscha)* (Kirillov 1972, 1976, Levanidov 1951, 1959). In such cases, seasonal occurrence is involved. In the taimen, too, the "usual" diet should be considered to contain mainly small cyprinid fishes, such as roach, minnow *(Phoxinus phoxinus* and *P. lagowskii)*, dace *(Leuciscus leuciscus baicalensis)* and, in places, perch and ruffe *(Gymnocephalus cernuus)* (Berezovskii 1924, Pravdin 1949, Kirillov l.c., Levanidov l.c., Misharin and Shutilo 1971, Bukirev 1967).

Even though *H. hucho* is predominantly a territorial predator, as will be shown below, foraging migrations may occur. Thus Luk'yanchikov (1967b) reports that the concentration of taimen in the Kheta river (a tributary of the Khatanga river) increases in autumn due to migration from the tributaries to the main stream where the taimen prey on *Coregonus sardinella* which spawn there in masses at that time.

Besides fishes, the stomachs of *H. hucho* were found to contain amphibians, reptiles, small mammals and birds (Herman 1887, Sabaneev 1911, Kulmatycki 1931c, Ivaška 1951, Kirillov 1976, Misharin and Shutilo 1971, etc.). However, one may conclude from the interesting report by Sigunov (1972) that in periods of overcrowded populations and subsequent migrations of lemmings *(Lemmus)*, Siberian chipmunks *(Eutamias)* and squirrels *(Sciurus)*, these mammals, crossing rivers on their migrations, often become the only, although short-term and seasonal, prey of the taimen.

The representation of individual species of fishes in the diet of the taimen also depends on their abundance and availability. While Teplov (1951) writes that in the Ilych river taimen prefer salmon fingerlings to those of grayling, this observation being based on their opposite ratios in taimen stomachs and in hauls obtained with seine nets, it is our assumption that smolt salmon are more readily available prey than the stationary grayling young.

PREY SIZE

The size of the prey of *H. hucho* varies in dependence on the age and size of the predator.

Levanidov (1951) reported that young taimen 23–26 mm long *(Fl)* preyed on organisms whose size was 8–10% of its own length. The average size of fishes consumed by somewhat bigger young taimen (50–190 mm *Fl*) varied between 11 and 40%. The data of Nagy (1976) indicate that the size of the prey of young huchen varies within the range 22–34% (Table 29). With bigger fish the relative prey size is smaller although it may be quite large absolutely. In Slovak huchen this is 13–45% (Table 31), in those from the Transcarpathian region of the Ukraine 15–29%, in the taimen from the Amur river system 10–30% on an average (Nagy l.c., Shnarevich and Moshuk 1957, Levanidov 1951). The prey size varies not only with the age of the fish (Table 32) but also according to season, as emphasized by Levanidov; this author found that young-of-the-year taimen preyed on relatively bigger prey in autumn (20–43%) than in summer (10–25%) or in the next spring (15–22%). The same holds for sexually mature taimen, in which the sizes of prey fish increased from 10–25% to 20–43% in autumn and dropped to 10–18% in winter.

Naturally, the absolute prey size increases with increasing age and size of the predator. In huchen from the Dunajec river, 267–1093 mm in *Tl*, the correlation between the size of prey *(y)* and predator *(x)* is described by the equation $y = -1.99 + 0.29\,x$ ($r = 0.75$; Witkowski and Kowalewski 1984). Although the predominant part of the population feeds on small cyprinid fishes, large fish may prey on bigger prey species, as demonstrated by Levanidov (1951) on concrete examples: in the stomach of a taimen 115 cm in length *(Fl)* and 13.3 kg in weight (age 10+), he found two specimens of *Erythroculter mongolicus*, totalling 850 g in weight; a taimen 1470 mm long and weighing 24 kg contained two specimens of lenok *(Brachymystax lenok)* 1.2 kg in total weight; and the stomach of a taimen 172 cm long and weighing

TABLE 32

Relationship between the size of huchen and that of its prey (after Shnarevich and Moshuk 1957)

Age group	*Fl* (cm)	Average prey size (cm)	Average relative prey size (per cent predator's size)
0 — I	33	7.5	22
II — III	38—45	9	29—20
IV — V	53—57	10.7	20—18
VI	67	11	15.9
IX —X	78—84	13	16—15

32 kg was found to contain a female *Oncorhynchus keta* weighing 3 kg. According to Teplov (1951), stone loach and Siberian sculpin occur in the stomach of taimen up to 2 kg in weight, whereas grayling and young salmon (both 12–15 cm in length) were not found in the stomachs of taimen until 4–6 kg in weight. Additional concrete cases have been described by other authors; let us mention three more of these: Kirillov (1972) reported that a muskrat *(Ondatra zibethica)* weighing 1 kg was found in the stomach of a taimen 118 cm long *(Fl)* from the Lena river. Sigunov (1972) mentions a taimen 5 kg in weight, found in the Vilyui river, choked to death when swallowing a pike equal in weight. Lastly, Kvet (1961) describes a case in which a huchen 74 cm long, observed in the Váh river, seized a two-day-old dog cub, but choked to death when attempting to swallow it.

Big and heavy huchen in pond cultures, depending on food supplied by man, also seize large prey (according to Ivaška 1951, a male huchen 20 kg in weight consumed fish over 1 kg in weight; according to Skácel, up to 3 kg in weight). In the wild, however, such cases are exceptional rather than regular.

AMOUNT OF FOOD CONSUMED; FOOD COEFFICIENTS

In the stomachs of huchen from Slovak waters, Nagy (1976) found 1–7 individuals of various invertebrates in young-of-the-year and yearlings, and 1–3 individuals of invertebrates or 1–2 fish in huchen older than 2 years. In the stomachs of huchen 1–5 years old, Shnarevich and Moshuk (1956) found an average of one, in older specimens of two fish. In huchen from the Cheremosh river, Kulmatycki (1931c) found 1–5 fish; in those from the Dunajec river, Witkowski and Kowalewski (1984) found 1–9 fish, the higher numbers coinciding with smaller dimensions of the prey. Teplov (1951) reports that in taimen from the Ilych river sculpin and grayling (the latter up to the size of 15 cm) occurred in twos, stone loach in threes, and young salmon (12–15 cm in size) in fours in their stomachs. The index of filling in the huchen and the taimen varies in dependence on the season and size of the fish. In huchen from the Dunajec river, the mean values of this index are 0.9% in summer, 2.07% in autumn, 5.09% in winter and 0.1–3.35% in spring (Witkowski and Kowalewski 1984). Similarly, in taimen from the Khor river (the Amur river system), the index of filling was lowest in summer (0.9%) and highest in autumn and winter (3 and 5%, respectively). In young-of-the-year from that river, which fed on invertebrates, the index of filling was 2–2.8%, but in summerlings and yearlings feeding on fish it was 2–5.5% (Levanidov 1959).

The amount of food consumed during one feeding bout may be considerable, as demonstrated by the following examples: a huchen 5 kg in weight, caught in the stilling basin below the Orava reservoir in 1956, contained in its stomach 205 individuals of sunbleak *(Leucaspius delineatus)* (Nieslanik); in the Kama river, Bukirev (1967) found, in taimen examined by him, up to 40 individuals of minnow *(Phoxinus phoxinus)* and up to 15 individuals of stone loach *(Orthrias barbatulus)*. In a huchen 720 mm long from the Turiec river, Skácel found three individuals of brown trout *(Salmo trutta* m. *fario)*, two individuals of grayling *(Thymallus thymallus)*, one pike *(Esox lucius)* and one sculpin *(Cottus gobio)*. A huchen from the Drava river, 20 kg in weight, contained 7 burbot *(Lota lota)* 2.4 kg in total weight (Schulz 1985). Sigunov (1972) describes a case where, on 2 June 1961, when a migrating herd of lemmings *(Lemmus)* had crossed the Anabar river, he caught a taimen which contained 42 individuals of this rodent, and three coregonids and 15 lemmings in the stomach of another taimen. In the autumn of 1962, squirrels swam across the Severnaya river, and in the evening of that day waves of the river washed ashore a dead taimen containing 14 squirrels in its stomach; the fish choked to death when swallowing the last one. However, the author makes no mention of the size of these taimen.

The food coefficient of *H. hucho*, i.e. the amount of food required to produce 1 kg of weight of the predator, has not been the subject of special investigations. Jungwirth (1978) states that it varies between 4 and 5, but this value appears to be too low when compared with other predatory fishes (Fortunatova and Popova 1973). According to Schulz (1985), who sets out from the data of Jungwirth (1980) and of Carinthian breeders, the food coefficient of huchen in the second year of life is 5; by the 5th year it increases to 10, and at the age of 20 years its value is 19. According to Schulz's calculations, it requires 500 kg fish to produce a weight of 34 kg (in the course of 20 years). Nothing is known about the food ration, i.e. the amount of fish consumed by a predator per day, month or year (thus the daily, monthly or annual ration). According to Ivaška (1951) and Skácel, in huchen pond cultures it is necessary to reckon with an amount of 10 kg fodder fish per kg of sexually mature huchen per annum (Chapter 24). This is obviously overestimated, since according to many years' rigorous observations and experiments by Fortunatova and Popova (l.c.), the annual ration is 2.3–2.4 for the pike, 1.7–2.2 for the wels, 2.0–2.4 for the pike perch, and 1.5–1.7 for the asp. Pavlík (1955) states that in young pond-reared huchen, the average food consumption over 90 days is 35–40 young cyprinid fish per individual, i.e. one fodder fish is enough for a huchen per 2.5 days on average. Within these 90 days the huchen will grow from 4–6 cm to 15–18 cm *(Tl)*.

The time required for food digestion is unknown, except for a single case (Sigunov 1972), when a taimen 6 kg in weight digested 6 individual arctic grayling *(Thymallus thymallus arcticus)* within three days.

SEASONAL AND DIURNAL DYNAMICS OF FEEDING

The seasonal dynamics of feeding of the nominate form are still not known exactly. It may be inferred from the experience of practical fishermen that huchen prey most actively in autumn and winter, but do not ingest food during long frosts (Robida 1902). According to Nieslanik, huchen take food most actively after spawning, i.e. in spring, and also before the onset of winter. They ingest food even during the spawning period. It is evident from the papers of several authors (Levanidov 1951, Nikol'skii 1956, Bukirev 1967, Luk'yanchikov 1967b) that taimen ingest food most intensively in autumn and winter, least so in summer when food intake may even be stopped entirely. Misharin and Shutilo (1971) are the only authors to mention decreased food intake in winter; most probably this pertains to periods of heavy frosts when the food is scarce and simply not available to the fish.

H. hucho feeds not only during the spawning period but even at the spawning site (Pirozhnikov 1955, Nieslanik), during the day and even at night (Pravdin 1949, Berezovskii 1924, Bukirev 1967, Nieslanik). Some authors (Nikol'skii 1956, Misharin and Shutilo 1971), however, deny any food intake during the spawning period.

MANNER OF TAKING FOOD

Ivaška (1951) states that the huchen ingests its prey head first as a rule, but if its stomach already contains fish the subsequent prey is seized and swallowed tail first. According to Bogyi (1962), the huchen seizes its prey from above, by its back, then turns it in its mouth and swallows it head first. His observation has been confirmed by Nieslanik who adds that only small prey fish are swallowed regardless of the position of their heads.

Hucho hucho actively searches for prey and often pursues it so eagerly that it is stranded in shallow water (Sigunov 1972). Berezovskii (1924) observed a taimen which, lying near the bottom, slowly moved its tail from time to time to raise the fine bottom sediment, whereby it attracted minnows, which it then seized, four times within 30 minutes on average. Sigunov (1972) states that taimen often hunt in groups, one part of the group lying in ambush, the other driving the prey actively.

Besides live prey, which it prefers, *H. hucho* also takes dead food, although only rarely and very cautiously (Sigunov 1972); however, fish kept in ponds soon learn to take dead food and later they take fresh dead fish quite readily (Nieslanik).

It is evident that from early youth *H. hucho* is a predator. Its orientation is primarily visual, as follows from the observations of Nieslanik where young huchen preyed on cladocerans; the fish took notice of their prey only if it was moving actively.

DIFFERENCES BETWEEN SEXES IN FOOD INTAKE

Literature is so far lacking on differences in quality, amount or rate of food intake between male and female *H. hucho*. The sex ratio among fish caught with hook and line suggests, however, that females are more voracious than males, and hence are more frequently caught, which is the main cause of the uneven sex ratio, as also assumed by Nieslanik. It may be a merely seasonal phenomenon, based on the fact that in Czechoslovakia huchen fishing is permitted only during winter months, in which period the development of female gonads requires a greater supply of energy than that of male gonads. This assumption requires increased attention or even experimental verification in future. Should it prove to be true, it would be necessary to shift the open season to a period in which the feeding intensity (and thus also the catchability of huchen with hook and line) is equal for both sexes.

FOOD COMPETITORS; RELATIONS TO OTHER SPECIES

Until the change to predatory (piscivorous) habits, the range of food items taken by *Hucho hucho* greatly resembles that of brown trout (Straškraba et al. 1966, Tuša 1968), as pointed out by Nagy (1976). This, however, is only a temporary similarity, as the competition for food with trout disappears due to the fast growth of *H. hucho*. In the eastern part of its range, *H. hucho* competes for food with inconnu *(Stenodus leucichthys nelma)*, burbot *(Lota lota)*, pike *(Esox lucius)*, lenok *(Brachymystax lenok)*, and grayling *(Thymallus baicalensis)*, as indicated in the papers by Pravdin (1949) and Tugarina and Kupchinskaya (1977). The trophic relations were specially studied by Levanidov (1959) in the lower reaches of the Khor river. The author found that they were very complicated, particularly as regards the parallel occurrence of chum salmon *(Oncorhynchus keta)*. On the one hand, one- or two-year-old taimen act as predators, consuming young of chum salmon, and on the other taimen of these and older year classes intensively prey on the major competitors of chum salmon *(Phoxinus lagowskii, Gobio gobio cynocephalus, Pseudorasbora parva, Brachymystax lenok)*, thus indirectly securing food for chum salmon and, moreover, removing its direct enemies.

In the middle part of the Irkutsk reservoir, Tugarina and Kupchinskaya (1977) found a high percentage dietary overlap between burbot, pike, lenok and grayling, amounting to 71.13 between taimen and pike plus burbot, and to 62.87 between taimen and perch. The authors concluded, however, that the competition pressure decreases due to low numbers of predators in various localities.

CANNIBALISM

Cannibalism occurs among captive *H. hucho* (Ivaška 1951, Nieslanik) as well as in the wild, the latter less frequently (Misharin and Shutilo 1971). Pirozhnikov (1955) believes that predation by taimen on members of its own species is a reflex action and only exceptional, e.g., when confined to a diminishing space by a seine net being hauled. Levanidov (1951) found cannibalism among taimen in the Bir river (the Amur river system) and correlated it with low numbers of other fish species, stating that in this case it was self-regulation of the numbers of their own species. In the wild manifestations of cannibalism are rare as, on the one hand, *H. hucho* can find enough available food and, on the other, young fish are spatially segregated from adult ones (Chapter 8).

12. Predation

Although Eurasian huchen is practically the only big specialized predatory fish species in the submontane zone and, like any predator, influences populations of other species, concrete data on the extent of this influence are almost completely lacking, particularly as regards its quantification. The only data in this point were obtained after the introduction of *H. hucho* into the Usses river (the Rhône river system, Upper Savoy, France) and published by Vivier et al. (1964) and later adopted in a slightly altered form by Svetina (1967). Prior to the introduction in 1956, control fishing in a limited section of the river yielded 8 tonnes of fish; the catch was dominated by the nase carp *(Chondrostoma nasus)* and there were occasional chub *(Leuciscus cephalus)* and barbel *(Barbus barbus)*. In the same section and manner, only 2.8 tonnes of fish were caught in 1961 (4 years after the introduction), 1.9 tonnes in 1962 and 1.3 tonnes in 1963. It is of interest that the space vacated by the nase carp and other, secondary and accompanying fish species was gradually occupied by brown trout *(Salmo trutta* m. *fario)* whose numbers increased, particularly in weight groups 100–200 g. At complete variance with this report are opinions of anglers who assert, as we were informed by Laurent (in a letter of 11 Aug. 1977), that the introduction of huchen was followed by a decrease in the numbers of brown trout (allegedly as a result of their destroying trout redds!!!) and nase carp and, on the contrary, by an increase in the numbers of barbel and chub. It is merely possible to add that if the numbers of brown trout actually decreased, this must have taken place only after the numbers of nase carp and other cyprinids had decreased to such an extent that they became relatively less easily available to the huchen, whereby the relative availability of brown trout was increased.

Novák (1971) correlates the increasing numbers of nase carp in the Turiec river with the decreasing numbers of huchen in that river. He states that until 1945 the numbers of nase carp in the Turiec were not very high, and anglers even caught nase carp and transported them over the weir near the paper mill in order to increase their numbers upstream of Martin. The same author also mentions a record, found in the files of the former Anglers' Club at Martin from 1941, stating a good stock of huchen but a bad one of "white fish". The author furthermore notes that after a drop in the numbers of huchen in 1950–1960, the numbers of nase carp increased to the extent that this species became dominant in the Turiec river. Bastl et al. (1976) also believe that the decrease in the numbers of huchen in the Turiec river contributed considerably to the increase in the numbers of nase carp, chub and dace.

In a foregoing section, we mentioned the case of the Hornád river where, after a successful introduction of the huchen into the Hornád defile near Spišská Nová Ves, the numbers of this species were later gradually lowered, as it allegedly exerted a negative influence on brown trout and grayling populations. Skácel (1976), who briefly mentions this case, states that "due to adaptation and pollution of the stream, huchen migrated up to the trout zone". It follows from this statement that, in fact, the huchen were deprived of their food sources in the foothill section of the stream and were compelled to ascend to the submontane brook zone. Skácel (l.c.) mentions another case of forced elimination of huchen in the Rajčianka river, but gives no further particulars.

The numbers of huchen have been limited in other streams as well. Thus an anonymous author (d. 1970) reports that in the lower reaches of the Ammer river it is necessary to limit the numbers of huchen in view of those of grayling and trout, similarly as in the upper reaches of the Isar river.

In the Tormes river, Spain, no negative influence on the population of the endemic species, *Chondrostoma polylepis,* was observed after the introduction and acclimatization of huchen (Lobón-Cerviá in a letter of Nov. 1979). This is probably because the numbers of huchen are small or because they populate only a short stretch of the stream.

Indirect and often uncertain reports only are available on the predation pressure of the taimen. Thus Pirozhnikov (1955) reports that since the numbers of taimen are larger in the Lena than in the Enisei, the hauls of fish are smaller, particularly of species which constitute their prey *(Coregonus sardinella, C. tugun,* small individuals of *C. autumnalis migratorius).* For that reason, the author recommends increasing the harvest of taimen to 150–200 tonnes annually. In the Vilyui river (Kirillov 1962), within the limits of the Suntar region, between the Kuchchugui Khan rapids and the mouth of the Ogogut river, the local fish fauna lacks *Coregonus muksun, Rutilus rutilus* and *Leuciscus idus,* allegedly due to a large number of predatory fish species, predominantly taimen (besides *Esox lucius, Lota lota* and *Perca fluviatilis*).

Tugarina and Kupchinskaya (1977) characterize predatory fishes (including the taimen) as biological ameliorators, as they feed on less valuable fish species.

It is obvious that the predation pressure exerted by *Hucho hucho* may not be uniform; it may vary from one case to another according to the character of the stream, the species composition of its fish fauna, and the abundance of individual species. Knowledge of the predation pressure is very important for a correct estimation of the economic importance of *H. hucho* in a particular stream, and it must rank first when acclimatization experiments are considered. An idea of the predation pressure of *Hucho hucho* can be derived from the data of Gritsenko and Churikov (1977) on the related species, *Hucho perryi,* in Sakhalin. The authors report that in the Nyiskii Bay, at a daily ration (i.e. amount of fish consumed per day, related to the weight of the predator) of 1.84%, a population of *H. perryi* 1270 kg in total weight consumed, during June to August, 16.5% of the standing stock of *Osmerus mordax dentex,* 7.9% of *Hypomessus olidus,* 17.1% of *Eleginus gracilis,* and 1.3% of *Zoarces viviparus elongatus.* The total weight of fish consumed within 90 days was 1677 kg (including over 1 tonne of herring, *Clupea harengus pallasi,* whose population density was not estimated, as it was a case of mass spawning migration). The authors emphasize that in spite of its low abundance, *H. perryi* exerts a considerable pressure on the numbers of certain fish species. Apparently, similar pressure may be assumed to be exerted by *Hucho hucho,* and it is the problem of the feeding ration and its corollary the determination of optimum numbers of this predator, which will have to be resolved in the near future.

13. Predators and Enemies

Predators which are dangerous for the laid eggs, newly hatched young and older juveniles of the huchen include several species of fishes (nase carp, grayling, brown trout, sculpin, burbot, chub, pike), amphibians (frogs, newts), reptiles (grass snake), birds (grebe, kingfisher, heron, cormorant, gulls, stork) and mammals (water-shrew, otter) as well as aquatic insects and their larvae (water-beetle, backswimmer) (Robida 1902, Vladykov 1929, Hollý 1934b, Ivaška 1951, Prawocheński and Kolder 1968, Novák 1971, Nieslanik). Ivaška (l.c.) denotes the nase carp, grayling and brown trout as "the greatest enemies" of the huchen, since "they directly accompany the female going to spawn, to get possession of her eggs to such an extent that they destroy as much as 95% of them". It should be remarked that this statement is unsubstantiated; the fish species enumerated by Ivaška, as well as other species feeding on fish eggs, are unable to do any harm to huchen eggs or to those of other brood-hiding species, as they can only seize those eggs which were not covered in the redd and thus are doomed to annihilation. Moreover, the male huchen will drive away from the redd members of its own species as well as other fish species (Chapter 9).

Pavlík (1955) states that huchen 6–9 cm in size fall prey to pike 15–18 cm, chub 18–22 cm, burbot 18–20 cm and other huchen 21–23 cm in size. Beyond doubt, these data are rough estimates, as relating the sizes to that of the predator results in exaggerated percentage values compared with data determined in an exact way (Popova 1967).

Young huchen reared in ponds in Spain for the purpose of introduction were consumed by grey herons *(Ardea cinerea)* from a nearby breeding colony, and by grass snakes *(Natrix maura)* (Lobón-Cerviá in a letter of Nov. 1979).

The predators of *Hucho hucho taimen* include, in addition to most of the species mentioned above, lenok, perch, pike and the members of the same species, as distinct predators of young, while adult taimen may fall prey to polar foxes, bears, wolverines and seals (in Lake Baikal) and, of birds, eagles (Kirillov 1976, Sigunov 1972).

Almost all authors emphasize, however, that adult huchen and adult taimen have no enemies except man and his interference with river ecosystems.

14. Population

No exhaustive data are available on the population of *Hucho hucho* and its dynamics, as these important aspects have not been the subject of special investigations. In the literature only fragmentary data are available, pertaining to bag sizes obtained from certain streams, their age structure, or relative population density.

AGE STRUCTURE

Only the data published by Sabioncello et al. (1970) are available, pertaining to the age structure of the population of the nominate form in the Dobra and Kupa rivers, Yugoslavia. Together with information on the probable age structure of the huchen "population" in the Sava and Ljubljanica rivers, these data are given in Table 33. As to the latter "population", it is necessary to add that the information was derived from over twenty years' records made by Jagoditsch and Munda between 1900 and 1925 and published by Munda (1925). This information, however, pertains to body weights which were converted to standard lengths and analysed by Andraščík (1980) with regard to age. Sabioncello's and Andraščík's data differ mainly in that the former pertain to an actual population (sampled by means of an electrical shocker), whereas the latter pertain to anglers' bags. The comparison shows that at the beginning of the century the bulk of the bags consisted of sexually immature fish or those that had spawned for the first time. Beyond doubt, the rather low representation of fish 1–2 years old in the Kupa and Dobra rivers is due to the fact that individuals of these age groups predominantly inhabit the upper reaches of streams.

Information on the structure of taimen populations is provided by several papers, all of which are based on analyses of samples stemming from net-caught bags. In the

TABLE 33

Percentage age structure of huchen populations in selected river drainage basins

Basin	Age group 1	2	3	4	5	6	7	8	9	10	11	12	13
Kupa[1]	4.8	9.7	35.5	24.2	6.5	6.5	8.0	3.2	—	1.6			
Dobra[1]	4.0	4.0	30.0	8.0	16.0	18.0	4.0	4.0	10.0	2.0			
Sava, Ljubljanica[2]	—	—	—	39.8	27.7	17.4	1.2	0.1	6.6	1.3	4.0	0.1	1.8

[1]Sabioncello et al. (1970); [2]Andraščík (1980).

Frolikha river, but in other rivers as well, around 70% were fish 4–6 years old, the rest were 7–20 years old (Misharin and Shutilo 1971). In the Kama river, fish 3–5 years old constitute the bulk, whereas older ones occur singly (Bukirev 1967). The ages of the taimen in the Khatanga river range between 4 and 18 years (Luk'yanchikov 1967b), in the Angara river (in the area of the future Ust'-Ilimsk reservoir) fish up to 15 years are caught (Olifer 1977). From the data on the length ranges of the taimen in commercial catches of fishermen on the Lena river (in which fish 81–120 cm *Fl* predominate, those 21–50 cm *Fl* making up 20% of the catch), it may be estimated that the bulk of the catches is fish 7–13 years old. Kalashnikov (1978) reports that catches of taimen in the Vitim river are predominantly composed of fish younger than 7 years, which make up 68.1%.

Perhaps it should be added to the above information that by the average weight of the huchen in the waters of Slovakia (Table 37), their age can be estimated at 4–15 years.

Data on the size structure of the population or bags of the huchen and the taimen are given in Chapter 10.

ABSOLUTE AND RELATIVE ABUNDANCE AND BIOMASS

Data on population density and biomass of *Hucho hucho* are practically lacking, as no special investigations have been made in this respect; however, some idea may be obtained from several data on the number of fish caught per unit of angling effort, or on bags obtained from a certain stream or its part.

According to Kulmatycki (1931a), individual anglers caught up to 180 huchen (1–2 kg in average body weight) annually in the middle reaches of the Cheremosh river. Hollý (1934) reports that in 1906 it was possible to catch 10–15 huchen (10–30 kg in body weight) within 2–4 hours of spinning in the Váh river near Ružomberok. According to the same author, seven huchen (two weighing 10 kg each, two 4 kg each and three smaller) were obtained in a single haul with a seine net in the Váh river near Vlašky. Angling in the Váh river (near Ružomberok) in 1907, Baroness Révay and her son caught the following numbers of huchen: on 2 November, 8 fish 5–12 kg in weight in 5 hours of angling; on 22 November, 10 fish 3–18 kg in weight in 6 hours; on 2 December, 10 fish 3–11 kg in weight in 3 hours

(Stork 1908, also showing a photograph). Novák (1971) reports that in 1925, an angler named Ďurkovič caught with rod and line 120 huchen in the Turiec river within 4 months, and Samo Ivaška caught 28 huchen in the same river in a single summer evening of 1929. In 1933, Nieslanik caught 14 huchen 5–14 kg in weight within one afternoon of angling in a section about 5 km long on the left bank of the Váh river near Kraľovany. It is therefore no exaggeration when Hollý (1934) states that prior to 1918, on the Váh river near Ružomberok, it was not uncommon for an angler to take away his whole day's huchen bag on a cart! On the basis of data on bags obtained in the Turiec river in successive years of 1926–1931, Bastl et al. (1975) estimated that about 1540 huchen of age groups I–VI were present per hectare of water surface. While this value is very high, it does not seem to be unreal in view of the data stated above. Additional data, pertaining to the situation after World War II, were recorded by Nieslanik: on 12 December 1946, in the Váh river near Vlašky, an angler named Kuna caught 11 huchen 107 kg in total weight (i.e. two fish 5 kg each, two 8 kg each, two 10 kg each, one 16 kg, three 12 kg each, and one 9 kg in weight). The same angler caught 30–40 huchen within one season; he maintained that in a pool in the Váh river near Vlašky it was possible to catch 12–15 huchen a day.

At present, the density of the huchen population is generally much lower. Shnarevich and Moshuk (1957) estimated that in the Cheremosh river there are only 2–8 huchen per kilometre of the stream (the authors themselves stating that the population has been decimated). According to Kux and Weisz (1964), the numerical ratio of yearling huchen to the remaining fish species in the Turiec river was 1:100 to 1:200. The low abundance of the species is also evident from the paper by Kosorić et al. (1975); according to these authors, in the catches obtained with gill nets (15 × 15 or 20 × 20 mm mesh size) from the Zlatarske reservoir on the Uvac river, Yugoslavia, the huchen was represented by only 0.96%, ranking after *Chondrostoma nasus, Leuciscus cephalus, Barbus barbus* and *B. meridionalis petenyi, Rutilus pigus, Alburnoides bipunctatus* and *Gobio gobio.* At present, a very low representation of huchen is observed in the Poprad river: Kirka (1975) found the relative abundance varying from 0.45 to 1.45% and biomass from 0.42 to 1.83% in the river section between Spišská Belá and Čirč.

Still fewer data are available on the population density of *H. h. taimen.* In the Kama river, in a place called Govorlitskaya Yama, a single haul with a seine net 150 m long yielded 63 taimen, half of which were over 10 kg in individual body weight (Bukirev 1967). The author does not specify the circumstances under which this catch was obtained; it could have been a pre-spawning grouping. As a matter of fact, Pirozhnikov (1955) states that "even in regions where the annual bag attains 350–380 metric centners, one or two taimen can be obtained in a haul of a seine net as a rule". Additional information, pertaining to commercial catches of taimen, is given in Chapter 19.

SURVIVAL AND MORTALITY

In this connection we can only mention our calculations based on data on the age structure of huchen populations in the Dobra and Kupa rivers, contained in the paper by Sabioncello et al. (1970) and in that by Andraščík (1980), pertaining to the Sava and Ljubljanica rivers. The respective values of percentage representation of individual age groups, after being smoothed and plotted in a co-ordinate system, provided a basis for constructing a catch curve (Fig. 30). It is evident that the curve showing the decrease in the numbers in age groups 5–8 in the Sava and Ljubljanica rivers has a very steep decline; the average annual survival rate is only 51.2%, which is considerably less than in the same age groups in the Dobra and Kupa rivers where the respective survival rates are 77.2 and 65.8%. Taking into account the values of the age groups on the whole declining part of the curve and calculating the survival rates *(S)*, mortality rates *(A)* and the instantaneous mortality rate *(Z)*, the resulting values for the three streams are as follows: Dobra $S = 0.783$, $A = 0.217$, $Z = 0.245$; Kupa $S = 0.701$, $A = 0.299$, $Z = 0.355$; Sava and Ljubljanica $S = 0.560$, $A = 0.440$, $Z = 0.580$. Hence it follows that the mortality rate in the present populations (in the Dobra and Kupa rivers) is higher than in those 50–70 years ago (in the Sava and Ljubljanica rivers). There are two possible explanations: either the huchen populations were more numerous in the past than they are at present (see the facts mentioned in the preceding chapter), or it is a reflection of their incorrect management in the past, when the legal size was too small and the open season too long, 9 months a year.

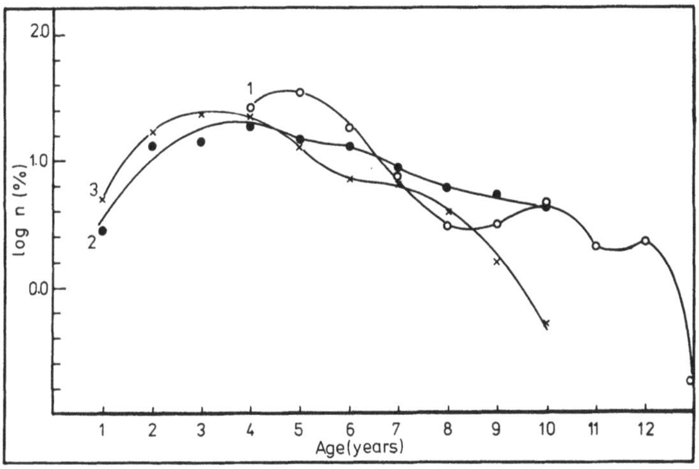

Fig. 30. Catch curves for huchen populations in (1) the Sava and Ljubljanica rivers, (2) the Dobra and (3) the Kupa rivers. See text for explanation.

As for the taimen, we can only present values calculated from the data published by Misharin and Shutilo (1971) and concerning a catch of 500 taimen in the Angara river near Dubinino in 1959. The catch contained fish 50–130 cm in length *(Fl)* and

4–20 years old, but a continuous series comprises only fish 4–15 years old. The respective parameters for a non-smoothed and a smoothed catch curve are $S = 0.645$, $A = 0.355$, $Z = 0.44$, and $S = 0.607$, $A = 0.393$, $Z = 0.50$. Compared to that ascertained in the huchen, the survival rate of the taimen is evidently higher. However, considering only the survival rate of age groups 5–8, the resulting value of $S = 0.508$, which is almost identical to that of the huchen in the Sava and Ljubljanica rivers at the beginning of this century. This fact tends to indicate that the taimen populations are also managed incorrectly, like those of the huchen in the past. Actually, there are no limitations on taimen fishing so far and therefore the catches predominantly contain sexually immature individuals, as explicity pointed out by Misharin and Shutilo (1971).

With the present status of knowledge, it is impossible to decide on the distribution, in the cases mentioned above, of natural and so-called fishing mortality. Per analogiam with other fish species, the fishing mortality should include losses due to mortality of wounded fish which succeeded in freeing themselves from the hook, or those wounded by seine fishing, and subsequently dying as a result of fungal disease, shock, or inability to ingest food. Svetina (1970) states that the number of fishes being twice lower in the length group 71–80 cm (*Tl;* beyond doubt the author refers to Munda's data from the Sava and Ljubljanica rivers) is due to a high mortality rate, resulting from wounds suffered by fishes that freed themselves from the hook; unfortunately, the author presents no evidence supporting this statement.

15. Migration and Territory

MIGRATIONS

Hucho hucho can be described as a stationary species whose members, under normal conditions, occupy certain limited territories. In general, huchen young keep to places where they hatched, but with advancing growth in length and with increasing habitat and food requirements the young fish gradually move downstream until they finally settle in a definite territory; they stay in it more or less permanently, except for sudden changes occurring in their environment and for the spawning period.

Apart from spawning migrations, as mentioned in Chapter 9, the most frequent migrations are caused by such changes in the habitat as variation in water level, water turbidity, increased water temperature, disturbance by angling, hydraulic works, and water pollution. In such cases of temporary environmental changes the huchen withdraw into tributaries or lateral branches, moving both upstream and downstream. As soon as any of the factors mentioned above ceases to affect the environment huchen will return to the original place. Krauss (1933) found that migration may also be induced by tagging and releasing the fish; in that case younger individuals migrated upstream and older ones downstream. According to Krauss (l.c.), such migrations are short, not exceeding 1–3 km from the release point. Vladykov (1926) reports that huchen keep to a certain place throughout the summer, but as autumn approaches big and sexually mature individuals migrate downstream to the mouths of rivers where they had spawned, whereas young and sexually immature ones stay in the same places. Shnarevich and Moshuk (1956), however, state that the huchen migrate into submontane river sections in late October and return in December and January. As also indicated by data concerning the taimen, such migrations can obviously be classified as foraging migrations, induced by increased energy requirements during gonad development. The distance of such migrations is not known, but apparently depends on the places of concentration of fish species preyed upon by the huchen.

Young huchen also migrate to lower reaches of streams at the approach of winter, as observed by Nieslanik. In that case the fish, one or two years old, start migrating at the onset of first frosts and select deeper pools, which do not freeze solid even during the heaviest frosts. According to this author, frosts are also a signal for bigger huchen to migrate away from shallower sections liable to become frozen to considerable depths.

Enforced migrations caused by long spells of hot weather which result in overheating of water and a decreased content of dissolved oxygen, are known to have occurred in both the Orava and Tisza rivers. Dyk and Nieslanik (personal communications) mention that in such periods (with frequent mass mortality of huchen) the huchen aggregated in the mouths of tributaries with cooler water. Thus in the summer of 1950 a considerable number of huchen from the Orava river aggregated in the mouth of the Studený potok Brook, flowing down from the Roháče Mts. At that time the huchen were so numerous in that place that the Forest Department at Oravský Podzámok had to have the section guarded by rangers. In the Tisza river basin, this phenomenon was observed by Dyk and Kryštof: during the hottest spell the huchen of the Teresva river near Usťchorna ascended to the cooler waters of the Brusturanka stream, but never stayed there permanently nor spawned in that stream.

Some data are available on the migrations of the taimen. Even in this case, apart from spawning migrations, mainly foraging migrations have been mentioned. In the Kama river (Bukirev 1967) the extent of the migrations increases in autumn, the taimen migrating to the lower reaches of streams. According to Levanidov (1959), taimen leave the Khor river in autumn for the Ussuri and Amur, where they feed intensively. The concentration of taimen in the Kheta river increases in autumn as they migrate there to feed on *Coregonus sardinella*. A similar phenomenon occurs in the Lena river (Pirozhnikov 1955) where the concentration of taimen varies in relation to the population densities of *Coregonus autumnalis migratorius* and *C. sardinella,* the chief prey species of taimen.

In addition to foraging migrations, those induced by changes in water temperature have been mentioned in connection with the taimen (Bukirev 1967, Pirozhnikov 1955, Sigunov 1972, Wu 1979). In addition, migrations connected with changes in water salinity have been reported for taimen populations living in the deltas of rivers beyond the Polar Circle, such as the Lena, Kotui, Kotuikan and Voevorlikhan rivers (Pirozhnikov l.c., Sigunov l.c.). Increased salinity levels cause the taimen to emigrate upstream and return to the original habitats when the salinity level has decreased again.

Besides longitudinal, there are also lateral migrations, i.e. movements towards banks and back to the stream line. Such movements are generally of a foraging character and are mostly diurnal. This is so in the Lena (Pirozhnikov l.c.), Vilyui and Olenek rivers (Kirillov 1962, 1972) where taimen largely occur inshore at night, staying in the stream line in daytime.

Migrations connected with variations in water level were observed by Luk'yanchikov (1967a) who reported on taimen migrating upstream at the time when the Bratsk reservoir on the Angara river was being filled with water. Wu (1979) also correlated migrations of taimen into bigger and deeper waters in winter with the decrease in both water temperature and level in that season of the year.

TERRITORY

An adult *Hucho hucho* is a territorial fish. According to observations made by Ivaška (1951), a huchen will defend its territory against intruding individuals of the same species. Svetina's (1962) statement, that a huchen's hunting territory is 1–5 km long, is unsubstantiated and is probably based on an erroneous interpretation of tagging results. As a rule, a sizable and fairly deep pool is the territory of a big huchen. A very extensive pool may be occupied by several, or even tens of individuals. Thus Nieslanik observed, in 1954, 35 big huchen standing almost motionless in the middle of a large pool in the Orava river below the Čurilovská skala Rock around midday. According to the author, the strongest individuals usually control the upper part of a pool and the inflow to it, whereas the smaller ones occupy the central and lower parts. The same is true of the taimen. Nikol'skii (1956) states that, contrary to the wels *(Silurus glanis)*, several big taimen may stay in one and the same pool. Sigunov's (1972) observations on the spatial distribution of taimen according to size are similar to those of Nieslanik on the huchen, and tend to confirm the specific identity of the two forms. Sigunov reports that taimen usually stay in deep pools below reefs, with grayling *(Thymallus thymallus arcticus)* standing in the "head" of the reef, and big taimen close below. Medium-sized individuals occupy the lower part of a pool, whereas small, sexually immature ones inhabit its lowest part. In this connection, Sigunov mentions an experiment carried out by Kuteinikov, a geologist, on the Molodo river: when the biggest taimen were fished out, their place was occupied by medium-sized ones, and the vacated places were invaded by young taimen immigrating from sections lying lower downstream, but they were soon displaced by pike. Obviously, this habitat allocation has a prominent trophic aspect in both huchen and taimen.

As in other salmonids, the territory of the huchen is apparently delimited by obstacles on the bottom or by the morphology of banks, as indicated by the observations made by Ivaška (1951). Skácel's observations over more than ten years suggest that, except in the spawning period, a particular huchen will invariably occur in one and the same place.

Thus an adult huchen is not a solitary fish, but occurs in small groups, as mentioned above. With such a big and, moreover, predatory fish species, this is surprising to some extent. The number of individuals in a group is probably limited by the size of the territory, and indirectly also by angling intensity. The latter is apparently the cause of the fact that nowadays huchen occur singly even in fairly big pools in the waters of Slovakia, although such suitable habitats were shared by several big huchen not very long ago (Rudolf 1964).

16. Diseases, Parasites, Abnormalities and Injuries

DISEASES

The present knowledge of the diseases of Eurasian huchen is very fragmentary and insufficient, particularly as regards wild populations. According to Ivaška (1951), a huchen wounded when handled without care runs the risk of being infested with parasitic fungi of the genera *Saprolegnia* and *Achlya* and dying as a consequence (Pl. 8.1). According to Havelka and Ivaška (1954), the annual losses due to fungal diseases amount to 3–5% of the breeding stock in pond cultures. More detrimental and usually lethal is the infestation of pectoral fins and eyes, from which the fungus rapidly spreads to the gill apparatus. Since males are more frequently handled when hand-stripped, it is this sex that succumbs to this disease – the authors state that males make up 75% of the total number of huchen dying from this cause. But this is not due to the handling itself, as believed by Havelka and Ivaška, rather to the fact that at spawning time the skin of males differs in structure from that of females. It is a known fact (Pickering 1977, Roberts and Shepherd 1974, Wilson 1976) that at spawning time the skin of salmonid fishes is generally susceptible to infection, in particular that of males. Its epidermis lacks the mucous layer which acts as a protective barrier, preventing potential pathogens in the environment from penetrating into the skin. Nieslanik states that the introduction of anaesthetics into the hand-stripping procedure (Chapter 25) was followed by a considerable decrease in the number of fish infected with fungi especially among parent fish caught in the wild (the anaesthetic decreases the risk of these big and heavy fish being wounded during handling).

The fungi mentioned above, together with the flagellate, *Ichthyobodo necatrix*, have caused mass mortality and subsequent failure of experiments aimed at feeding young huchen with different kinds of food (Jungwirth 1977b).

Ivaška (1951) described mass mortality of huchen due to the so-called "plague" which he and Havelka later identified as a furunculosis caused by the microorganism, *Aeromonas salmonicida (Bacterium salmonicida)* (Havelka and Ivaška 1954). An infection introduced by inflowing water from earth works being carried out higher upstream, caused acute enteritis. Within two weeks, only 22 out of 63 huchen (parent fish) were still alive, i.e. the mortality rate was 65%. According to the authors, the symptoms of the disease were not typical of furunculosis, as no typical muscle furuncles occurred. The symptoms included mould developing on the body surface, and even blindness in several cases. At that time, Ivaška (1951) considered the fungal disease most important, being in his opinion the major cause of the failure of rearing huchen in captivity. At present, ulcerative dermal necrosis (UDN) appears to be the most dangerous disease, causing an almost total mortality of all salmonids; in 1978, this disease was the cause of disastrous mortality of huchen on the farm at Podsuchá near Ružomberok (Peňáz and Příhoda 1981). According to Skácel's observations, this disease manifests itself more, both in the wild and in captivity, at the time of ripening of sexual products, during and after spawning. The losses usually amount to 60% but the fish that had once overcome the disease are then resistant to it. Besides this disease, the cause of which is still unknown, great losses in huchen hatcheries in Bavaria have been caused by viral diseases, viz., viral haemorrhagic septicaemia (VHS) and infectious pancreatic necrosis (IPN) (Bohl 1979, Harsányi 1982). A bacterial disease of the gill apparatus, caused by bacteria of the genus *Myxobacterium,* is also known to occur (Harsányi 1982).

Mortality of young huchen can also be caused by dropsy of the yolk sac (Ivaška l.c.), as well as by the occurrence of air bubbles in the intestine (Balon 1956), resulting from a high air content of water supplied to the hatching apparatus.

PARASITES

The most complete data so far on the parasites of the huchen were published by Žitňan (1976), who found a total of 11 parasitic species in huchen from the Váh, Poprad and Hron river systems. In his opinion, the parasites of great epizootological importance include the nematodes, *Raphidascaris acus* and *Cystidicola farionis* and the copepod, *Basanistes huchonis* in free waters, and the flagellate, *Ichthyobodo necatrix,* the infusorian, *Ichthyophthirius multifiliis* and the copepod, *Basanistes huchonis* in pond cultures. Žitňan points out the risk of transmitting parasites both by introducing the huchen and by supplementing the breeding stock with huchen from free waters. In the degestive tract of huchen from the Ljubljanica river, Brglez (1966) also found three trematode species of the genus *Asymphylodora* not reported by previous authors. Of the parasites listed in Table 34, attention should be paid to the nematode, *Rhabdochona denudata,* so far reported only by Zakhvatkin (1951) as occurring in huchen from the Tisza river in the Transcarpathian region of the Ukraine. Ergens et al. (1975) doubt the correct identification of this species. Similarly, Moravec (1975) points out the fact that cyprinid fishes are the main

TABLE 34

List of species parasitizing *Hucho hucho* (after Schulze 1890, Zakhvatkin 1951, Bykhovskii 1962, Kirillov 1962, Strelkov and Schul'man 1964, Brglez 1966, Roitman 1967, Moravec and Ergens 1970, Ergens et al. 1975, Moravec 1975, Lukin 1976, Žitňan 1976 and Harsányi 1982)

	H. h. hucho	*H. h. taimen*
Protozoa		
Chilodonella cyprini	+	
Ichthyobodo necatrix	+	
Ichthyophthirius multifiliis	+	
Myxosoma cerebralis	+	
Platyhelminthes		
Asymphylodora imitans	+	
Asymphylodora markewitschi	+	
Asymphylodora tincae	+	
Azygia robusta		+
Azygia lucii	+	+
Cyathocephalus truncatus	+	
Diphyllobothrium strictum		+
Eubothrium crassum	+	+
Eubothrium salvelini	+	+
Nanophyetus schikhobalovi		+
Orientocreadium pseudobagri		+
Sanguinicola sp.		+
Scolex pleuronectis		+
Tetraonchus huchonis		+
Tetraonchus skrjabini		+
Tetraonchus spasskyi		+
Triaenophorus nodulosus	+	
Aschelminthes		
Camallanus lacustris		+
Capillaria brevispicula		+
Cucullanus truttae	+	+
Cystidicola farionis	+	+
Cystidicoloides ephemeridarum	+	+
Hysterothylacium aduncum		+
Pseudocapillaria salvelini		+
Rhabdochona denudata (?)	+	
Raphidascaris acus	+	+
Acanthocephala		
Echinorhynchus cryophilus		+
Echinorhynchus salmonis		+
Neoechinorhynchus rutili	+	+
Paracanthocephalus tenuirostris		+
Pomphorhynchus laevis	+	
Hirudinea		
Cystobranchus respirans	+	
Limnotrachelobdella taimeni		+
Taimenobdella amurensis		+

TABLE 34

(Continued)

	H. h. hucho	H. h. taimen
Arthropoda		
Argulus coregoni	+	+
Basanistes enodis		+
Basanistes huchonis	+	
Basanistes woskoboinikovi		+
Salmincola stellata (?)		+

Species marked with a query (?) are probaly misidentified.

hosts of this parasite, findings in other fish species being rare and requiring verification.

A specific and, according to the observations so far, most important parasite of huchen is *Basanistes huchonis* (Pl. 11.2) which aggregates in large numbers on the inside of opercles and on the palate of the fish. The parasite itself is not considered too harmful. But by injuring surface tissues of its host it enables parasitic fungi to penetrate (Volf and Havelka 1958). Witkowski and Blachuta (1980) observed this parasite invading mainly young and sexually immature individuals.

Important losses among pond-reared young are also caused by *Myxosoma cerebralis,* causing the so-called whirling disease, or myxosomatosis which, together with ichthyophthiriosis, was responsible for heavy losses among young reared on the farm of the Slovak Anglers' Union in Banská Bystrica in 1973–1977 (the average survival rate of young from spring to autumn was only 29.7%).

Few data are available on the parasites of the taimen. The most frequently occurring parasites are thought by Kirillov (1962) to be the digenetic leech, *Azygia lucii,* and the copepod, *Basanistes enodis* (the latter usually infesting the inconnu, *Stenodus leucichthys nelma;* Bykhovskii 1962). Of sixteen taimen from the Vilyui and Olenek rivers, he found parasites in fourteen, i.e. 88.5%; nine of these contained *Azygia robusta* and *Basanistes enodis,* six contained cestodes, four contained nematodes, and in one case there was an acanthocephalan; the author does not mention their identification as to species. In taimen from the Amur river the most important parasite was the larvae of the digenetic trematode, *Nanophyetus schikhobalovi,* with extensity of invasion up to several hundreds of larvae (Strelkov and Shul'man 1964), whose adult stages have been found even in humans (Podyapol'skaya and Kapustin 1950).

It is evident from Table 34 that 43 species have been found to parasitize Eurasian huchen. However, the number cannot be considered final, since the hitherto parasitological investigations have mostly been accidental. The two forms of *H. hucho* have 9 parasites in common; besides them, additional 21 were ascertained in the taimen and 13 in the huchen. The smaller number of species parasitizing the huchen is due to the smaller range inhabited, a smaller number of fish species contacted, and even to its having received less attention.

ABNORMALITIES AND INJURIES

Kirka (1958) described several fin malformations occurring in young huchen in artificial cultures. The most frequently occurring malformity was an anomalous anal fin, lower or doubled at the base. Sometimes rudimentary pectoral fins with undulated rays occurred. In some cases the author even found an additional minute fin with 2–3 rays, situated in front of the ventral fins, the length of which sometimes attained that of the true ventral fins.

Linka (1934) mentions a huchen from the Hron river bearing a white dorsal spot, probably caused by an earlier injury (spearing). Holčík (1970) mentions a huchen from the Klíčava reservoir, which had a healed but open throat perforation, probably caused by an earlier hooking. The skin of male huchen is often covered with scars from healed wounds inflicted during duels in the spawning period (Pl. 10.2 and 11.3), especially on the flanks behind the head and in the region of the caudal fin. Mutilated pectoral, ventral and caudal fins also often occur. Nieslanik noticed that when handling big huchen (and especially when carrying them in a scoop net), the cornea of the eyes may be injured by the net and infected with fungus.

17. Hardiness

On the basis of egg morphology as well as of the practical experience of breeders, Peňáz and Příhoda (1981) conclude that, within the group of salmonid fishes, the eggs of the huchen are among those relatively less susceptible to shocks. Thus they can bear handling as well as transportation, especially in the so-called "eyed egg" stage (at the end of the 9th embryonic step).

Having partly absorbed the yolk sac, young huchen can tolerate temperatures up to 20° C, provided that the water is sufficiently saturated with oxygen (Chapter 30). According to Prawocheński and Kolder (1968), however, young huchen kept in an aquarium lost orientation at a water temperature of 16° C.

Huchen are very sensitive to external lesions, which are most frequently caused by handling the fish on farms, or inflicted in duels at spawning time; the lesions enable infections with fungus and microorganisms to penetrate. It is most probable that in such cases a regular water exchange plays an important positive part. Jungwirth (1978) mentions that in his feeding experiments with young huchen the water in the containers was exchanged every 10 minutes, and he recommends not to prolong that interval when feeding the fish with minced meat or fish, "since the dissolved food particles increase the risk of gill infection". The positive role of streaming water is also indicated by observations made by Nieslanik; according to him, huchen infested with fungus can heal themselves by staying at the water inlet to a pond, exposing the infested parts of their body to the stream of falling water.

The huchen can be considered a coldwater species; according to Prawocheński and Kolder (1968) it can tolerate water temperatures up to 20° C, given sufficient saturation with oxygen (8−9 mg l^{-1}), which is confirmed by the observations of Nieslanik (see footnote 8 at Chapter 8).

At higher water temperatures both the huchen and the taimen withdraw into deep holes and pools or stand in front of the mouths of tributaries with cold water, below waterfalls, flow-offs, cascades and riffles (Bukirev 1967, Kirillov 1972, Sigunov 1972). At very high water temperatures they may become somewhat torpid (Kirillov 1972). At water temperatures higher that 20−22° C and provided that the fish cannot retreat to deeper places or withdraw to the mouths of tributaries with cold water, they may die. Such a case was observed by Nieslanik in the section of the Orava river between Podbiel and Kraľovany in the summer of 1952 (Pl. 12.2). In that summer, due to a long spell of drought and high air temperatures (over 30° C), the water level in the river dropped considerably and mass mortality of huchen occurred. Kirillov (1976) considers the taimen an oxyphilous form, which is also true of the nominate form, *H. hucho*.

Prawocheński and Kolder (1968) state that the huchen is very sensitive to water pollution, which does not wholly correspond to reality, however. Even Hollý (1934b) and Ivaška (1951) noted that the huchen can withstand pollution with industrial waste water relatively well, and that it is more resistant than grayling. They even suggest that huchen be stocked in waters from which grayling had disappeared due to pollution. The present, though rare, occurrence of huchen in the Váh river near Trenčín, Ilava, Bytča and Sučany, and also in the polluted Czechoslovak section of the Danube, documented by yearly catches, even within Bratislava (Blahák 1972), and in the Isar river in Munich itself, a city with a million inhabitants (d. 1970), confirms quite clearly that the huchen, and apparently *Hucho hucho* in general, is in fact a fairly resistant species.[9] The resistance of the huchen is also evidenced by the occurrence of this species in the Hornád river, in Košice itself and in the section below the town (as documented by several catches of fishes weighing 2.5−6.5 kg), in spite of the fact that this section of the Hornád suffers from disastrous pollution waves with phenol waste waters from the East Slovakian Ironworks twice a year on average (a written report by J. Zubrický from March, 1982). Hollý (1934a) even assumes that huchen are capable of reproducing in polluted waters, although he doubts that the eggs or young could possibly survive. However, the term "polluted water" is so broad that a detailed investigation is required to assess the resistance of the huchen to different concentrations of various toxic substances contained in such waters, or to their mixtures. According to information submitted by Nieslanik, huchen occurred in the polluted section of the Váh river below Ružomberok in the late fifties, but only until July; with the onset of

[9] The Czechoslovak section of the Danube is known to be classified in water purity class IV, the total of waste waters in the profile above the Ipeľ mouth representing a full fourth part of the tributaries to the Danube (at Q_{355} = 894 m^3 s^{-1}), as pointed out by Rothschein (1976).

decreased water discharge and, hence, increased concentration of waste waters, huchen gradually withdrew to lower sections. In a heavily polluted section of the Turiec river below Martin, huchen (similarly as other fish species) did not occur for several decades. But when the release of waste waters was stopped (after the paper mill at Martin was oriented to different production) the huchen almost immediately repopulated the section (since March, 1980).

Nieslanik describes an interesting case which confirms the admirable hardiness of the huchen. The case involved a male huchen, 5.5 kg in weight, which had been in a dead branch of the Belá river near Liptovský Hrádok for three or four years. The branch served as a dump; it was gradually filled, became shallower (up to 60 cm deep) and was largely overgrown with grass. The huchen probably stayed in the branch even after it was closed when the canalization works were finished. Finally, it should be mentioned that young huchen have thrived rather well even in a silted and heavily overgrown carp fishpond (Dyk 1938).

The huchen is very sensitive to the effects of electric current, as evidenced by our own and foreign (Prawocheński and Kolder l.c.) experience with fishing by means of an electrical shocking gear.

18. Present Status

The decline in the numbers of huchen is apparently longstanding, as can be derived from reports of various authors and from suggestions for the protection of the species, published in various European countries in the period between World Wars I and II (Robida 1902, Mahen 1926, Kafka 1924, Šivic and Bajkov 1925, Z. Dyk 1926a, b, V. Dyk 1937b, Vladykov 1926, 1929, Kulmatycki 1931b, Hollý 1934b, Polášek 1936, Munda 1935, Seez 1939). Data in Chapter 15 may offer at least partial picture of the numbers of huchen in relatively recent times, which sharply contrast with the very low numbers at present. According to an investigation carried out by Bako in 1946, the numbers of huchen existing at that time were estimated at about 10% of their previous amount. The author stated that the fish were mostly small, up to 5 kg in weight, as the bigger ones were depleted when the front of World War II passed through Slovakia. Novák (1971) states that the present number of huchen in the Turiec river is approximately 1/20 of their original amount. In spite of these data being mere approximations and rough estimates, a decline in the numbers of huchen is evident in many places (Skácel 1976, Budaj 1976, Randík 1976, Bastl et al. 1976).

Besides the numbers, also the range of the species has considerably decreased, as illustrated by the map in Fig. 19, compiled from various sources of data. It is seen that previously the huchen populated considerably longer stream sections than they do at present. Moreover, the present occurrence of the huchen is no longer continuous in

TABLE 35

The present approximate status of distribution of huchen in individual drainage basins of the Danube river system

Drainage basin	km				%		
	Common	Rare	Extinct	Total	Common	Rare	Extinct
Danube	—	333	625	958	—	34.8	65.2
Left tributaries up to							
Morava	—	—	350	350	—	—	100.0
Morava	—	—	275	275	—	—	100.0
Váh	363	75	138	576	63.0	13.0	24.0
Hron	50	50	88	188	26.6	26.6	46.8
Tisza	313	63	95	471	66.4	13.4	20.2
Mureş	—	—	295	295	—	—	100.0
Timiş	—	—	50	50	—	—	100.0
Cerna	—	—	50	50	—	—	100.0
Oltul	—	—	150	150	—	—	100.0
Argeş	—	—	300	300	—	—	100.0
Ialomiţa	—	—	100	100	—	—	100.0
Siretul	170	400	38	608	28.0	65.8	6.2
Prut	288	—	—	288	100.0	—	—
Iller	125	—	—	125	100.0	—	—
Lech	145	—	150	295	49.2	—	50.8
Isar	175	43	143	361	48.8	11.9	39.6
Inn	203	195	53	451	45.0	43.2	11.8
Right tributaries from							
Traun to Traisen incl.	53	120	395	568	9.3	21.1	69.5
Drava	350	588	283	1221	28.7	48.9	23.2
Sava	1320	825	775	2920	45.2	28.3	26.5
Velika Morava	163	363	—	526	31.0	69.0	—
Total	3718	3055	4353	11126	33.4	27.5	40.2

many places, as it was in the past, and these are often but isolated localities. A similar situation to that in Czechoslovakia is also in other countries (Bănărescu 1964, Frank 1972, Jungwirth 1978, Vlasova 1959). Tables 35 and 36 present numerical data on changes in the occurrence of the huchen in different river basins and countries. It is evident that the species has completely disappeared from the systems of some streams. For example, it is absent from the left tributaries of the Danube down to the Morava river, from the Morava itself and from the Iller river; of the tributaries south of the Danube, the huchen is absent from almost all Rumanian tributaries, beginning with the Mureş and ending with the Ialomiţa. Its numbers are variable in the remaining streams, only the Prut and Velika Morava appearing to be least affected. Considering the total lengths (in km) of the streams, the total disappearance of the huchen has affected 39% of the length of all streams in which the huchen occurred in the past; the huchen is common in 33% and rare in less than 28% of the total length of the streams. Rumania is the most affected, the huchen having disappeared from almost 62% of the length of the streams which it populated

TABLE 36

The present approximate status of distribution of huchen in individual countries along the Danube

Country	km				%		
	Common	Rare	Extinct	Total	Common	Rare	Extinct
Austria and FRG	938	613	1809	3360	27.9	18.2	53.8
Czechoslovakia	413	125	501	1039	39.7	12.0	48.2
Yugoslavia	1470	1576	928	3974	37.0	39.7	23.3
Rumania	208	400	986	1594	13.0	25.1	61.9
USSR	564	341	254	1159	48.7	29.4	21.9

in the past; then follows Austria, together with the Federal Republic of Germany, and Czechoslovakia (although the endeavour to protect the huchen is most intensive and the care most thorough in this country!); the best situation is in Yugoslavia. It should be added to the two tables as well as to the map in Fig. 19 that the data in them are of different validity and time limitation, mostly reflecting the status before and after World War II.

The causes of the continuous decline in the numbers of huchen and decrease in the length of stream sections in which they occurred have received the attention of investigators ever since the beginning of this century. Robida (1902) was already aware of the fact that the main causes of the decline in the numbers of huchen are to be associated with modern industry (release of waste waters), canalization of rivers, and construction of dams, i.e. a complex of factors which now act with greater intensity. As a result of their influence, a rapid retreat of huchen has been recorded, especially during the last 20–30 years (Randík 1978). The influence of these factors causes a change in the natural character of submontane rivers and in their hydrological, hydrobiological and hydrochemical regimens. As a consequence, the possibility of reproduction is decreased (by silting of spawning sites), the mortality of eggs and young is increased (due to impaired oxygen content of water, action of toxic substances, lack of suitable and available food) and the number of habitats or whole stretches inhabited by the huchen is decreased. Intensive large-scale agriculture is a comparatively new factor which exerts a strong negative influence on the quality of river water and intensifies the overall unfavourable influence of industrial waste waters and municipal sewage.

According to Šivic and Bajkov (1925), the major causes include canalization[10], pollution and war events with their consequences (use of explosives by poachers). The influence of pollution has also been pointed out by Z. Dyk (1926a, b), Bako (1930), Hollý (1934a, b), Janza (1958), and others. At the same time, however,

[10] However, Mahen (1926: 305) records an opposite effect of canalization: "According to local, old inhabitants, the Orava did not previously flow in such a deep bed as it does now after it has been canalized, and the huchen were less numerous, contrary to pike, which are now abundant in the Čierna Orava river. Here canalization has favoured the penetration of huchen, which is more than interesting."

voices are heard complaining of excessive fishing (Mahen (M.) 1924, Z. Dyk 1926a, b, Vladykov 1926). Shnarevich and Moshuķ (1956) mention the unfavourable influence of building impoundments; Vlasova (1959) and Frank (1972) also point out deforestation and its consequences on the water flow in streams. The latter author includes among the causes even the biological properties of the huchen (long reproductive cycle, rapid growth, spawning in spring, and susceptibility to certain environmental factors); these, however, cannot be considered as causes but rather as circumstances which may accelerate the influence of the actual causes. It is worth noting that, as reported by Vlasova (1959), in the Transcarpathian region of the Ukraine they have not succeeded in suppressing poaching and illegal fishing, mentioned as early as 30 years ago by almost all authors who reported on the status of the huchen in that region (e.g., Z. Dyk 1926a, b, V. Dyk 1937b, Vladykov 1926, Komárek 1955).

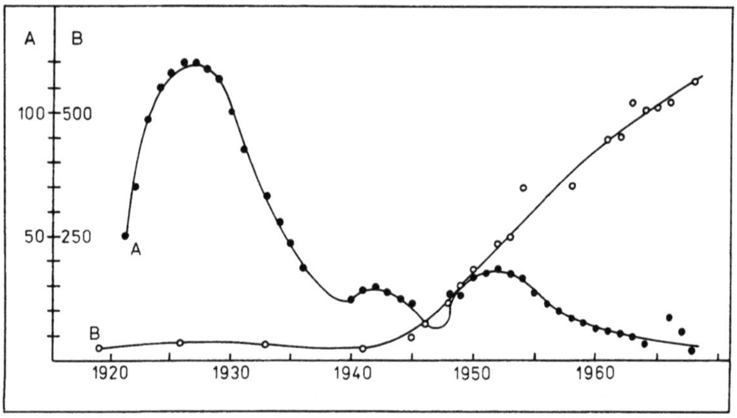

Fig. 31. Correlation between numbers of huchen caught in the Turiec river (A and the left ordinate) and numbers of members of the Anglers' Organization in Martin (B and the right ordinate) (after Bastl et al.1976).

Although a mutual and sometimes even synchronous interaction of the factors mentioned above is involved in almost all cases (Homei 1956), in some of them it is possible to determine the major factor which causes the decline in the numbers of huchen or their isolated distribution. Thus in the Turiec river the local huchen population was separated from that in the Váh river by the construction of a paper mill in Martin in 1904 and by the erection of weirs preventing the fish from migrating from the Váh river (Novák 1969). In subsequent years, the numbers of that population were decreased by excessive fishing (Bastl et al. 1975, 1976), mainly due to increasing numbers of anglers (Fig. 31). The erection of the Orava dam in 1954 divided the previously continuous range of the huchen in the Orava river into two mutually isolated parts, the lower section suffering mainly from variations in water discharge and an unsteady thermal regimen. Even in 1958–1959 the numbers of small huchen in that section were considered low compared with those prior to the erection of the dam (Ivaška 1959b). The artificial variation in water discharge, with sudden impacts and repeated sudden drops, was emphasized as the major cause of the decline in the numbers of huchen in the Orava river below the dam by Bastl and Kirka (1959). In the section affected, decreased not only the numbers of huchen but also those of grayling and nase carp (Skácel 1976). A similar negative influence resulted from the erection of the Váh river cascade, where excessive water pollution in the section below Ružomberok also played a major role. The limitation of the occurrence of the huchen in the Hron river or its confinement to the upper reaches of that river was due to continuous pollution downstream of Dubová and frequent toxic waves of disastrous proportions (one of which is also reported by Ivaška 1951). Apparently a similar situation is found in the Inn, where extensive fish poisoning occurred in 1970, the huchen being among its victims (Schm. 1970).

One of the causes of an almost complete extermination of the huchen population in the Rumanian part of the Tisza river system was a disastrous flood in the spring of 1970, which virtually washed out the huchen to the lower reaches, from which they did not find their way back and had not returned by the autumn of that year (Frank 1971).

Seez (1939) paid careful attention to the influence of river canalization and unreasonable fishing, which in his opinion are the very causes of the decline of huchen in Austrian and German waters. According to him, canalization decreases stream diversity, thereby depriving the huchen, and particularly the young, of shelter, food and spawning sites. By unreasonable fishing he understands an incorrectly fixed legal size (50–55 cm *Tl* at the time) which made it possible to catch and remove sexually immature individuals. According to his findings, almost 63% of huchen caught with rod and line in the Enns and Drava rivers and examined by himself were sexually immature. Even Kulmatycki (1931c), Vladykov (1926) and Shnarevich and Moshuk (1956) point out excessive removal of sexually immature young fish.

In our opinion, unreasonable management can also be seen in the fact that in Czechoslovakia, as well as in most other countries, huchen fishing is permitted during the autumn and winter months. In that period (Chapter 11) the huchen feed very intensely, the females apparently more than the males due to higher energy requirements, so that the females can be caught much more easily. For example, from the catch cards of the local organization of the Slovak Anglers' Union in Ružomberok for 1972–1973, one can learn that of 44 huchen caught, 67% were females and 33% males. The same sex ratio was found in a series of 78 sexually mature huchen caught in various streams in Slovakia and elsewhere. In a natural huchen population, undisturbed by fishing, a balanced sex ratio or even an excess of males may be expected to occur, as can be concluded from the information by Ivaška (1951); according to this author, catches of huchen in the Orava and Váh rivers contained only 26% females and 74% males in 1949 (after World War II, huchen fishing was generally prohibited and thus the above data may be considered as reflecting the actual normal sex ratio in a huchen population). It follows from the above that an excessive removal of females must inevitably result in a decrease in the number of spawning pairs and consequently in a decline in the numbers in the population. Increased attention will have to be paid to this problem in the nearest future.

The situation of the taimen is no better, though it would seem at first sight that with regard to the vast range inhabited by this form, sparse human population and relatively low level of industrialization the present status has not changed compared to the past. Unfortunately, the contrary is true, and Syroechkovskii and Rogacheva (1975: 76) state explicity that "... the freshwater species, *Hucho taimen* and the rare *Hucho hucho,* and the lenok, *Brachymystax lenok*... were previously common in all rivers in the north of Eurasia but are almost extinct in many places". An idea of the extent and degree of the decline of taimen populations can be obtained from the data presented by Soviet authors.

Nikol'skii reported in 1956 that taimen were gradually declining in the Amur river system. He stated that whereas the catch of this species in the whole Amur river was 204 tonnes in 1891, it dropped to only 28.6 tonnes on an annual average in 1940–1947. Nikol'skii considers the taimen to be almost exterminated in some of the tributaries of the Amur. A rapid decline of taimen in the Angara river is pointed out by Luk'yanchikov (1967a), according to whom 24.6 tonnes of taimen were harvested in that river in 1946 but no more than 215 kg (!) in 1950. The same author (Luk'yanchikov 1967b) reports that in the Khatanga river system, 2.6 tonnes of taimen were harvested in 1952 but only 100 kg (!) in 1964. The catch of taimen in the Krasnoyarskoe reservoir dropped from 129 kg in 1967 to 29 kg in 1968, and the species no longer occurred in the hauls in 1969–1974 (Ol'shanskaya et al. 1977). The catches of taimen decreased even in the Bratsk reservoir: 241 tonnes in 1963, 0.96 in 1964, 0.23 in 1967, 0.12 in 1968 and only occasional taimen were caught in subsequent years (Pushkina 1977). As described by Olifer (1977), the section of the Angara river below that reservoir suffered especially from the circumstance that after the dam had been erected, the fishing in that section concentrated on mass aggregations of taimen stopped on their migrations. In 1973–1974, 5–8 times fewer taimen were caught there than in 1970–1972. Consequently, the numbers of young taimen in the mouth of the Ilim river dropped from 2–3 individuals per haul with a fine-mesh seine net in 1970–1971 to zero in subsequent years. A similar situation was in the Kama (Bukirev 1967) and in most Yakutian rivers (Kirillov 1972). According to the data presented by Kirillov (l.c.), the average annual catches of taimen in the Lena river system were as follows (in tonnes) 100.4 in 1941–1945, 52.5 in 1946–1950, 27.2 in 1951–1952, 28.1 in 1956–1960, 30.3 in 1961–1965, and 12.8 in 1966–1970. Misharin and Shutilo (1971) estimated that while the commercial catches of taimen in the Siberian rivers and in the Amur amounted to around 200 tonnes on an annual average prior to 1950, they dropped to only 100 tonnes in 1960. In general, the main cause of the decline is seen in incorrect management: lack of legal size, excessive removal of sexually immature fish, fishing during the spawning period, and fishing by illegal means. From the example of the Angara river it follows, however, that canalization of rivers plays a negative part even in the case of the taimen. Beyond doubt, the rapid decline in the numbers of taimen is also partly due to the fact that at first this species was considered as noxious in the economic respect, and that intensive fishing of taimen was therefore recommended (Teplov 1951, Pirozhnikov 1955).

Judging the above facts soberly, and bearing in mind the huge range of *Hucho hucho,* one must admit that Kasecker (1970) is right in assuming that no other fish species has been decimated by man to such an extent as this very species. In our opinion, Eurasian huchen must be considered a species which, in spite of its considerable adaptability, is threatened with inevitable extinction unless effective conservation measures are soon introduced.

Part IV

UTILIZATION

19. Economic Importance and Exploitation

ECONOMIC IMPORTANCE

The huchen may by considered one of the most valuable species of freshwater fishes. It is highly appreciated by anglers as a trophy species which offers unusual fishing experience (Pl. 14); in a broader aspect, the huchen is of importance as a very effective and, in many places, the only biological ameliorator of submontane streams. This was suggested by Kulmatycki (1931a, b) and Ivaška (1951) who pointed out that with its rapid growth, the huchen improves the value of less valuable coarse fishes. In spite of the fact that the range of the huchen has decreased in comparison with that in the past and that even its numbers have dropped substantially, its importance has increased rather than decreased, as indicated by the increased interest in huchen farming, e.g., in Austria (according to Jungwirth in litt.). In Austria, 20–25 hatcheries are engaged in rearing huchen at present, although huchen eggs and young were still imported into Austria from Yugoslavia or Czechoslovakia in the early seventies. Interest in the huchen is also shown by anglers, whose numbers are rising continually, interest in the huchen being shown even in such countries where it has never occurred (Belgium, Denmark, Finland, Canada, Morocco, Spain, Sweden); this is also evidenced by the export of fertilized huchen eggs into these countries from Czechoslovakia, amounting to 150–200 thousand annually (Budaj 1976). While the aspect of sport angling is the predominant one at present, attention should also be paid to the bio-amelioration importance of the huchen. The latter is already underlined by the fact that in the foothill zone the huchen is practically the only autochthonous species with piscivorous habits from early youth. Stocking huchen to foothill streams has been recommended, e.g., by Svetina (1970), who argues that the huchen prevents invasion of cyprinid fishes. Sufficiently high numbers of huchen in foothill reservoirs may show a positive effect in the subsequent decrease in the numbers of accompanying species, not utilized in fishing, which constitute (relatively and absolutely) the most numerous component of fish communities in such waters (Wajdowicz 1964, Holčík 1966a, 1977a, Kirka et al. 1976). In the final effect, this leads to decreased food competition among non-predatory species and to improved growth of economically preferable and secondary species. The amelioration influence of the huchen could be successfully utilized especially in water supply reservoirs, where an overcrowding of undesirable fish species (bleak, roach, white bream, perch) may lead to impaired crude water quality and consequently increase treatment costs (Lusk and Vostradovský 1978, Straškraba and Straškrabová 1980, Holčík 1978). It may also be of use to pay serious attention to pond rearing of marketable huchen, as suggested in their time by Kostomarov (1937a, b) and Bastl (1958). The authors, like Kulmatycki (1931a, b), noticed that the growth of huchen could compete with that of carp in pond cultures. In that case, huchen pond cultures should be situated in the vicinity of large river dam reservoirs to secure supply of fresh fodder fish. Last but not least, the huchen is an important object of breeding experiments. Being an evolutionarily polyploid form, *H. hucho* may be expected to have considerable ecological plasticity coded in its genome. Even hybridization experiments may be of importance and may yield surprising results that could be utilized in both theory and practice.

The above holds true in its full extent for the taimen. However, this form of *H. hucho* is even of commercial importance, as it is still being fished on a commercial scale, by means of mass fishing techniques, and local inhabitants utilize taimen either fresh or salted (Berezovskii 1924, Pravdin 1949, Pirozhnikov 1955, Nikol'skii 1956, Bukirev 1957, Misharin and Shutilo 1971, Olifer 1977, Kalashnikov 1978). Most authors evaluate the taimen as a species of economic importance or as a very important one; Pravdin (1949), Bukirev (1967) and Wu (1979) even point out that it may be a valuable object of farming. Voices claiming the noxiousness of the taimen (Teplov 1951, Pirozhnikov 1955) are exceptional, the taimen is predominantly evaluated from a broader ecological and economic viewpoint (Bukirev 1967).

Utility Value

Very scarce data are available on the utility value of huchen or taimen. According to Skalin (1976), the edible part of huchen (around 4 kg in weight) is 57.8–58.2% of total weight. Of this, 28.7–31.3% is the head and skin, 7.0–7,7% the esophagus, stomach and intestines, 0.7% the testes, 3.6% the ovaries, 0.6–1.0% the liver, 0.6–1.0% the kidneys and 0.2% the heart. As for the taimen, some such data can be found in the tables published by Budagyan (1961). According to these data, the edible part of taimen is 57%, of which 74.7% is water, 18.9% proteins, 4.8% fat and 1.6% ash. The energy value of 100 g of the edible part is 510.8 kJ (= 122 kcal), and this amount of the edible part contains 316 mg potassium, 21 mg calcium, 25 mg magnesium, 217 mg phosphorus and 1 mg iron. The liver of taimen contains a high level of vitamin A; Higashi (1961), drawing on Soviet authors, states that 1 g of taimen liver contains 2910–3800 international units (I.U.) (i.e. 0.87–1.14 mg), and 1 g of liver oil up to 55 thousand I.U. (i.e. 16.5 mg) of vitamin A. The same author says that cases of hypervitaminosis were even reported from the region of the Enisei, caused by ingestion (apparently long-term one, note by J.H.) of taimen and inconnu *(Stenodus leucichthys nelma)* liver.

In general, the quality of huchen and taimen flesh is valued very highly (Herman 1887, Munda 1926, Pravdin 1949, Ivaška 1951, V. Dyk 1956), and the flesh is characterized as yellowish-pink or golden pink, fragrant and moderately fatty. Ivaška (l.c.) states that the quality of the flesh depends, among other things, on the quality of the water inhabited by the huchen, being inferior in polluted waters and at low water levels.

Until recently, the skin of taimen (together with that of other species of fishes) served as material from which garments and footwear were produced by natives of certain northern regions of the USSR (Tsepkin 1980b).

Fishing Methods

At present, huchen are predominantly caught with rod and line, viz. by trolling, most frequently on a small dead fish (Pl. 13.1). The rod used for this purpose is around 2 m long, fitted with a spinning reel (in the past, a single-action flycasting reel), a line 0.35–0.50 mm in diameter. The bait is sewn to two or three big triple hooks. Less frequently an artificial lure is used (spoon, devon, wobbler, artificial fish) (Ivaška 1951, Šimek and Kopáčik 1959, Sedlár 1970, Hartlieb 1948, 1969, 1970) (Pl. 13. 2). According to the season, either the so-called summer system (i.e. surface trolling) or the winter system (trolling at various depths) is used. Besides the authors mentioned above, a synopsis of the methods used in huchen fishing was presented by Robida (1902), Heintz (1920), Thaller (1932), Tejčka (1934), Šimek (1946, 1954), Wendland (1965) and Harsányi (1982).

Exceptionally, huchen are even caught with nets (in the Danube and Orava reservoirs), viz. seine nets and gill nets. In such cases the huchen are caught as part of the bag, i.e. not on purpose.

In the past, when huchen were more abundant, they were also fished by spearing (especially at night with lights, by wading, or from small boats), with dip-nets, haul nets (even beneath the ice in winter), in fish pots, snares, by pounding, with explosives and plant poisons as well as by special, now obsolete, means and implements (Herman 1887, Vladykov 1926, Hollý 1934a, b, Tejčka 1934, Ivaška 1951, Komárek 1955, Andreska 1974), or simply caught by hand (Kulmatycki 1931a).

If a farm does not rear huchen in a so-called closed cycle, or if it is necessary to supplement a shoal of parent fish, the parent fish are caught with rod and line, traps and the haul seine nets. The use of an electric shocking-gear has proved unsatisfactory (low effectiveness of the usually used types, and risk of damaging or killing the fish with stronger ones).

Taimen are mainly caught with seine and gill nets, but also with drift nets, in fish-pots, by spearing, with night lines and, in recent times, more and more frequently by anglers with rod and line (Pravdin 1949, Kirillov 1972, Misharin and Shutilo 1971, Kalashnikov 1978). Certain special fishing methods are described by Sabaneev (1911), Berezovskii (1924), Egorov and Meier (1958), Vasil'ev (1960) and Sigunov (1972).

Major Fishing Areas

In essence, they are the same as the areas in which huchen occur; naturally, those yielding the best bags are the most important. In the past, good "huchen waters" in Slovakia included the Váh river between Ružomberok and Nové Mesto nad Váhom (in particular the environs of Kraľovany, Vlašky, Žilina, the Strečnianska úžina Defile, the mouth of the Turiec river), the Orava river in the section Tvrdošín–Oravský Podzámok, the Turiec river above Martin up to Príbovce or Kláštor pod Znievom, the Hron river between Brezno and Žiar nad Hronom (Mednyansky 1844, Kornhuber 1861, Ivaška 1951, Novák in litt.). At present, according to the data presented by Skácel (1976) and with regard to the numbers of caught huchen, the most important huchen streams include the Váh river in the area of Ružomberok, the Orava river between Trstená and Dolný Kubín, the Hron river near Podbrezová, the Poprad river in the section between Stará Ľubovňa and Spišská Stará Ves and, prior to the elimination of huchen in that stream, also the Hornád river near Spišská Nová Ves. In Austria, judging from the map published by Jungwirth (1978) and from the news occurring in the Oesterreichische Fischerei Zeitung, most huchen are caught in the Drava river between Lienz and Rosegg, in the Mur river between Murau and Judenburg, and in the Pielach river between Kirchberg and the mouth into the Danube. As for Yugoslavia, we lack exact information on the present situation of the major fishing areas but, apparently, both in the past and at present most huchen have been caught in the Sava, Ljubljanica, Kupa and Drina rivers (Munda 1925, 1926, Ivaška 1951, Sabioncello et al. 1970).

Most taimen have been caught in the Amur river in the areas of Elabuga, Daierga, Novoil'inovka and Sukhanovka (Nikol'skii 1956); in the Angara river, prior to the construction of the reservoir, it was the area of Bratsk (Olifer 1977); in the Lena river, mainly in its lower reaches (Pirozhnikov 1955).

Seasons and Conditions for Fishing

Experience obtained in huchen fishing has been summarized by Robida (1902), H-ý (1934), Ivaška (1951, 1959c), Šimek and Kopáčik (1959), Hartlieb (1969, 1970) and Sedlár (1970). According to these authors, the autumn and winter months are most satisfactory for huchen fishing, beginning in October and ending in February, but H-ý (l.c) writes that huchen actively take any kind of bait even in May and June.

On the basis of his own experience, Komárek (1955) also demonstrates that huchen can be fished even in summer. The fishing is most rewarding on cloudy or misty days with rain or snow showers, when the water is at normal level or falling and slightly turbid, the barometric pressure is medium high or falling, the air temperature is stable and a westerly or northwesterly wind is blowing. The best fishing is in October and November, when the first frosts set in and the water edges begin to freeze. Then huchen bite all day long, most intensively around midday and before sunset; in the morning they bite only after dark nights. When the frosts last several days in succession the huchen cease to bite, sometimes for several days, but the chance of a catch increases when the weather becomes settled. Huchen bite even at the beginning of the thaw, before the ice has started to move and the water is still not turbid. They do not bite when the water is clear, its level is too low or too high, on clear and moonlit nights, on days with extreme air temperatures, and when northerly or easterly winds are blowing. Naturally, local conditions are of no less importance.

Being mostly fished by industrial means, taimen mostly occur in catches in spring, and then not until the autumn or winter months, when sexually mature fish aggregate to form bigger shoals (Pravdin 1949). According to Nikol'skii (1956), in the Amur river in the area of Elabuga in 1941 to 1946 the highest percentage of catches was on December (24.2% of the total annual catch of taimen), January (28.5%), March (12.8%), and May (13.1%).

Catches

There seem to exist practically no regular statistics on the catches of huchen, since the genus *Hucho* is not mentioned separately in available statistical records (e.g., in the Yearbooks of Fishery Statistics, published by the FAO, Rome). A synoptic summary of the catches of huchen in Slovakia, compiled on the basis of statistical records of the Central Committee of the Slovak Anglers' Union in Žilina, is given in Table 37. Even this summary is incomplete, however, as it only contains anglers' bags, and as we know from our own experience not all catches are reported and recorded. Since huchen fishing requires special permission, the absolute catches obtained in successive years are quite stable, except in the last years, when a decreasing trend is apparent. Due to increasing numbers of anglers the percentage by weight of huchen in the all-Slovakian catch of all species caught with rod and line is continuously decreasing, being only 0.18% on the annual average. Considering that huchen are caught only in those stream sections in which they commonly occur (Table 36), the long-term average catch in Slovakia is 0.33 huchen per km of stream, or 1.78 kg km^{-1}. No summary information on huchen catch is available for other European countries, except for two data. Verce (1973) reports that in Slovenia, Yugoslavia, 148 huchen were caught in 1966, totalling 635 kg in weight (avg. 4.28 kg per fish; percentage by weight, 0.45% of the total catch of fish); in 1971 it was 128 huchen 704 kg in total weight (avg. 5.49 kg per fish; 0.40% of total catch by

TABLE 37

Review of catches of huchen in Slovakia in 1954—1980 (compiled from the statistics of the Central Committee of the Slovak Anglers' Union)

Year	Number	Weight (kg)		Percentage by weight of catch of all species
		Total	Average	
1954	118	614	5.20	0.56
1955	100	562	5.62	0.37
1956	288	935	3.25	0.46
1957	71	372	5.24	0.16
1958	66	371	5.62	0.17
1959	176	535	3.04	0.18
1960	106	617	5.82	0.24
1961	201	782	4.34	0.32
1962	312	1120	3.59	0.36
1963	106	715	6.75	0.14
1964	134	976	7.28	0.13
1965	148	801	5.41	0.17
1966	176	900	5.11	0.15
1967	101	805	7.97	0.13
1968	144	1045	7.26	0.17
1969	213	976	4.58	0.20
1970	104	749	7.20	0.13
1971	108	564	5.22	0.07
1972	107	549	5.13	0.08
1973	118	610	5.17	0.09
1974	124	634	5.11	0.10
1975	142	717	5.05	0.08
1976	112	705	6.29	0.07
1977	105	745	7.03	0.07
1978	85	615	7.24	0.06
1979	73	462	6.33	0.04
1980	73	442	6.05	0.03
Total	3610	19.008		
Mean	136.7	704. 00	5.63	0.18

weight). According to Vlasova (1959), the huchen stock in the streams of the Transcarpathian region of the Ukraine is very low, only 2–10 kg per kilometre of stream. From the data reported by Schulz (1985) one can deduce that an annual average of 11.3 huchen was caught on a 30 km long stretch of the Drava river between Villach and Mauthbrücken in 1971–1981, i.e. an average of 0.38 fish km^{-1}.

Catches of taimen have been partly reported in Chapter 18. It may be added to the numerical data contained in that chapter that the annual catches obtained in Lake Baikal were estimated at 5–6 tonnes and in the Enisei river at 25–30 tonnes (Pravdin 1949). In the rivers of Yakutia, the annual catches in 1940–1972 varied from 8.6 (1967) to 175 (1972), averaging 41.2 tonnes (Kirillov 1972). In the Lena river itself, the average annual catch of taimen was 83.8 tonnes in 1942–1946, most of which came from the lower reaches and the delta of the river, where 25% of the catch was obtained (Pirozhnikov 1955). It can be estimated from the data presented by Kalashnikov (1978) that in the Vitim river system the annual catch of taimen is around 6.2 tonnes (in 1958–1968). The percentage representation of taimen in the total catch of fish is variable. According to Pirozhnikov (l.c), taimen made up 0.8% of the total catch in the Lena river system in 1945. It can be calculated from the data presented by Kirillov (1972) that the catch of taimen obtained from the same river in 1941–1970 averaged 1.12% of the total catch, or 2.51% after the subtraction of diadromous fishes. The percentage representation of taimen is lower in the Krasnoyarsk reservoir on the Enisei river (0.2–3.9%), in the Bratsk reservoir (0.02–2.1%), in the Baunt lakes (0.12%) and in the Vitim river north of the Yuzhnomuiskii Khrebet Mountain Range 0.7–0.8%), as can be found by recalculating the data presented by Ol'shanskaya et al. (1977), Pushkina (1977) and Kalashnikov (1978). Locally, however, the percentage representation of taimen may be very high. Thus, Olifer (1977) writes that in the Bratsk region the catches of taimen made up 29% in the Angara river and 10% in the Nizhne-Ilimsk region in 1934, whereas in the Vitim, Chita region, taimen comprised 11.2% of the total catch in 1962–1967.

The generally low share of taimen (or, under normal conditions, also huchen) in catches (and in fish communities) is altogether understandable, when bearing in mind that they are top fish predators. Their lower representation in lakes and reservoirs, as compared to rivers, is due to the fact that *Hucho hucho* is a riverine and not a lacustrine species, i.e. a species ecologically adapted to life in running waters.

In concluding this chapter, let us present a rough estimate of the total catch of *Hucho hucho* in their entire range. Assuming that the catches of huchen come only from those streams in which the species normally occurs (Table 36) and that the catch per kilometre of stream is identical with that in Slovakia in 1954–1979 (i. e. 0.33 fish or 1.73 kg per km of stream), then the total catch of this species is around 309 individuals, 1622 kg in total weight, in Austria, 484 individuals, 2541 kg in total weight, in Yugoslavia, and about 68 individuals, around 360 kg in total weight, in Rumania. It follows that around 1000 huchen about 5.2 tonnes in total weight are caught annually in the rivers of the original European range. Including Poland, the Transcarpathian part of the Ukraine, and Soviet Bucovina, and assuming that illegal and unreported catches amount to approximately the same quantity, the resulting amount is 2000–2500 huchen 11–14 tonnes in total weight caught annually in Europe.

With regard to the lack of available statistics, an estimate of the catches of taimen must be based on that of Misharin and Shutilo (1971), according to which the industrial catch of taimen in all Siberian rivers and in the Amur river is around 100 tonnes. But at least the same amount is made up by the catch of taimen in other regions (the Kama and Pechora rivers; Mongolia; China), where catches of this particular species are not recorded, and by those of subsistence fishermen. It should be mentioned in this connection that according to Nikol'skii (1956), 50.9% of taimen caught in the Amur river were handed over by non-professional fishermen;

in the Vitim river, as stated by Kalashnikov (1978), it is about 60%. Thus the total annual catch of taimen is around 200 tonnes. Assuming that the average weight of a taimen in that catch is similar to the average weight of a huchen in Slovakia (which is not improbable, see Chapter 10), i.e. 5.61 kg, this amounts to roughly 36 thousand taimen.

20. *Protection and Management*

LEGAL REGULATIONS

Bag Limit

In Slovakia, the permitted catch of huchen was unlimited until 1949, except for local regulations of anglers' clubs. According to the first local fishing regulation of the Anglers' Club in Martin, dating from 1926, it was permitted to catch and keep two huchen per angling trip, and the number of angling trips was set at two weeks annually. In 1940, the same club issued special permission for huchen angling, being part of a fishing permit (Novák 1971 and in litt.). In 1949, the former Slovak Mandate of Agriculture permitted huchen angling to some of the anglers' clubs, but a special permit for each individual fish was valid for one day only. Since the administration of the fishing rights was put under the control of the state by the Act No. 62/1952, huchen fishing in Slovakia has depended on a special permit which entitles an angler to keep only one huchen a year.

In Poland it is permitted to catch and keep a huchen a week (Polski Związek Wędkarski 1977). In Rumania, huchen fishing is subject to a special permit (Buşnită and Alexandrescu 1963, Papadopol in litt.). In Yugoslavia, huchen fishing is regulated by a special rule: each angler must have special equipment, pay a special fee, and may keep one huchen a day.

We have no information on the situation in the remaining countries of the original range of the huchen.

No limitations pertain to taimen fishing (Misharin and Shutilo 1971).

Close Season

Originally (Hungarian Fishing Act from 1888, Article 19), huchen in Slovakia (and whole Hungary) were protected only in the period from 1 March to 31 May. In 1948–1949 they were protected throughout the year, and from 1949 on huchen fishing was permitted from 1 October to 15 December. At present, the huchen is protected in Czechoslovakia from 1 January to 31 October; permits for huchen

TABLE 38

Close season and legal size (*Tl*) of huchen in various countries

Country	Close season	Legal size (cm)
Czechoslovakia	1 Jan.—31 Oct.	70
Dunajec and Poprad rivers	1 Jan.—31 Oct.	60
Yugoslavia[1]		
Croatia (selected rivers)	15 Feb.—1 Oct.	70
Slovenia (selected rivers)	1 Jan.—30 Sept.	70
Remaining waters	1 Feb.—31 May	50
Federal Republic of Germany[2]		
Bavaria	15 Feb.—15 May	70
Baden-Württemberg	15 Feb.—15 May	70
Austria[3]		
Tirol	1 Mar.—31 Mar.	50
Carinthia	1 Mar.—15 May	75
Salzburg	16 Mar.—30 Apr.	60
Lower Austria	1 Mar.—15 May	75
Vienna	1 Mar.—8 May	75
Upper Austria	1 Mar.—15. May	55
Styria	1 Mar.—30 June	75
Rumania[4]	1 Mar.—31 May	80
Poland[5]	1 Mar.—31 May	60
France[6]	28 Sept.—11 June	70
Switzerland[7]	?	70

[1]Ribiška Zveza Slovenija (1966); [2]Kasecker (1970); [3]Schindler (1953) and Jungwirth (in litt.); [4]Busniţă and Alexandrescu (1963) and Papadopol (in litt.); [5]Polski Związek Wędkarski (1977); [6]Laurent (in litt.); [7]Service Cantonal de la Chasse et de la Pêche, Fribourg (1977).

angling are issued only for the period from 1 November to 31 December. In other countries the close seasons differ and often are regulated separately for different regions (Table 38).

General Protection

Except in Czechoslovakia, where the huchen (spawning shoals in their spawning sites) is included in the list of protected animal species (Intimation No. 125 of the Presidium of the Slovak National Council, dated 6 November 1965), the species is not protected in any other countries.

Legal Size

Originally, the legal size of huchen was 54 cm (or only 50 cm in Moravia) in total lenght *(Tl)*. After 1925 it was increased to 65 cm, and since 1979 the valid legal size has been 70 cm *Tl* (or 60 cm in the Poprad and Dunajec rivers). The legal sizes valid for other countries are given in Table 38. No legal size has been set for taimen; Misharin and Shutilo (1971) suggest 70 cm *Fl*, i.e. about 79 cm *Tl*. Vlasova (1959) suggested a similar length (legal size) for huchen in the Transcarpathian region of the Ukraine.

It should be pointed out here that a legal size below 70 cm *Tl* is too small, as it allows the taking of even those females which have not spawned. Skácel's observations show that only about 50% of huchen 650 mm in *Tl* can be considered sexually mature and capable of spawning. Krauss (1933), Seez (1939), Ivaška (1951) and recently also Rothschein (1980) stressed the consequences of too small a legal size. In Yugoslavia, for instance, the legal size of 70 cm was introduced as early as 1922, and was further increased to 75 cm by the tenants of some of the fishing grounds, which was favourably reflected in increased numbers as well as catches of huchen in subsequent years (Munda 1925). According to Rothschein (l.c.), the legal size of 70 cm is insufficient to utilize the growth capacity of the huchen and does not allow maximum yield from a given number of stocked fish. The author recommends a legal size of 80 cm, which would ensure obtaining the same yield even at low fishing efficiency, and a still higher one at a fishing efficiency of $F = 50\%$. Svetina (1970) suggests that the legal size be at least 85 cm, as at least two spawnings would be obtained per female in this way.

Reserves

To our knowledge, there is only one protected huchen locality throughout its range, viz. in the Turiec river, Czechoslovakia. This reserve has been established by the Intimation of the Ministry of Education and Culture, No. 6527/1966, dated 10 July 1966. The protected locality comprises a section of the Turiec river 21 km long and 62 ha in approximate area. Here the fishing rights are regulated by a special statute which is renewed from time to time. Among other things, this statute forbids huchen fishing for other than breeding or research purposes. Unfortunately, as stated in an Opinion of the Institute of Fishery Research and Hydrobiology in Bratislava, dated 3 November 1980, the statute is not being observed. Moreover, the very existence of this reserve is endangered by the planned canalization of the Turiec river and by the construction of a water supply reservoir in its upper reaches.

STOCKING

Stocking of streams with huchen is an old practice, dating back to the late 19th century, when the first experimental rearing aimed at this very purpose was begun. By the beginning of the 20th century, huchen stocking had reached a considerable extent. Munda (1935), for example, writes that in 1911–1934, several streams (especially the Savinja, Ljubljanica, Sava and Drava rivers, but also the Bosna and other streams) were stocked with a total of over 440 thousand huchen alevins and 9351 fingerlings.

At present stocking is of decisive importance in the efforts to increase the numbers of huchen. The positive influence of stocking was pointed out by Skácel (1976), who stated that huchen had not disappeared completely from streams due not only to protective measures taken, but also to the new experience obtained in huchen farming. Systematic though not quite sufficient (in terms of numbers) stocking has helped to regenerate the numbers of huchen in the Turiec river (Bastl et al. 1975, Kirka and Bastl 1980) alhough even the present status is far from resembling the situation existing 20 years ago. Unfortunately, the production of huchen for stocking in Slovakia has experienced a considerable decline in recent years due to a decrease in egg production by the Slovak Anglers' Union: the average annual egg production was 388,000 in 1968–1972, but the production dropped to only 60,000 eggs in 1974–1977, so that it was even necessary to stop exporting the eggs. Prior to that period, part of the egg production was exported: in 1968, 100,000 eggs to Yugoslavia, 30,000 to the Federal Republic of Germany, and 20,000 to Finland; in 1969, 30,000 eggs to the FRG, 20,000 to Switzerland, and a 1000 to Denmark; in 1970, 100,000 eggs to Poland, 7000 to the FRG, and 2000 to Switzerland; in 1971, 5000 eggs to Austria and 100,000 to the FRG; finally in 1972, 100,000 huchen eggs were exported to the FRG, 100,000 to Austria, and 5000 to Luxemburg.

In Slovakia, the Váh, Turiec and Orava rivers are more or less regularly stocked with huchen. Ivaška (1951) mentions that the best effect is obtained from stocking with fingerlings, and not with alevins. For that reason, older huchen young are increasingly being released in recent years, and not alevins as in the past. According to the statistics of the Central Committee of the Slovak Anglers' Union, the numbers of huchen yearlings and summerlings released in Slovakian waters during successive five-year periods, were as follows (annual averages in thousands of individuals): 1965–1969, 19.8; 1970–1974, 20.6; 1975–1979, 45.8. The numbers and age of huchen released in other countries are unknown to us except for Poland. Witkowski and Kowalewski (1980) give a statistical review indicating that in Poland an annual average of 27,294 huchen eggs was produced beginning in 1955, yielding almost 25,000 young which were used for stocking various streams. In Austria, as it seems, only alevins are still used for stocking (Jungwirth 1978).

Stocking huchen are released not only directly in those streams which are inhabited by the species but even in small tributaries in which huchen do not occur permanently and which serve as so-called rearing streams. From such streams older and bigger young huchen gradually move downstream into bigger streams or are fished out and moved elsewhere if needed. This method was developed by Czechoslovak breeders. It was first used by Ivaška (1951, 1959a) and later by others (Matyas 1954, Pavlík 1955, Režný 1968). According to Pavlík (1955), this method is not suitable, as the losses amount to 90–95%. The same author recommends releasing summerlings in shallower parts of streams where young of other fish species are present in sufficient numbers. The rate of stocking is 40–60 summerlings per kilometre of stream.

It follows from various parts of this chapter that management of a huchen population can only be understood in a very broad sense of that word. Essentially, the old method is still practised, involving occasional emendations of the close season and legal size, and stocking waters with older or younger stock (according to what is available and not what is needed). The ratio of numbers of harvested: released fish is not taken into account, nor reproduction success in successive years, nor the population density in a particular stream; permits for huchen fishing are issued mechanically and not in accord with population analyses carried out in the preceding years; the legal sizes are mostly valid for extensive regions and not for individual streams, etc. This is doubly true of taimen, where not even the old-fashioned management mentioned above is applied.

With regard to the critical and endangered situation of the species *Hucho hucho* as a whole, it will be necessary to alter the ways of management radically. In our opinion, the open season should be shifted to summer months, when the huchen are not so voracious and when males and females are approximately equally vulnerable to angling. The legal size should be fixed separately for separate populations (streams) and the quota of fish harvested (number of permits issued) should be determined from year to year according to objective indices (growth rate, abundance, reproduction success, recruitment). Before stocking, the success or failure of natural reproduction, the numbers in a population, and the size of the bags should be considered.

21. *Introduction and Acclimatization*[11]

21.1. RESULTS OF EXPERIMENTS WITH INTRODUCTION AND ACCLIMATIZATION

Attempts to introduce and acclimatize the huchen outside its original range are about hundred years old and are closely connected with the first attemps to rear this species and to stock streams with it. Naturally, they were at first of an elementary character, but later they were better and better planned and learned from previously obtained experience. In the chapter to follow, they are intentionally described at greater length: on the one hand, the acclimatization experiments show the

[11] This chapter was published separately in a more condensed form in 1984 (Holčík, J., 1984: Review and experiments with introduction and acclimatization of the huchen, *Hucho hucho* (Linnaeus, 1758) (Salmonidae). EIFAC Tech. Pap. 42, Suppl. Vol. 2: 290–298.

adaptability of *Hucho hucho* and on the other they are a valuable source of information which is immediately connected with the protective measures to be taken in order to conserve the species. All experiments carried out so far pertain exclusively to the nominate form, the huchen; in the available literature, we have found no mention of experiments aimed at the introduction and acclimatization of the taimen.

Czechoslovakia

While Vutskits (1918) reports on huchen occurring even in the Poprad river (a specimen being kept in the Hungarian National Museum), we are justified in assuming that it was introduced into the system of that river somewhere at the turn of the century. Indirect evidence of the absence of huchen from the Vistula river system can be found in its geographic distribution (Chapter 7) as well as in the fact that archaeological findings in the Vistula river basin contain no remains of huchen (Dr. Urbanowicz, personal communication, March 1980). Allegedly huchen vertebrae found in a ritual well near Gánovce, dating from the early Bronze Age (in this case 1465 B.C.), and described by Ambros (1959), were those of salmon, *Salmo salar* (Holčík and Lepiksaar 1980). On the one hand this indicates that the fish were caught in the nearby Poprad river and on the other it is indirect evidence of the fact that the huchen never occurred in that river. At present huchen are released in the Poprad river on the territory of Czechoslovakia and in the Dunajec river in Poland (Žitňan 1974, Witkowski, personal communication, Witkowski and Kowalewski 1980). In both these rivers, huchen have become acclimatized and are reproducing.

Huchen have also been introduced into the Hornád river. A detailed report on this introduction has been sent to us by J. Zubrický from Košice. According to that report, huchen were introduced here twice. The first introductions were in the nineteen sixties, when huchen were released in the environs of Margecany. The introduction was successful, as show the present catches of huchen 7–9 kg in weight from natural reproduction. From that period we also have verified (L. Skácel) reports that huchen were also released in the Hornád defile above Spišská Nová Ves, whence they were later removed by systematic fishing (including the use of an electrical shocking-gear), because the huchen allegedly decreased the numbers of brown trout and grayling. At present, huchen occur in the area of Slovenský raj, in the environs of Smižany and Čingov (Kriššák 1985). Since 1977, huchen have been released, in the form of summerlings, below the equalization basin of the Ružín reservoir, where around 11 thousand individuals have been stocked so far. Huchen fishing is not yet permitted in that river section but records are available on accidental catches, sometimes reported as "big trout". According to reliable and verified information, several huchen were caught in the stretch of the Hornád river between Malá Lodina and Kysak in 1980–1981. Above Košice, near Ťahanovce, two huchen, 82 and 85 cm in respective lengths *(Tl)* and 4.6 and 5.7 kg in weight, were caught in 1981; in the same places, two additional huchen 2.3 kg in weight were caught in 1981. In Košice, below the weir in the Anička locality, two huchen 2.5–3 kg in weight were caught in 1980. Below Košice, at Krásna nad Hornádom, a huchen weighing 6.5 kg was caught as early as in 1976 (this individual apparently came from the progeny of the huchen introduced into the area of Margecany; note by J.H.) another 4 huchen weighing 2.2–2.5 kg were caught there in 1980–1981. While there are no reports on huchen caught down to the boundary section, in 1981 a report was published in a regional newspaper in Hungary on the catching of two fish around 3 kg in weight, "resembling a pike", in the section of the Hornád river some 8 km below the Czechoslovak–Hungarian border. It may be of interest to find out whether or not huchen have become acclimatized in the section of the Hornád below Ružín, the water of which is far from being the best quality, and whether or not they will even reproduce.

In 1962, huchen were also introduced into the Topľa river, when 2000 yearlings around 120 mm in *Tl* were released in the lower reaches of the Zborovský potok Brook. They became established in their new environment and even spawned in 1966–1968. Young huchen from natural spawnings were then caught several times in the Topľa – they were sexually immature individuals 250–300 mm in *Tl.* In 1967–1974, several huchen 1.6–8.0 kg in weight were caught with rod and line in the stretch between Tarnov and Dubinné. After 1974, however, there are no reports on the occurrence of huchen in the Topľa and the following are the probable causes of its disappearance: (1) stocking was not repeated, (2) poaching, and ignorance of anglers, who mistook huchen for trout, (3) gradual deterioration of the environment (dredging, pollution). V. Křička, who has sent the data in a letter, believes that the Topľa river is suitable for huchen to occur, especially in the section downstream of Bardejov.

From 1938, huchen were introduced into the Rajčianka river, in which they allegedly never occurred before. The huchen became acclimatized and even reproduced in that stream (Matyas 1954) but later they were removed, as in the Hornád river and for the same reasons. Besides the Rajčianka, Ivaška (1959a) also mentions the Blatnický potok Brook, Malý Čepčín and Žarnovica (tributaries of the Turiec), where huchen did not become established. Skácel (1967) even mentions the Kysuca river but huchen are apparently autochthonous in that stream, since Mišík (1959) reports that in 1957, at spawning time, huchen migrated upstream the Čierňanka, a tributary of the Kysuca, up to the Polish territory. Huchen were even introduced into the Muráň and Slaná rivers (a letter from the local organization of the Slovak Anglers' Union at Jelšava of November 1970, and a letter from A. Makara of November 1982). In the Muráň river, huchen attained a length of 820 mm *(Fl)* and a weight of 5.2 kg. No reports are available to us on the fate of the huchen in the Slaná river; it is only known that through ignorance the local anglers took the huchen for "strange big trout" and fished them out gradually. At present, huchen do not occur in the two streams any more.

More exact data are available on the introduction of huchen into the Nitra river, presented here on the basis of reports submitted by Kattoš (1966, and in litt.) and

Sedlár (1969, and in litt.). While Tejčka (1934) writes that huchen were already introduced into the Nitra river in 1931, any more detailed information or any confirmation of that report is lacking. Hearsay evidence indicates that the first huchen were brought into the Nitra (between the Handlovský potok Brook and the mouth of the Chvojnický potok Brook) in the number of 100 yearlings in 1943. In 1948 a huchen weighing 4.5 kg was caught above Prievidza; still earlier in 1945, after the front had passed, two huchen were allegedly found there, weighing over 1 kg each. Systematic stocking with huchen was begun in 1952, when 2000 yearlings were released in the Nitrianske Rudno reservoir on the Nitrica river. A further stock, totalling 2500 yearlings, was released in the Nitra river in 1962, 1963 and 1965. The huchen became acclimatized in both the Nitra and Nitrica rivers and reproduced normally. Very good growth was recorded therein (a huchen 8 years old weighed 12 kg) and fish weighing 5–10 kg were caught several times; in 1977, a huchen was found killed by a dredge, measuring 120 cm in *Tl* and weighing 16 kg. At present, the Nitra river is canalized for the most part, and huchen occur there no more. The huchen in the Nitrica river deserve attention. Since this is a small and rather shallow stream, which becomes overheated during spells of hot summer weather (in the summer of 1959, huchen were dying there of heat, several fish weighing 6–12 kg being found in the mouth of the stream into the reservoir), huchen mostly stay in the reservoir, moving from it into the Nitrica river only to spawn.

Attempts to introduce the huchen into the streams of Bohemia and Moravia were begun before World War I. Probably the first attempt to acclimatize the huchen was made in the Olše river (the Odra river system) where the former German Anglers' Union for Silesia at Český Těšín released 800 huchen yearlings in 1888–1891 (Ehrler 1935). Nothing is known about the outcome, however. Later, huchen got again into the Olše river in 1937, both unintentionally, when the fish escaped from Nieslanik's hatchery during a flood in 1937, and as a result of intentional stocking. In 1943, spawning of these fish was observed, but in spite of this, huchen did not survive in the Olše river due to continuously deteriorating water purity.

In 1909 and 1910, Šivic attempted to stock the Bečva river with young huchen from eggs bought in Austria and hatched at Valašské Meziříčí. The fish were released in the Bránecký potok Brook from which the huchen moved into the Bečva river a year later. Even in 1914 several huchen were observed spawning below the Jurinský weir downstream of Valašské Meziříčí (Polášek 1936). In 1950–1953, about 400 huchen yearlings were released in the Bečva river above Valašské Meziříčí and the stocking was later continued until 1965. In the Bečva river, huchen have probably survived until the present, as in 1972, four dead huchen up to 60 cm in *Tl* were found on the confluence of the Vsetínská Bečva and Rožnovská Bečva rivers (Lusk 1976).

In 1936, huchen alevins were released in the Čtyřdvorský potok Brook (the Svratka river system) numbering 1300 individuals. Later, huchen were observed, and in 1940 two huchen weighing around 5 kg were even caught in that stream. The subsequent fate of the huchen in that stream is unknown, however; they have apparently not survived (Lusk 1976).

Attempts at stocking the Moravice river (the Odra river system) with huchen are described by V. Dyk (1956), Sýkora (1957) and Lusk (1976). In 1949–1950, the Salmonid Fish Breeders' Club at Bělá brought young huchen from Ivaška's hatchery at Martin and raised them up to 9–12 cm in *Tl*. Three hundred such huchen were then released in the Moravice river in the environs of Karlovec (the present Karlova Pláň). When the stream in that area was affected by heavily polluted waters of a tributary, the Černý potok Brook, the huchen gradually moved 10–15 km downstream. Further stocking was carried out in 1951, 1952 and 1953, when unstated numbers of huchen alevins, 2200 summerlings and 15 bigger huchen about two years old were released in the Moravice river. Young from natural spawning were already observed in 1954. In 1955, spawning was observed in the mouth of the Černý potok Brook; the huchen participating in that spawning were 70–80 cm in *Tl*. Spawning huchen were also observed in the Moravice itself, at Karlova Pláň, in 1970 and 1971. In those years, an angler caught a huchen weighing as much as 15 kg, which he turned loose again (Dyk, in a letter of 17 May 1981). Although scattered occurrence of huchen in the Moravice river cannot be excluded even at present, the species is practically absent from that stream.

In 1949–1953, attempts were made to re-acclimatize huchen in the Dyje river. First several hundreds of huchen young and summerlings were released in the Dyje river below the Vranov reservoir, and in 1963 also in the Dyje river above Znojmo. In the years to follow, huchen weighing 5–6 kg began occurring in anglers' bags, and in 1976 a huchen weighing 8.2 kg was found dead in the Dyje river near Dobšice. Lusk (1976) writes that at present huchen occur in the Dyje river, mainly in the area of Znojmo, both in the river itself and in the impoundment where fish around 100 cm in *Tl* and 10 kg in weight have been observed (Lusk. l.c.).

Into the system of the Labe (Elbe) river in Bohemia, huchen were first introduced in 1949 and 1952, when they were released in the Otava river in the environs of Žichovice and in the upper reaches of the Vltava river (Smíšek 1963). Most probably, Ivaška was an indirect initiator of the introduction of huchen in the Otava river, as in a lecture on huchen farming held in Prague in 1946, he expressed his opinion that the Otava river might be a suitable stream for huchen (Bardoun 1946). Huchen were also stocked into the Vltava river in 1958, 1961–1962 and 1973–1979, this time in the section between Prague and the Vrané reservoir, directly in the reservoir and in the environs of Český Krumlov (Vostradovský and Novák 1959, Havelka and Vostradovský 1974, Vostradovský et al. 1973, Skácel 1976, Anonymus 1986). In 1963 and 1965, huchen were released in the Ohře river near Klášterec nad Ohří and Kadaň, and in 1980 even below the Nechranice dam (Flasar and Flasarová 1981). Few particulars on these introductions are known, however. Huchen probably did not hold out in the Otava river, although they exhibited good growth (attaining 20–25 cm in length within one year, 60–70 cm and up to 1.8 kg in weight in the second year, after the Otava had been poisoned by the matchworks at Sušice in 1954, huchen weighing up to 8 kg were found; Dyk 1956, in a letter of 17 May 1981). Yet in the Vltava river huchen occur up to present, as confirmed not only by findings of dead huchen but also by occasional catches, reported directly as "huchen" or as "big trout" (Havelka and Vostradovský 1974, Michálek 1976), in the section between the Vrané reservoir and Prague and in the environs of Český Krumlov. The possibility of the huchen surviving in this part of the river is undoubtedly connected with the construction of the Vltava river cascade, thanks to which the water temperature decreased downstream, the processes of decay were slowed down and a good oxygen regimen was maintained, so that the section of the Vltava (between Vrané and the mouth of the Berounka river) was changed into a trout stream.

In 1957–1959 the Sázava river was stocked, first with alevins and later with fingerlings 4–6 cm in length. As early as 1959, in a control fishing, two huchen (54 and 72 cm in *Tl* and 2 and 3.5 kg in weight) and further seven huchen (varying from 34 to 72 cm in *Tl* and up to 3–3.5 kg in weight) were caught in 1960–1962 (Krechler 1963). At present, however, huchen do not seem to occur in the river any more.

In 1960, 101 huchen two years old, about 30 cm in length, were released in the Klíčava water supply reservoir (the Berounka river system). The aim of this introduction, together with other measures, was to reduce the numbers of coarse fish, particularly perch. In 1962 and 1964, two huchen were caught there (Holčík 1970). At present, huchen occur there no more (Švátora, personal communication, July 1981).

In 1969, huchen eggs were brought to the hatchery of the Czech Anglers' Union (CAU) at Hynčice near Broumov and, at the same time, to the hatchery of the CAU at Hradec Králové–Malšovice. From there the reared yearlings, or alevins, of huchen penetrated into the Stěnava river (the Odra river system) and into the Orlice river (the Labe river system), respectively. The outcome of this unintentional introduction into the Orlice river is unknown, but in the Stěnava river an unstated number of huchen under 1 kg in weight is caught every year (Lohniský 1973, 1984). No huchen have been found in the Polish section of the Stěnava river (A. Witkowski, in a letter of 20 April 1979) and thus it appears that the huchen is only distributed in our section and that it probably even reproduces in it.

According to Poupě (in a letter of 7 Sept. 1977), huchen are said to occur even in the Berounka, Kamenice and Nisa rivers, but no particulars on stocking or on the outcome are known to us.

A large number of countries not only in Europe but also in Africa and North America is interested in the acclimatization of the huchen. According to Skácel (1976), fertilized huchen eggs from Slovakia have been almost regularly exported not only into countries of autochthonous occurrence of the huchen (Austria, the FRG, Yugoslavia) but also into countries where the species had never occurred before (the GDR, Poland, Finland, France, Bulgaria[12], Denmark, Spain, Belgium, Morocco and Canada, Israel is also said to be interested in its import). It is a known fact, too, that besides Czechoslovakia, Yugoslavia also exported huchen eggs,

---

[12] According to Živkov (in a letter of 13 May 1977), however, the huchen has never been imported into Bulgaria.

whereas it was Austria in the past. Unfortunately, very scanty information is available on the outcome of the introduction and acclimatization of huchen in those countries, much of which is scattered in local, poorly accessible magazines, and thus evades attention. For example, Vooren (1972) only mentions the introduction of huchen into England and France, and his data were later adopted by Karpevich (1975). According to these authors, the results of the introduction of huchen into France are unknown, although the contrary is true, as will be shown later. Since the problems associated with the introduction and acclimatization of huchen in other countries are interesting and important from both theoretical and practical points of view, we mention them here in more detail, both on the basis of published data and according to those obtained by several specialists, our foreign colleagues, in their letters.

England

England was probably the first country to import huchen. Here they were to replace salmon, whose migrations into English rivers were becoming increasingly scarcer due to their increasing pollution. We are informed on the introduction of huchen in England from several sources (Anonymus 1905, Pölzl 1910, 1929, Wheeler 1969, 1979, Wheeler and Maitland 1973, Cacutt 1979). At the request of The Thames Salmon Association, F. Pölzl, a well-known Austrian fish breeder, imported 20,000 fertilized huchen eggs into England in 1905. One part of young huchen reared from them was released in the upper reaches of Thames in 1906. The second part was then raised at Denham by some Mr. Gilbey. He succeeded in spawning them in 1910, when the fish were 5 years old and weighed 3 kg. Young huchen from this spawning were also placed in the Thames at Hedsor, Taplow and Teddington. The stocking made in 1914, mentioned by Cacutt (1979) probably came also from the same brood stock at Denham. However, the fate of huchen from these implantations is not certainly established. In spite of numerous reports on huchen up to 20 kg in weight, allegedly seen and even hooked by anglers, and in spite of a special netting of the Thames arranged, no huchen was ever landed and positively identified. While local anglers and some authors (e.g. Marlborough 1963) say the huchen is still in the Thames river perhaps unrecognized, the authorities (Wheeler 1969, 1979; Wheeler and Maitland 1973) say that attempts to introduce huchen have failed completely.

Switzerland

We are informed on the introduction of huchen in Switzerland from several sources (Steinmann 1948, Ivaška 1951, Brodbeck 1953, Svetina 1964, Renz 1969, Anonymus 1969, Service cantonal 1977, Bél in his letters of 27 Feb. and 18 May 1978). All attempts were made in the Rhine river system, either directly in the Rhine or in the Sarine (Saane) river, a left tributary of the Aare river. The introduction was preceded and accompanied by a discussion of the suitability of huchen for Swiss waters: it follows from a contribution by Steinmann (l.c.) that the introduction was backed by angling circles on the grounds that the huchen might replace salmon, which had disappeared from the Rhine, whereas scientific circles opposed the acclimatization.

The first import took place in summer 1949, when 15 thousand fertilized eggs were imported from Czechoslovakia in exchange for those of *Coregonus lavaretus* and *Salmo trutta* m. *lacustris,* and hatched in August Hohler's hatchery at Augst. The hatched alevins, numbering 11,000, were released in the Rhine near Kaistenbach below the Laufenberg dam. The rest, fingerlings 6–8 cm long, were released in the Rhine below the waterworks at Ryburg-Schworstadt and Rheinfelden. A further import of 50,000 fertilized eggs was realized in 1950, of which 7500 fingerlings were released in the Rhine in the area of Basel. The Rhine was later stocked with additional 700 yearlings reared from eggs, again imported from Czechoslovakia, in 1955. The huchen did not become acclimatized in the Rhine, however: the total catch comprised 4 huchen–three about 30 cm in *Tl* were caught in the Rhine near Rheinfelden in 1950, and one was caught in a net by a professional fisherman in the Rhine near the mouth of the Erholz stream in 1953 (the fish was 60 cm in *Tl* and weighed 2 kg).

In contrast to the Rhine, where the huchen introductions may be characterized as uncontrolled, the acclimatization of this species in the Sarine river was well-thought-out. Here, Professor M. Svetina, of the University of Ljubljana, carried out an investigation in 1964, on the basis of which he recommended introducing the huchen into the section between Fribourg and the Gruyère power plant, which he characterized as the "lower part of a grayling zone or the upper part of a barbel zone". Besides grayling, barbel and brown trout, there occurred additional species, predominantly cyprinids, with a marked predominance of nase carp. The introduction was aimed at restoring the allegedly disturbed equilibrium in the local fish community. In 1965, the first 20,000 eggs were imported from Czechoslovakia, the most numerous batch, numbering 10,000 alevins, was released in the Sarine in summer, about 6700 fingerlings were released in the autumn of the same year and 500 yearlings in the spring of the next year. The stocking was repeated in 1966 and 1967, when 20,000 young, again originating in Czechoslovakia, were released in the same section. The stocked section was under special supervision, anglers were specially informed on the presence of huchen, and were obliged to report all accidental catches. Besides, occasional control fishings were made. In this way, starting in 1966, records were obtained on several tens of huchen, showing that the new environment was suitable: one-year-old fish were 20–30 cm in average *Tl;* two-year-old ones 42–46, three-year-old ones 52–56 cm. In 1969, fish as big as 70 cm were caught. However, the huchen did not stay in the section appointed but occurred even lower – below as well as in the lakes Perolles and Schiffennen. Later huchen fishing was permitted, except in the periods from mid-April to the end of May and from 30 September to 1 March. The legal size was set at 30 cm of *Tl,* and the bag size at one fish per day during the open season. No spawning was observed, however. It follows from the contribution by Renz (1969), who presents the above particulars, that they intended to start their own huchen breeding, based on fish caught in the Sarine, and to use their progeny for further stocking. According to a report of the Cantonal Fishing and Hunting Management in Fribourg of 3 Nov. 1977, the expected results, i.e. a reduction in the numbers of cyprinid fishes in the Sarine, was not quite achieved although no final conclusions can be drawn because of the short duration of the experiment. Since 1968, the streams are preferably stocked with brown trout and grayling which are more valued by anglers, allegedly because it is difficult to catch huchen with rod and line. The above report also mentions difficulties encountered in the protection of trout and grayling in winter when, on the contrary, it is permitted to catch huchen. At present it seems that interest in the huchen has died away.

Morocco

Huchen were imported to Morocco twice: 30,000 fertilized eggs in 1953, and further 20,000 in 1959, in both cases from Czechoslovakia (Preudhomme 1955, 1959a, H.M. 1962, Skácel 1963). The eggs were hatched in the D'Azrou hatchery and fingerlings were stocked in several streams in the Central Atlas Mts. (the Fellate, Umm-ar-Rbia, Gígú). While certain reports were very optimistic in describing the fate of the huchen in Morocco, mentioning even natural spawning (H.M. 1962), the actual situation was different, as follows from the reports published by Preudhomme (1959b, 1961). According to this author, the huchen showed a very rapid growth in Morocco (fish 15–16 months old having attained 46–57 cm in *Tl*, but a huchen caught in the Gígú in 1960, evidently 7 years old, was only 68 cm long *(Tl)* and weighed only 3 kg) but never reproduced in the wild. Attempts at rearing huchen were successful but once, when fish 80–90 cm long were reared from some 30 captive fish. However, there was only one female among them, which yielded 700–800 eggs by hand-stripping. It is unknown whether or not any young were obtained from the eggs. The above author is very sceptical as to the possibility of establishing the huchen in Morocco, pointing out that the local streams almost lack a foothill zone with satisfactory water temperatures. Moreover, the water in those streams is permanently turbid. According to the information obtained from Morocco and kindly submitted by P. Laurent of Thonon (in a letter of 24 Aug. 1977), huchen do not occur there at present.

Belgium

Dr. J. A. Timmermans (in a letter of 3 May 1977) kindly informed us on attempts at acclimatizing huchen in Belgium: Fertilized huchen eggs of Yugoslavian origin, totalling over 22,000, were imported in 1954, 1960, 1961 and 1962. Hatched young were reared in fishponds until they attained a length of 10–18 cm. Over 6000 fingerlings were stocked in the Lesse river, and an unknown but smaller number in the Semois river (the Meuse river system). Huchen did not become acclimatized in either river although they were observed and several times also caught with rod and line in the three years following the introduction.

Poland

Originally, huchen did not occur in Poland except in the Čierna Orava river. The first attempts to introduce the species into the Poprad and Dunajec rivers began as early as the end of the 19th century, when transporting and acclimatizing fishes was fashionable among the fishermen and anglers of that time (Witkowski and Kowalewski 1980). However, particulars are no longer available on those attempts. Further experiments were not begun until later, in 1955, when the progeny of huchen coming from the Čierna Orava river was released in form of summerlings in that stream again. The experiment was carried out by the Polish Anglers' Union (Polski Związek Wędkarski, PZW) in Cracow; the eggs were hatched and the stocking fish were reared in a hatchery at Łopuszna (Matlak 1958, Kolder 1964). Further eggs were also obtained from fish caught in the Čierna Orava, and young hatched from them became the basis of a later breeding shoal numbering 4–13 fish. Besides, Polish anglers contacted the Slovak Anglers' Union and imported fertilized huchen eggs, alevins and summerlings in 1964 and 1970. At present, Poland seems to have a prosperous closed-cycle huchen farming. It is centred in the hatchery of the PZW at Łopuszna near Nowy Targ where over half a million alevins and 20,000 summerlings or yearlings were produced within 25 years, with an average annual production of 6–60 thousand individuals. Besides Łopuszna, attempts at huchen breeding are also being made in a hatchery of the PZW at Rumia, in a branch of the Institute of Freshwater Fisheries at Oliwia near Gdansk, in the Fish Stocking Centre of the PZW at Czarci Jarz, and in the Institute of Ichthyobiology and Fisheries in Olsztyn. Huchen are stocked, besides the Poprad and Dunajec rivers where most of the reared fish were released, in other streams as well. Thus in August 1968, 2000 summerlings reared from eggs imported from Czechoslovakia were released in the Nysa Kłodzka (the Odra river system) and in recent years (as reported by A. Witkowski in a letter of 20 Apr. 1979), huchen have been introduced even into the Tresna reservoir on the Sola river, into the Rabski potok Brook (the San river system) and into the Skawica river below the Babia Góra Mts. In 1977, 500 huchen young were released in the Drwęca river (about 30 km south of Olsztyn; J. Guziur in a letter of 15 Jan. 1978). However, except for the Dunajec and Poprad rivers, where huchen have become naturalized, the outcome of the above introductions is not as effective as desired; huchen are only occasionally caught (and are often believed to be "big trout") or observed, as for example, in the Nysa Kłodzka river. A. Witkowski believes that the cause lies both in the small number of young stocked and in the fact that the stocking is being carried out but once. The stocking procedure is as follows: huchen young are released in smaller streams where no predatory fishes occur, such streams being fished out with the use of an electrical shocking-gear from time to time to remove bigger trout and sculpin. The objective of introducing huchen into Poland is to compensate for salmon and sea trout, whose numbers are in steady decline in Polish rivers.

France

In 1957, an attempt was made to introduce and acclimatize the huchen in France. In this case, for the first time in the history of huchen acclimatization, the attempt was preceded by careful selection of a suitable locality and by ichthyological investigation. A left tributary of the Rhône river in Upper Savoy, called the Usses river, viz. its section populated by brown trout in the upper part and by barbel in the lower, was selected for the introduction. The nase carp was the dominant species there. The stocking procedure and results are reported by Vivier et al. (1964), Svetina (1967) and Laurent (in letters of 8 and 11 Aug. 1977). A year before the introduction (1956), the numbers of nase carp were reduced by electro-fishing and in each of the two subsequent years (1957 and 1958), 25,000 fertilized huchen eggs were imported from Yugoslavia. After partial rearing, the young hatched from these eggs were stocked in the section selected. The stocking was repeated in 1959 and 1960, when 10,000 eggs were imported on each occasion, again from Yugoslavia. Check fishings, carried out in August 1959, revealed the presence of huchen coming from the previous introductions. It was found that young-of-the-year attained a length of 10–12 cm, yearlings 30–35 cm and two-year-old fish 45–48 cm *(Tl)*. A check fishing in May 1960 yielded 5 huchen, among them one 57 cm and another 49 cm long. No more stocking was done in the years to follow, but observations were made in order to ascertain incidental natural spawning. In June 1961, two sexually mature individuals were caught, measuring 65 and 67 cm in *Tl*. In May and October 1962, the fish already attained 77–87 cm in *Tl*. Young from a natural spawning were ascertained in the spring and summer of 1963 when huchen 19–27 cm long were caught, and in July of that year several individuals measuring around 10 cm were obtained. Reproduction was directly observed in 1976 some 20 km upstream of the release site; at the spawning site there was a male about 90 cm long and a female 120 cm long. Huchen apparently spawn there every year, as can be judged from the varying lengths of the huchen caught. Svetina (1967) writes that the first season for huchen fishing was opened in the Usses river in 1965, and that each year several big huchen among them some weighing over 16 kg, and one even 25 kg, have been caught there ever since. However, neither Vivier (P.V. 1968) nor Laurent (l.c.) mention such catches. The decrease in the numbers of nase carp under predation pressure of huchen has already been mentioned in Chapter 12.

From the Usses river the huchen moved into the Rhône, where an individual is said to have been caught about 400 m below the mouth of the Usses.

The last check fishing in 1973 and also reports of anglers have shown that huchen still occur in the Usses but, as written by Laurent in a letter of 11 Aug. 1977, their numbers are probably very small. Laurent believes this to be the result of the huchen being mistaken for trout by anglers, and of the great angling vulnerability of huchen in shallows during the spawning period. Neither the legal size nor the open season set for the huchen are observed, as anglers have lost interest in the huchen. They believe that the Usses is too small a stream for such a big species and maintain that huchen destroy spawning redds of brown trout (!), this being the cause of the decrease in the numbers of trout that followed the introduction of huchen.

Sweden

In 1963, an attempt to acclimatize huchen was made in Sweden, when 8000 fertilized huchen eggs were imported from Yugoslavia. Hatched young were reared in troughs and small ponds and released in the Rexforsån rivulet and in the Indalsälven reservoir in 1964. The experiment was a total failure: no catches or observations have been recorded (Nilsson and Svärdson 1963, Nilsson in a letter of 5 May 1977). Nevertheless, Sweden is seriously interested in introducing huchen, particularly into submontane man-made lakes and possibilities and ways of introducing it are being studied intensively, as follows from the information obtained from J. Henricson of the Fishery Board of Sweden in 1978.

Spain

The acclimatization of the huchen seems to have been successful in Spain, where the species was introduced for angling purposes. Besides published sources (Anonymus 1969, 1973) which, however, are at second hand, reliable information has kindly been submitted by J. Lobón-Cerviá of Madrid (in a letter of 2 Nov. 1979). The most recent information has been submitted by C. Mondejar-Reyna of Madrid in the form of an unpublished paper presented to the 1st Congress of Spanish and Portuguese Ichthyologists in Leon in May 1981 (Mondejar-Reyna 1981). According to that information, fertilized huchen eggs from Czechoslovakia (not from Switzerland, as stated by Anonymus 1969), numbering 20,000, arrived in Spain in May 1968. One group (10,000 eggs) was placed on fish farms in La Coruña (northwestern Spain). After hatching, the fish were kept in ponds, and after a few years some of them were released in nearby streams, where they apparently did not became established, as no further information is available on catches or direct observations of huchen there. The rest of the group still lives in fishponds; the fish are annually used for artificial reproduction (in spite of obviously insufficient feeding, as according to the report of Lobón-Cerviá the huchen look like "monsters", having big heads and small bodies).

Some months later, the other group, comprising young-of-the-year, was released in the Tormes river (a left tributary of the Duero river) near Salamanca. In 1972, i.e.

4 years later, natural spawning was observed in that river. That this spawning was successful is indicated by catches of two-year-old fish around 1.8 kg in average weight in 1974. In the same year a sexually mature female 93 cm in total length and 10.5 kg in weight was caught, containing 16,900 eggs in its abdominal cavity (J. Lobón-Cerviá kindly sent us a colour photograph of the fish to confirm this information). In 1976 a huchen was caught weighing 14.5 kg, in 1981 another one 1060 mm in total length and 11 kg in weight was caught along with another 30 fish, 9 of which were over 700 mm long (Corredera-Martin 1981). The above data indicate that the huchen in that stream exhibit conspicuously rapid growth, exceeding the values usual for this species. According to dissections carried out in several huchen caught in the Tormes river, they feed on nase carp, *Chondrostoma polylepis,* the most numerous local fish species. At present, the density of the huchen population in the Tormes river is very low as a result of mass mortality of fish that occurred there in 1976 due to artificial draining of a reservoir lying upstream. The rest of the huchen population, estimated at some 30 fish, comprises individuals weighing 7–10 kg.

Other Countries

Attempts were also made to introduce huchen into North America. Since the data published on this are very brief (Crossman 1968, Lachner et al. 1970), we state here particulars kindly communicated by J. F. Bergeron (in letters of 15 Oct. and 4 Dec. 1985). In 1967 and 1968, Professor V. D. Vladykov obtained 25,000 and 34,000 fertilized huchen eggs from Czechoslovakia; he submitted them to be hatched in the Baldwins-Mills hatchery owned by the Government of Quebec, Canada. In 1968 and 1969, several rivers (the St. Lawrence River and the Richelieu River) and lakes (Lake Saint-Louis, Lake Mandeville, Lac Deligny) in the environs of Montreal were stocked with 11,000 young, 100 fingerlings and 869 yearlings; in 1973, with an additional 900 yearlings of huchen. The latter came from the progeny of the huchen reared from the imported eggs. There is a single record of a catch of a huchen 718 mm in *Tl* and 3685 g in weight, caught with rod and line in the Deligny brook on 4 July 1973 (Bergeron and Brousseau 1981, Bergeron in a letter of 4 Dec. 1985), but Bergeron (in a letter of 10 Oct. 1985) admits that "...there is a possibility that anglers confuse it with other salmonids like the recent arrivals in the Quebec waters of two species of Pacific salmon *Oncorhynchus tschawytscha* and *Oncorhynchus kisutch,* coming from the Great Lakes at the headwaters of the St. Lawrence River". An interesting experimental hybridization was made also in Canada: in April 1976, eggs of eight female rainbow trout *(Salmo gairdneri)* were fertilized with milt of three male huchen in the Baldwins-Mills hatchery. The total of 19,656 fertilized eggs yielded, in late June, 300 fingerlings of a hybrid, 99 of which were released in the Rapides de Lachine (the St. Lawrence River). Particulars of the hybrid are unknown, as with the subsequent fate of huchen in Canada, but it appears that it has not become established. Some time ago, interest was also shown in introducing huchen into the USA, but the intention has never been implemented (Buss in a letter of 9 Jan. 1977, Willoughby in a letter of 9 Jan. 1978).

21.2 PRINCIPLES OF HUCHEN INTRODUCTION

In concluding this chapter, we suppose it necessary to deal here with the principles which in our opinion, make it possible to acclimatize or re-acclimatize the huchen. It should be noted on this occasion that the principles governing the introduction and acclimatization of the species were formulated by Matyas (1954) on the basis of his experience obtained in introducing the huchen into the Rajčianka river. The principles are still topical as will be seen below.

According to Matyas, huchen cannot be introduced just anywhere, without due planning and a sound consideration, and the following procedure should be observed:

(a) Stocking should be regular, year after year, at least for 4–6 years, until the huchen introduced begin to reproduce naturally;

(b) As soon as the first stock is found to have attained over 1 kg in weight, the section in which it thrives best and in which the largest numbers of huchen occur should be closed and declared a long-term reserve;

(c) Fishing of whatever kind should be prohibited in a section populated by the huchen (which, being a predator, when in motion can be deceived by any kind of bait, be it a fly, a worm, pastry, or fruit);

(d) Wherever possible, huchen summerlings or yearlings should be used for stocking, as stocking with fry does not warrant a hundred percent success.

It is evident from this quotation which, except for minor stylistic emendations, is given in its full extent, that the principles of successful introduction were known in Czechoslovakia over 30 years ago, and that the failure of attempts to acclimatize the huchen outside its original range has been caused by insufficient knowledge of its biology, by using insufficiently developed stocking fish, and by the fact that in most cases the stocking was made once and for all.

On the basis of the present knowledge of the bionomics and ecology of the huchen, as well as on experience obtained in attempts to transplant the species into other waters, the following main principles of introduction can be formulated:

(1) The huchen can be released in streams of the foothill zone, with rocky and gravelly or gravelly and sandy, sufficiently diversified bottom and banks, with water temperatures not exceeding 20° C during the summer months. Stream sections with broad and shallower riffles alternating with deep pools are of advantage, as well as places where islets divide the stream into lateral branches, where the banks are overgrown with trees and shrubs and are interrupted by mouths of tributaries. The section should be selected so as to join the foothill brook zone, offering suitable sites for natural spawning;

(2) The introduction must be preceded by a limnological investigation aimed

primarily at the qualitative and quantitative composition of the benthic and fish fauna of the selected stream;

(3) Before releasing the huchen stock for the first time it is necessary to reduce the original fish fauna, especially the numbers of fishes which occur in masses and are bigger in size (such as chub, burbot, brown trout, pike). This reduction should be selective: the young of the above species should remain in the section selected;

(4) For stocking, huchen summerlings or yearlings, not alevins must be used on principle. If only alevins are available, they may be released in the tributaries of the selected stream or in its upper reaches; in this case, however, the original fish stock of the whole tributary or the selected section must be removed by electro-fishing repeated several times. It is of advantage when the selected stream section is separated from the adjoining upper and lower sections by natural or artificial barriers preventing other fish species from migrating into it;

(5) The stocking must be repeated every year during the first 4–6 years until the huchen from the first introduction begin reproducing naturally;

(6) The respective stream section must be excluded from normal fishing exploitation for the whole period of stocking. Fishing by any means whatever must be forbidden, except for check fishing;

(7) Throughout the period of introduction, the respective section and the whole stream must be subject to special supervision. Regular investigations must be carried out in the stream to monitor the numbers, growth, physical condition, food supply and sexual maturation of the stocked fish, their longitudinal and transversal migrations and, towards the end of the period, to establish the places and time of natural spawning.

Part V

FARMING

22. History

According to a brief historical review published by Ivaška (1951), the huchen farming has been going on for more than 100 years. The review, supplemented by other data, both published (Novák 1949, Holčík 1977 b) and unpublished (Novák in a letter of 7 Feb. 1978, Papadopol in a letter of 25 Aug. 1977, Bél in a letter of 18 May 1978, Jungwirth in a letter of 15 May 1978), provide the basis for an outline of the history of huchen farming which, however, is still incomplete, as it mostly covers Czechoslovakia and, to a limited extent only, the other countries. To our knowledge, taimen breeding has not so far exceeded experimental dimensions (Kifa and Vdovchenko 1976).

It is a historical fact that it was in Slovakia that huchen farming was begun. The first note, not about huchen farming but on keeping them in fishponds, is found in the land register of Budatín, dating from 1690 (Martinka 1930). This document, written in Slovak, states that in the gardens of Budatín Castle there are two fishponds and a pool "for sustenance on the days of fasting, which contain carp, huchen and pike... which to feed with small fish is the duty of fishermen Powinski and Radolsky".

It follows from the subsequent lines of the document that the fishermen and millers of the Budatín estate were obliged to send all huchen they caught to these ponds. However, proper huchen farming did not begin until 1876 when J. Guber and W. Rowland, employees of the then Orava compossessorship, hand-stripped a pair of huchen caught in the Orava river in the spawning period. Of the 12,000 eggs obtained, only 4,000 were fertilized, and five weeks later 1680 young hatched from these. In the summer of the next year, two fishponds were constructed for rearing huchen, supplied with water partly from a spring and partly from a brook. In December of that same year, 12 parent huchen captured in the Orava river were placed in the ponds. In April 1878 it was found, however, that the females were not mature enough. The fish later succumbed to a fungus infection (Rowland 1881).

In 1880, E. Giebner attempted to breed huchen in Silesia. The attempt was apparently unsuccessful, as the young supplied by Giebner were stocked in the Vistula and Odra rivers by the Silesian Anglers' Union.

In 1880–1920, many attempts to breed huchen were made in Austria, which wholly monopolized this field. The best known and best managed was the hatchery of F. Pölzl at Traismauer, which not only stocked many Austrian streams (the Ybbs, Erlau, Pielach, Melk, and Danube) with huchen but even exported fertilized huchen eggs. Every year hundreds of parent huchen were spawned there, obtained during their spawning migrations from the Danube into the Traisen river (including fish up to 35 kg in weight). Besides huchen eggs and young, the hatchery also produced more advanced stocking fish reared on live food (bloodworms, or chironomid larvae). Besides this hatchery there were tens of others, such as the hatchery of A. Weeder at Puchheim, H. Köttl at Redl-Zipf, or H. Gerson at Bachleithen; Austria also had a state hatchery for huchen breeding at Ilidža near Sarajevo. The breeding results were very modest, however, as less than 37,000 huchen (very probably alevins) were released in Austrian waters in 1883–1929, and the huchen made up only 0.36% of the total of all fish species, or only 0.65% of the total number of salmonid species stocked. According to a review published in the Oesterreichische Fischerei Zeitung in 1930, the following numbers of huchen were released in Austrian waters in individual years: 500 in 1883, 1000 in 1884, 10,500 in 1914, 10,000 in 1915, 5000 in 1923, and 9600 in 1929. Evidently the breeding was not regular; spawning was apparently carried out only when they succeeded in obtaining parent fish. Attempts to rear huchen in ponds were also made by Šusta (1898) but they were unsuccessful.

In 1913, R. Köttl tried to breed huchen at Kláštor pod Znievom, using huchen caught in the Orava river. Hand-stripping was repeated several times, using even parent fish from the Váh river, but without any major success (Köttl and Kafka 1927).

After World War I, attempts to breed huchen continued with increasing intensity. The centre of this activity was shifted from Austria to Yugoslavia and later to Czechoslovakia.

In Yugoslavia, huchen breeding began as early as 1903 when Professor Franke of Ljubljana obtained 25,000 young from eggs of huchen caught in the Ljubljanica river; some of the young were released back in the river and some were kept in a fishpond until the 4th year of life of the fish. In 1924–1934, at least 1.1 million fertilized huchen eggs were obtained in Yugoslavia; of these, more than 93% in 1925–1934; 25,000 eggs were exported to foreign countries (Munda 1935). Although based on eggs obtained from parent fish caught at natural spawning sites, the breeding attained quite a high level: according to Munda (l.c.), losses at hatching amounted to 5–10%, which is just slightly higher than at present. However, no solution was found for rearing older stocking fish: according to the same author, a total of half a million alevins but only slightly more than 9000 older stocking fish were released in Yugoslavian rivers in 1911–1934.

In Czechoslovakia, efforts to breed huchen began, or continued, with some delay, and apparently in complete isolation from what was happening abroad. The attempts at huchen breeding were revived by gradual deterioration of water purity in our streams, increasing intensity and extent of stream canalization, and fading away of the events of World War I (spreading poaching, use of explosives in illegal fishing). On authorization from the former Agricultural Council in Bratislava, Kafka examined, in 1924, the possibility of breeding huchen artificially and of utilizing the hatcheries at Kláštor pod Znievom and at Handlová for the purpose. Fertilized eggs were to be obtained directly in the wild, first of all in the Orava and Váh rivers. A year later, on 29 June 1925, Šivic and Bajkov submitted to the then Ministry of Agriculture in Prague a memorandum in which they pointed out the depressing situation of huchen and its causes, and asked that a hatchery with an annual capacity of half a million eggs be created to supply huchen young to Slovakia, then Ruthenia (Carpatho-Ukraine), and Moravia. The application involves a proposal that the hatchery be supervised by an expert and managed by a board of trustees consisting of specialists in breeding salmonids. Already at that time, possibilities for exporting huchen eggs were examined, as indicated by one part of the memorandum mentioned, asking for special permission for this purpose. The memorandum remained unanswered, however.

A year before, in 1924, R. Köttl had tried again to breed huchen at Kláštor pod Znievom with the aim to obtain parent fish capable of spawning. He was quite successful when all of his five huchen, 12 cm in initial length, reached sexual maturity six years later and tried to spawn with added rainbow trout. However, Köttl does not seem to have obtained progeny of these huchen.

In 1927, Vladykov attempted to breed huchen (Vladykov 1929). From several huchen captured in the Teresva river (with rod and line, or with a spear) he obtained sexual products and transferred fertilized eggs to a hatchery of the State Forests at Buština. Due to improper handling during transportation, however, all alevins died.

Upon suggestion by R. Köttl (Köttl and Kafka 1927), the Anglers' Club at Dolný Kubín established a hatchery for huchen breeding at Oravský Podzámok in 1928. The hatchery was run by Š. Butkovský, assisted by J. Novák, a graduate of the Secondary School of Fisheries at Vodňany (Butkovský 1930, Novák 1932). After a failure at its start, the hatchery began to prosper, but ceased to exist shortly thereafter. The cause or causes of its abolition are unknown but it may have been financial difficulties, in spite of subsidies granted from the so-called "huchen fund" established by the Union of Anglers' Clubs and Associations in Slovakia, as well as of benefits received from "officials and devoted members" (Hollý 1934b).

In the years to follow the breeders' efforts slackened visibly, which can be attributed not only to failure due to often distorted (Kafka 1924) and quite insufficient knowledge of the biology of the huchen (Kostomarov 1937) but also to the worldwide crisis which also affected Czechoslovakia's economy. The establishment of another "huchen fund" by the editors of the magazine "Rybář" (The Fisher) in 1934 was of no avail either (Rousek 1934). The suggestion of Hollý (1935a, b) (who, upon authorization by the Agricultural Council for Slovakia, investigated the occurrence of huchen in the Hron river) to utilize the hatchery on the Tajovský potok Brook for huchen breeding and to create a new, specialized one on the Ľupčica brook at Slovenská Ľupča, did not materialize either. Similarly, no success was obtained in Ruthenia although the provincial authorities organized a conference in Uzhgorod in 1936 where a number of papers were presented, illustrating the situation of the huchen, and where several suggestions were made not only for the protection but also for breeding the huchen.

In 1935–1937, however, a turn for the better took place. At that time, huchen breeding was taken over by Samo Ivaška (Ivaška 1935) who, although an amateur, established at his own expense and risk a hatchery in Martin (Pl. 27), existing up to present, and realized the first successful artificial spawning of huchen in 1937. After many years of experimental work, Samo Ivaška successfully solved several problems associated with keeping parent fish in captivity, thus laying the foundations for breeding huchen in a closed cycle. In 1937, Ivaška submitted part of his young to J. Nieslanik, a trout breeder at Jablunkov in Silesia, who obtained from them viable progeny in 1943, and then bred another three huchen generations in succession. It is worth noting that in 1936–1943, Ivaška reared around half a million huchen young, and that in 1942 his breeding shoal numbered 46 huchen.

After World War II, huchen breeding began increasing at an unusual rate. Beside Ivaška, other hatcheries were also devoted to huchen breeding, such as the hatchery of the State Forests at Oravský Podzámok, or those of the Anglers' Associations at Demänová and in Ružomberok. Later the breeding was gradually concentrated on the hatchery of the Central Committee of the Slovak Anglers' Union at Dubná Skala, where all parent fish from Ivaška's hatchery in Martin were also transferred in 1964. On a smaller scale huchen were also bred by the State Forests at Oravský Podzámok. In its prime, the breeder shoal kept in the hatchery at Dubná Skala numbered as much as 95 huchen, and the annual production amounted to half a million fertilized eggs. Since 1970, huchen have again been bred in Ivaška's restored hatchery, which was purchased by the local organization of the Slovak Anglers' Union (SAU) in Martin, and the eggs produced there have been distributed to other fish-producing establishments of the SAU (at Považská Bystrica and Banská Bystrica). When it was necessary to close the hatchery at Dubná Skala, huchen breeding was undertaken on a farm at Podsuchá near Ružomberok, but without any major results obtained so far. At present, the centre of gravity of huchen breeding in Czechoslovakia is the farm of the local organization of the SAU in Martin. Experiments are also in progress on the farm of the State Fisheries at Kláštor pod Znievom. To be complete it is worth mentioning that since 1977 huchen have also been bred on the farm of the Czech Anglers' Union at Vysoký Potok. In 1980, a successful artificial spawning was also carried out by the local organization of the Czech Anglers' Union at Vsetín. From among the more important breeders, besides those mentioned above, J. Kocian, I. Bél, J. Andók, B. Lanka, D. Ivaška and P. Mizera should be mentioned.

At present huchen are bred apart from Czechoslovakia, also in the Federal Republic of Germany (some data on this were published by Bohl 1977, 1979),

Yugoslavia, Rumania (where the annual production is about 50,000 young) and especially in Austria. Judging from reports kindly submitted by Dr. M. Jungwirth, Austria is experiencing a renaissance of huchen breeding; at present 20–25 farms are engaged in this activity where scientific methods, including experimental injections of fish pituitary are applied. Unfortunately, we have not succeeded in obtaining any more particulars on huchen breeding in the countries mentioned above.

It is to be hoped that the present stagnation in huchen breeding in Czechoslovakia may be followed by an improvement, when the construction of a farm of the SAU at Príbovce, designed mainly for huchen farming, is completed.

23. Huchen Farm

Huchen farming has been carried out on farms which originally served for breeding brown trout and grayling. Gradually increasing experience has shown, however, that a specialized farm is necessary for rational huchen farming, as there are striking differences between farming huchen and the other salmonids, despite the fact that the problems and principles are very similar. The differences are mainly due to the fact that the huchen is a predatory species attaining considerable sizes.

SELECTION OF THE BUILDING SITE

Proper selection of the building site is one of the decisive factors to guarantee a successful huchen farming. To secure highly intensive and economically self-sustaining farming, a substantial and sound source of water must be available in a place with an advantageous hydraulic gradient. Besides, it is inevitable that the locality be situated on an impermeable geological substrate and have good roads.

The farming can be intensified by a closed cycle, i.e. by producing eggs, rearing young, stock and future parent fish from the farm's own breeding shoal. To secure economic independence, it is necessary to establish supplementary breeding of another fish species for marketable fish production (preferably rainbow trout), but this means increased demands in the number and area of ponds and elsewhere, including maximum possible mechanization of the work.

CLIMATIC AND HYDROLOGICAL CONDITIONS

Climatic conditions are among the major factors that govern especially the temperature and level of water. For huchen farming, suitable regions are those where extreme overheating or overcooling of water and especially extreme oscillations in water temperature do not occur in spring months. Freezing up of ponds for a long time is disadvantageous with regard of the need to check the health of the parent fish. Very low water temperatures also affect food intake. Cyprinid fishes, serving as live food for huchen during winter months, die rather soon in cold water. It is important that the water temperature does not drop below 4°C, especially during April and May, which could cause mortality of huchen young.

Hydrological conditions must satisfy two fundamental requirements. First of all, there must be a sufficient supply of water of good quality and innocuous in the sanitary respect. This condition is often underestimated, although long-standing experience in breeding salmonid fishes indicates unequivocally that water quality is one of the factors decisive of the breeding success or failure. The point is that the water should contain the least possible amount of organic compounds, which represent some kind of pollution exerting a negative influence particularly on the development of eggs and hatched young. Therefore, the water diversion must not be situated in places where there are major residential areas, farms or industrial plants and intensively cultivated land above the source, or where the headwater is passed by a road, etc. It is also unfavourable if the source is near extensive deciduous woodland from which considerable amounts of leaf litter get into the water in autumn. Any major organic pollution not only causes organic water turbidity (particularly during summer months), but also decreases the dissolved oxygen content, and may be a carrier of fish diseases.

Another fundamental requirement is that there must be a certain height difference between the source of water and the inlet, preferably at least 4–5 m. This is to ensure sufficient streaming of water and to simplify the supply, distribution and drainage of water in both the hatchery and the ponds.

On the basis of experience obtained so far, a farm specialized in huchen breeding may be situated, under the conditions of Czechoslovakia, at heights between 350 and 650 m above sea level. Water temperatures should not exceed 20°C in summer or drop below 1°C in winter, or below 5°C in spring (April and May). The water reaction should be neutral and should not drop below pH 6.5. Of advantage are waters springing from a limestone substrate. The minimum amount to supply a farm, about 4 ha in area, is $500\ l\ s^{-1}$. A hatchery requires $10-15\ l\ s^{-1}$ of spring water, or seepage water with the possibility to be mixed with spring water. The water supplying the ponds must have a sufficient hydraulic gradient, it must be pure, without any organic or inorganic impurities. If need be, effective filters or sedimentation pools must be constructed. The dissolved oxygen content must not drop below 80% even during critical periods. Surface waters which are turbid or show variation in their pH in periods of increased discharge, or which pass through extensive deciduous woodland or through intensively cultivated agricultural land, settlements or near industrial plants or major roads, or which are long fed with snow water during the spring thaw, must not be used as sources of water supplied to huchen farms.

PONDS

The size and number of ponds on a fish farm depends on their purpose, i.e. on the production required and on the species of fish reared. In the given case, we shall use a schematic outline of a specialized huchen farm.

Ponds for Parent Fish

A sufficient number of healthy parent fish is an essential condition for successful breeding. Therefore the ponds in which the parent fish are kept are very important parts of a huchen farm. The ponds should simulate natural conditions. It is important to keep individual age (size) groups of fish separately in separate ponds; for example, parent fish at the age of 7 years are kept in one, at the age of 8–10 years in another, and older fish in a third pond. The density of the stock in these ponds is round 200 fish per 1000 m^2 of area. If lack of space prevents construction of several ponds it is possible to keep the parent fish in one pond but there must be no fish younger than 6–7 years. Even in this case, the density of the stock must be as indicated above. The depth of water in such ponds must be between 1.0 and 1.8 m, and the water supply rate 10–25 $l\,s^{-1}$. The construction of the pond should prevent formation of so-called "dead" spaces, i.e. a continuous water exchange and current should cover the whole surface of the pond. The ponds must have a sufficient decline and their middle must be passed by a channel in which the fish can be concentrated by gradually draining the pond. The channel must be divided into several, around 4-m-long sections, so that these can be separated with lattices. The channel should be about 1.5 m wide and 40–50 cm deep, somewhat shallower in its upper part, made of concrete, with a smooth surface and rounded edges. An auxiliary pond simulating natural conditions should be constructed: about 200 m^2 in area, with a stronger water supply and a clean gravelly and sandy bottom, in which stripped parent fish are temporarily placed to finish spawning in the natural way. Such ponds, which simulate natural conditions to some extent, can also be used for other purposes, especially for "training" summerlings or older stocking fish for new living conditions before they are released in the wild.

Ponds for Rearing Alevins and Fingerlings

Huchen young, partly reared in the rearing plant, are transferred into outdoor installations, including rearing channels, ground pools and special hard-bottom ponds.

The channels are of the same type as those used for rearing grayling or brown trout, for example. Their bottom is fine-grained gravel and sand, the central collecting channel is concrete, but the sides are firm, reinforced with rocks. The ponds are rectangular, 125–300 m^2 in area. The water in them is exchanged three times a day.

The special ponds are hard-bottom reservoirs. Essentially they are concrete, rectangular or circular, smaller in dimensions. All types of containers serving for rearing alevins and fingerlings mentioned above, must be constructed so as to allow removal of the fish below their outlet. In all types the stock density is around 1050–1250 young per m^2 of area.

Ponds for Rearing Recruits

At least two ponds are necessary for this purpose. With a normal and uniform growth it is possible to keep together two successive age categories, e.g. fish 3 and 4 or 4 and 5 years old. The ponds for this purpose may be rectangular, 15 × 10 or 20 × 8 m. Their depth should be 0.7 and 1.2 m, respectively. Their central collecting channel must be 0.8 m wide and 0.4 m deep. The minimum water supply should be 5 $l\ s^{-1}$.

Auxiliary Ponds

The auxiliary ponds include especially those for keeping fodder fish, for rearing plankton, quarantine ponds and other auxiliary ponds. In addition, this group includes the distribution system, water supply and drains, as well as sedimentation or preheating pools. In size and construction, they are adapted to the character and extent of the main plant. In their construction certain principles must be observed: the water supply to the ponds (except the sedimentation and preheating pools) should not utilize pond water before it is passed through recirculation and regeneration equipment. It is important to remember that the whole distribution system must be constructed in sufficient capacity and so that it is to be easily cleaned, does not become blocked, and does not freeze in winter. The inlets to the ponds should be provided with a suitably designed cover which prevents huchen from jumping into them and from being injured or from escaping. The outlet should be led through a common double outlet box with screens or lattices of an appropriate size.

HATCHERY AND ITS EQUIPMENT

A huchen hatchery does not differ from any hatchery commonly used for hatching salmonid fishes. It may be modernized to various degrees, according to environmental conditions or according to the investor's wishes. The important principles for the construction of a hatchery are the following: The plant should be situated at an advantageous gradient in relation to the place of water diversion, i.e. the water feed must have sufficient pressure and discharge, and the possibility must be provided to mix the supplied water with that from a spring or from a hydrological bore. In view of the need for continuous watching, the plant should be situated near the dwellings or directly in the basement of such buildings. In some cases the hatchery and the rearing plant are built as one complex. Besides the hatchery proper and its equipment, there is usually a small laboratory, equipment for incidental adaptation of water temperature, food storage, etc. In our conditions the hatcheries are equipped with combination Rückel-Vacek hatching apparatus made of steel sheets, or with a battery-type apparatus. Water supply and adaptation and, in some cases, also filtering equipment are essential for a hatchery. The hatchery rooms must contain handling containers. In many cases, equipment of the rearing plant is used for this purpose in the period when hand-stripping is carried out.

AUXILIARY BUILDINGS

According to size, a huchen farm includes such auxiliary buildings as storage rooms, garages, a fodder preparation room, workshops, lavatories, a cold-storage plant and, of course, dwellings for the personnel.

WORKING APPARATUS, IMPLEMENTS AND EQUIPMENT

Besides the standard and permanent equipment of a hatchery and a rearing plant (such as incubators, troughs, water heaters, pumps, etc.) the work also requires the use of various implements in different seasons of the year. These include vessels to contain food and to carry fish, cradles for handling parent fish, graduated vessels for measuring eggs and young, various measuring apparatus for the hatchery and for checking water quality, wheelbarrows for the transportation of fodder, equipment for the laboratory and for disinfection, dip and other nets, screening apparatus, feeders of various kinds, etc. Besides, the staff must be equipped with rubber boots, gloves and aprons, simple mechanization, etc.

24. *Technology of Parent Fish Breeding*

OBTAINING PARENT FISH IN THE WILD

A sufficient number of healthy parent fish is the basis of successful huchen farming. It would seem that the most advantageous way of obtaining parent fish is catching them in their natural habitats. Under the present conditions, however, this can only be done exceptionally. Within the last decade, not only have natural spawning sites decreased in number, but also the numbers of huchen have fallen abruptly, mainly due to industrial water pollution. Thus catching parent fish in natural habitats is only possible in protected huchen localities; even in this case, however, it depends considerably on water levels. In spring the water discharge in streams shows marked variation and the accompanying water turbidity impedes fishing. As often happens at that time, only a small number of fish is caught, or of one sex only. In addition, at that time huchen are very sensitive to handling and as a result they are very often infested with fungi to such an extent that they die. For those reasons, obtaining parent fish from natural waters at the spawning time is the solution in exceptional cases only. If a breeder does depend on obtaining the parent fish from the wild, it is of advantage to get them in the course of the whole year, but particularly after the spawning period and in autumn. Since older fish are less adaptable to altered environmental conditions in a pond, it is best to concentrate on obtaining fish of younger age categories, i.e. those 3–4–5 years old. In those age groups acclimatization is practically no problem. The technique of obtaining parent fish in the wild varies according to environmental conditions. However, it is necessary to consider the fact that using gregarious fishing means, such as nets or electro-fishing (Pl. 28. 1), often results in injuries and subsequent infestation of the fish with fungus, frequently with lethal consequences. Even though it seems paradoxical, obtaining parent fish with rod and line has been found most satisfactory (Pl. 28. 2). With this fishing technique – particularly if used by an experienced angler – even a heavy huchen can be fatigued to such an extent that it does not resist being brought ashore and handled, which prevents the fish from being injured. Huchen can even be caught in a fish pot, preferably made of soft natural materials or of knotless netting to minimize the risk of surface lesions (of the skin and especially eyes) suffered by the fish.

Much care should be taken particularly when using an electric shocking-gear, as long expositions to electric current may cause serious injuries (usually a fracture of the backbone) or even kill the fish. This technique is the least suitable also because of the fact that huchen do not easily recover from electric shocks, particularly at higher water temperatures in summer. In general, regardless of the fishing technique and gear used, fishing for parent huchen must be carried out mainly on cool days.

After being caught, the fish is carried in a wetted dip net, canvas or polyethylene sheet (taking care that the surface does not freeze) and transported in suitable equipment. In the past (Ivaška 1951), first wooden and later aluminum boxes were used, fitted with a rubber lining and a cover, containing water saturated with oxygen if need be. They held 120 l water and could be used for transporting three huchen weighing 4–6 kg each to a distance up to 100 km. At present, such boxes have been replaced by thick-walled polyethylene bags filled partly with water and partly with oxygen. Their use almost excludes any injury or suffocation of the transported fish.

Before releasing, the fish must be subjected tó preventive treatment.

Huchen caught in the wild are shy and, when released in a pond, tend to jump out into the inlet or even ashore. For this reason it is necessary to lower the water level in the pond temporarily. Moreover, violent movements should be avoided when feeding the fish, during maintenance of the pond, etc.

If several huchen of different size are caught at once, they should be placed in separate ponds to minimize cannibalism or mutually inflicted injuries. Huchen obtained in this way must be fed live fish on principle.

BREEDING PARENT FISH IN A CLOSED CYCLE

As confirmed by experience gradually accumulated in practical breeding, successful breeding and production of young huchen depends, as in other salmonid species, on a sufficient number of captive parent fish. In such cases the breeder acts like natural selection by favouring the fittest and removing weaker fish. At last quality parent material is obtained for high egg production. Besides, the fish are under permanent veterinary control, protection and expert supervision. It is most advantageous for a breeder to rear his own parent fish directly from young. A minimum of 100 young is required to rear one parent fish. Emphasis must be placed on selecting for breeding purposes individuals with the highest growth rate and the best appearance. Selection should also be made during subsequent years of breeding so as to create a cohort of parent fish not differing too much in size. Slowly growing fish must be removed, and those showing fast growth must be transferred to separate pools.

Due attention must be paid to the quality and quantity of food given to the huchen; live food is necessary. The most suitable fodder fish are up to 70–80 mm long. Later they should be replaced by other, bigger fish, when the huchen attain the age of two.

It is important to examine regularly the health of the parent fish and to carry out prophylactic baths from time to time.

SUPPLEMENTATION OF THE PARENT SHOAL BY RECRUITS

Environmental conditions and adequate individual development, depending on a sufficient food supply, are among the factors that determine sexual maturation. Under normal conditions, males attain sexual maturity at the age of five years, females in the sixth, or exceptionally as late as the seventh year of life. Unsuitable or insufficient feeding results in extremely delayed sexual maturation and often even in sterility. For the farm's own egg production it is advantageous to use optimally effective parent fish, i.e. those showing the highest relative or absolute fecundity and the highest quality of sexual products. Such fish are in the age group of 6–12 years for males and 7–14 years for females. It is necessary to count on renewing the parent stock at least once every two years, at a rate of about 20% of the individuals of both sexes. It also is of great importance to check the quality of milt in males (a microscopic examination is very quick) and losses of eggs in females. If needed, unsuitable parent fish must be eliminated even earlier. In conditions where space or feeding of parent fish offer no problem a 1:1 sex ratio is ideal. Otherwise a ratio of 1:2 in favour of females should be maintained. Within one day males can be handled (hand-stripped) up to three times. If the females do not attain sexual maturity uniformly it may happen that when stripping the last females (within 10–20 days from the first strippings) the males may not produce any more milt. One of the breeder's principles is to consider carefully all circumstances before eliminating a parent fish. Although fish over 15 kg are less easy to handle, they may not always be too old. Their ripening is mostly very regular and by being handled repeatedly they have acquired resistance to fungus infection. They are well acquainted with being handled and fed by man. Thus it is advantageous to keep such fish in reserve. One should also take into account the fact that the costs of rearing parent fish during the first five to seven years are incomparably higher than the only slightly increased costs of feeding older fish plus the loss caused by their fecundity decreased by at most 20%. The above principles must be permanently observed in exchanging males; particular care must be paid to their weight, which must not exceed 12 kg.

FEEDING OF PARENT FISH

Sexually mature parent fish used for artificial spawning are fed with live fish over 10–15 cm in size. Their species makes no difference; generally, however, one should use cyprinid fishes, which are best taken by the huchen. When bigger perch or ruffe are offered they are sometimes only reluctantly taken by the huchen. If live fish are in short supply it is possible to offer even thawed fish previously stored in a freezer. These, however, must be offered singly and so that the huchen may catch the fish when it hits the water surface. Portioned fish can also be offered, even to younger huchen (with a temporary scarcity of live food). Whenever using dead fish, however, care should be taken that the fodder fish do not remain lying on the bottom unused; if so, they must be removed in time.

The voraciousness of huchen is governed by several factors. In general, they are less voracious before spawning and during hot summer spells, when the water temperature is increased, than after spawning. They take food most intensively in May and June and then in early autumn until the water surface begins to freeze. In winter, when the ponds are covered with ice, feeding should be limited and live coarse fish should be offered only from time to time.

Naturally, the predatory habits of huchen require that sufficient supply of fodder fish is available at any time, particularly in autumn, when the huchen begin to form sexual products, and after spawning when they recover. It is therefore of advantage for a huchen farm to have at its disposal pools suitable for keeping fodder fish, which can be fed suuplementary granulated feed from time to time. It is essential that the fodder fish are healthy, particularly those which are immediately given to the huchen. It is therefore of use to subject even the fodder fish to preventive baths before releasing them in the pond or before offering them to the huchen.

In compiling budgets of food supply and consumption, a feeding ratio around 10 (i.e. 10 kg food per kg of body weight of huchen) should be considered.

Parent fish are fed as long as they show interest in the food offered. Feeding must be interrupted whenever deviations from normal behaviour of the huchen or symptoms of disease are observed.

25. Artificial Reproduction

MATURATION OF HUCHEN, IDENTIFICATION OF SEXUAL MATURITY

Although the ripening of sexual products and the spawning of huchen depend on several factors they repeat every year with considerable regularity in Czechoslovakia. According to our records, huchen on the farms at Dubná Skala, Martin, Kláštor pod Znievom, Ružomberok, Podsuchá and Jablunkov attain sexual maturity and spawn within some two weeks, between the second week of April and mid-May, exceptionally until the end of May. The approaching spawning period is indicated on the one hand by certain phenological events (appearance of first leaves on alder trees, of buds on willow trees, and when 30–40 days have elapsed from the departure of ice) and on the other by the markedly changed behaviour of the huchen. Aggregation of females at the bottom below the inflowing water is the first sign. As early as the end of March, when the ice has disappeared from the ponds, the huchen can be observed circling around one another, the females often "yawning", i.e. gaping their white mouths. At the same time, the coloration of the fish begins to change visibly. The males gradually turn distinctly darker, especially in the posterior part of their dorsum between the bases of the dorsal and the caudal fins, their flanks turning copper red to purplish, therein differing at first sight from the paler, pinkish females whose ventral parts become distinctly thickset. The approaching spawning period can be deduced and sexes distinguished even by the increasing number of "scratches" on the heads of the males, inflicted during more and more frequent duels. Later the females begin to separate, assembling below the water inlet and often trying to jump into it. In a subsequent phase, individual females (first the biggest ones) descend to the bottom, find a suitable place for egg deposition and begin to clean it. The males, but occasionally even the females, attack one another with increasing frequency, they bite one another's flanks in front of the dorsal fin and around the caudal peduncle, or even tear one another's pectoral and caudal fins. The males attack one another with particular fierceness, circling round in larger or smaller circles. The last symptoms may follow shortly after one another and spontaneous spawning may occur on the same day or by next morning. It may happen, however, that this breeding behaviour is prolonged. In this period it is inevitable that the responsible staff take regular readings of water temperature; if it permanently drops below 6°C at night and mostly does not attain 8°C during the day the spawning period is delayed; otherwise it is approaching rapidly. Naturally, the best way to check the maturity of the fish is to see whether gentle pressure on the abdominal cavity of their body results in a release of sexual products. With such big fish as the huchen, however, any handling should be avoided or limited to the least possible extent, and not only at this time. With sufficient breeding experience the term of the spawning can be determined with an accuracy of one day. It should be emphasized again that this is one of the reasons why a breeder should keep detailed records of the temperatures and terms of spawning in successive years – an invaluable aid which enables one to forecast the spawning term and to carry out the necessary preparations in time. On the huchen farm at Dubná Skala, for example, spawning occurred at regular time for several successive years, falling on 9, 12, or 26 April.

As soon as the breeder observes that the females have selected their partners from among the duelling males and that they begin cleaning the redd, it is necessary immediately to decrease the water level in the pond, separate the fish by their sexes and degree of ripeness, and carry out the hand-stripping.

PREPARATIONS FOR HAND-STRIPPING

Since the huchen is a fish which on the one hand is rather ponderous and on the other is very sensitive to being handled, the process of hand-stripping must be organized and prepared as thoroughly as possible. The preparation for hand-stripping consists primarily in a perfect organization of the work, in the division of the respective duties, and in maintaining certain procedures and important principles. The latter include especially the following:

(a) The stripping is carried out directly in the pond in which the parent fish are kept;

(b) Depending on its size, the pond is drained so that the stripping can begin around 10 a.m.;

(c) The following numbers of persons are required to carry out the various duties: four to do the hand-stripping proper, two to carry and anaesthetize the fish, one to take measurements, mix, rinse and carry away the eggs, one to put down records on the fish and number of eggs produced and one to place the eggs in the incubating apparatus. The whole handling is organized and supervised either by one of the above persons or by a special supervisor. The number of persons involved in this procedure depends on the number of fish to be hand-stripped;

(d) This work requires sufficient equipment, such as rubber boots, aprons, sheets and gloves, cradles for carrying the fish, a stand and cradle to put away the spent fish, suitable towels or rags for holding the fish while being stripped, dishes to hold the eggs, a balance, anaesthetics, writing utensils, a tape measure, calibrated vessels, a vessel for anaesthetization, hand nets of suitable sizes to fish out the parent fish, a disinfectant solution for treating minor wounds to the fish, etc.

HAND-STRIPPING

Like other salmonids, the huchen does not exhibit high absolute or relative fecundity. In specialized huchen farming this means that on the one hand a farmer must keep a larger number of parent fish and on the other handling a large number of fish is rather difficult, particularly if their individual weights are high. This is another reason why a correct determination of peak ripeness, is of particular importance in huchen farming. In our experience, the oldest females are the first to attain ripeness and therefore they are used first. When the ripeness is determined correctly, sexual products can be obtained from almost all fish at the first hand-stripping. The fraction of fish that do not attain sexual maturity is sometimes only 5% but in extreme years, under unbalanced weather resulting in excessive variation in water temperature, this percentage may be considerably higher, up to 50%. These shortcomings might be removed by injections of fish pituitary. The first results obtained in Austria (Jungwirth 1979) were promising: all females matured within six days after the injections. Skalin (1983) attempted to induce ovulation by administering oxitocin, but his attempts were a failure as all fish treated with injections died. Experiments with carp pituitary (5 mg kg^{-1} live weight in a 0.65% solution of natrium chloride; divided into two doses for females) were successful (male sexual products were obtained after 84 hours, female after 132 hours, but the author does not mention their quality). The method, however, has not yet become fully established in practice because of problems associated with exact dosage (Bohl, personal communication).

On the basis of our own experience in handling a large number of parent fish, we would warn against checking the ripeness by squeezing out eggs from all females. If such a check must be made at all, it should involve only selected individuals. First of all, the degree of ripeness is checked in males. The males are easily identified by their coloration (Pl. 22). In our experience, colour inversion, i.e. simulation of the male breeding colours by a female, occurs in one out of 50 females kept in fishponds. When the males are found to discharge milt readily they are separated in a suitable handling space and the degree of maturity is checked in the females. For this purpose, females with a stout, voluminous, conspicuously white abdomen and with a more or less intumescent, pinkish genital pore are selected. By exerting gentle pressure on the abdomen in the direction from the pectoral fins towards the urogenital papilla, one can finger the retreating eggs which are then readily released through the genital pore as well. Then the female is stripped, the check is repeated in another two females, and the eggs thus obtained are fertilized with milt obtained from previously prepared males. Hand-stripping of all ripe parent fish should then begin on the next day at latest.

The following measures are carried out in the pond itself. After the water level in the pond has been lowered so that the water remains in its lower third and in the medial channel into which the water is continuously supplied, some of the parent fish to be hand-stripped are carefully driven towards the inlet and the individual compartments of the medial channel are separated with lattices. Then the fish are gradually segregated, placing males directly under the inlet and females in the next compartment. In the next phase, females are segregated: the ripe ones are placed in the neighbouring compartment, the unripe ones are transferred to another pond where they can mature. Spent males are added to this second group. With a small number of ripe and a majority of unripe females, the latter remain in the pond and the ripe ones are transferred to another.

In segregating the fish, the biggest ones are taken first, as these mature earlier than younger and smaller ones. Checking the ripeness of females requires certain experience. An unfailing sign of full ripeness is the stout and voluminous, white-coloured abdomen and the pinkish, slightly swollen genital pore protruding over the contour of the abdomen, as well as the eggs retreating under slight pressure of fingers on the ventral wall.

Suitable vessels for the anaesthetic bath should be placed near the handling channel.

Anaesthesia requires due care. On the one hand, it makes the work easier and on the other it decreases considerably the risk of injuring the parent fish or even of impairing the quality of the stripped eggs, which may be caused by violent movements of the handled fish (Pl. 12.2). Anaesthesia was first applied on the huchen farm at Dubná Skala, using the preparation MS 222 Sandoz (Skácel 1976). Anaesthesia occurs at a mean concentration of 8–10 g per 50 l of water, within 90–180 seconds after submerging the fish in the bath. It lasts from 3 to 6 minutes, which is ample time to strip the sexual products, to take measurements of the fish, and to remove ectoparasites (*Basanistes huchonis*). In running water, the fish will recover from anaesthesia within 1–6 minutes. Another successfully used anaesthetic is the Belgian preparation Propoxat, a white crystalline powder readily soluble in water, used in concentrations 1 : 500,000 to 1 : 300,000. The exposition time is 90–120 seconds. However, here the recovery time is considerably longer than with MS 222 Sandoz, about 15–30 minutes. In the Federal Republic of Germany, trichloromethylpropanol is also used at a concentration of 1.5 g per litre of water (Bohl 1979). In general, overdosing should be avoided in using any anaesthetic drug.

The following is the procedure of proper hand-stripping. The anaesthetized huchen is thoroughly rinsed with clean water and then transferred into the stripping cradle. (It is necessary to prevent the anaesthetic solution from contacting the sexual products.) Then one person takes hold of the fish by means of a collar made of fabric by placing it around the fish's body behind its pectoral fins. Another person lifts the fish so that it lies at an angle of about 30°, and a third person takes hold of it by its caudal peduncle, taking care not to press its spinal column, but holding it from both sides (otherwise the fish may be injured, its caudal part paralyzed, the spinal column deformed and the fish may even die). According to the size of the fish, a further person holds the genital pore of the fish to prevent spontaneous discharge of sexual products. As soon as the fish is in a suitable position, its genital orifice pointing towards a prepared dish held by the fourth person, the genital orifice is uncovered and the fourth person presses out the eggs into the dish, pressing his forearm or both hands against the abdominal cavity of the fish (Pl. 19.2). When all eggs have been stripped (i.e. when no more eggs are released by repeated stripping movements) the female is allowed to recover in a stream of water in an auxiliary pool (Pl. 20.1). When a sufficient amount of eggs has been obtained (15 to 20 thousand), milt is obtained from males in the same way.

When a male is to be used several (2–3) times within one day it is important to anaesthetize it for the first stripping only. For the second and third strippings it is better not to anaesthetize but to keep the male out of water (for about one minute), for it becomes tired much sooner now and recovers within a shorter period of time (Pl. 20.2).

Although good results can be obtained by repeatedly using one and the same male (according to Skácel's records made at Dubná Skala in 1964–1968, it was possible to obtain milt from a male three times within 90 minutes), we recommend the use of heterospermia by stripping two or three males at one time. When milt has been obtained by stripping the males, the milt, eggs and the ovarial fluid are carefully and thoroughly mixed with a feather. After 4–5 minutes water is added and, after the eggs had become partly swollen, suitable portions of the fertilized eggs are placed in an incubating apparatus (in some cases, eggs from individual females are incubated separately).

Records should be made of the origin of the eggs and for this purpose the individual females should be tagged. During stripping, the "obtained fecundity" should also be assessed for each individual female, if possible.

According to Nieslanik's experience, smaller fish (up to 10 kg in weight) can be stripped without previous anaesthesis, using a special stripping "muff" (or sleeve) which hangs on the chest of the person carrying out the stripping. An assistant inserts the fish from the right side into the muff head first so that its genital pore and anal fin protrude from a slit in the bottom right-hand side of the muff. Two assistants – one on the right and another on the left – hold the fish firmly by its head and tail and the man wearing the muff strips the fish of its sexual products, which are caught in a dish. The spent fish is then carried inside the muff to the handling channel or auxiliary pool and carefully released head first.

During hand-stripping, the quality of the sexual products obtained is checked regularly. Sound eggs are pinkish or orange, slightly sticky, clear and when fertilized and placed in the incubator, they do not change except for increasing in volume. Eggs that are hard when stripped, bouncing when released into a dish, and have a whitish spot in the middle of a paler, translucent field are over-ripe, incapable of being fertilized.

Sound milt resembles thin cream in appearance and consistency. Milt that coagulates immediately after stripping, contains blood or is too thin, is unsuitable for fertilization and must be eliminated. It is worth noting that newly stripped milt which has not contacted water can be transported in a cooled thermos (vacuum) flask, the spermatozoa retaining viability and fertilizing capacity for at least two hours.

When the eggs are mixed with milt, a small quantity of water is added to accelerate fertilization and induce swelling of the eggs. This swelling is quite rapid and intensive. According to Nieslanik's observations, the eggs increase by 10–15% in diameter and by 30–50% in volume within one hour after fertilization. The eggs harden and their surface becomes firm and resilient. After a short period of rest, the eggs are freed from the remains of milt by gently rinsing them with water, incidentally measuring their diameter, volume and number (using a measuring plate for 10–20 eggs, a calibrated tube 10–20 or 1000 ml in volume), and are placed in an incubating apparatus.

Before recovering from anaesthesia, the stripped fish are examined, ectoparasites are removed from their body surface and from beneath their opercular bones, and the fish are treated against fungal infection (Chapter 29).

After hand-stripping, a certain small number of eggs always remains in the abdominal cavity of the females, which they try to get rid of in the natural way – by spawning. For this reason, after being hand-stripped and treated against parasites and fungus, the fish are placed in a separate pool with a hard, gravelly bottom, with water 60–90 cm deep and an increased water supply. There, within a few hours after being transferred, the fish continue their breeding behaviour, prepare redds and spawn naturally shortly thereafter, the females getting rid of the remaining eggs. It has been found that such eggs are normal, of unimpaired viability and capable of producing normal alevins.

Beyond doubt, providing conditions for natural spawning for the hand-stripped fish is extremely good for their health, as it eliminates the negative effects of hand-stripping. The fish do not suffer from stress which could have an unfavourable effect, considering the size and repeated handling of the fish. Moreover, it prevents the formation of a so-called plug which sometimes develops from the residual eggs and impedes releasing eggs in the subsequent years.

If under exceptional circumstances one is compelled to strip huchen in natural conditions, utmost care must be taken in handling the fish to be stripped. They are kept separate in side branches or in slowly streaming water. After being stripped, they must be treated considerately and released carefully. After being rinsed with clean water, the activated eggs are transported into the hatchery in special transporting trays or in vacuum bottles (Chapter 30). Losses among eggs obtained in this way are usually not higher than otherwise.

26. Egg Incubation and Hatching

The technology used in the incubation of huchen eggs is basically identical to that used in rearing other salmonid species. Inseminated and rinsed eggs are placed in apparatus of the Rückel–Vacek type, in hatching trays or sometimes even in hatching troughs.

Since the eggs from huchen females of different sizes have quite different diameters, it is best not to blend eggs from different females, but to incubate them separately. As a rule, young huchen females produce smaller eggs, 4–4.5 m in diameter; eggs from bigger females vary from 5 to 5.5 mm in diameter; these values refer to swollen eggs, i.e. approximately two hours after insemination. The differences are still greater in terms of volume, since an egg 4 mm in diameter is 33.5 mm^3 in volume, while an egg 5.5 mm in diameter is 87.1 mm^3 in volume. For practical reasons of subsequent records of the numbers of young produced, an account is taken of the volume and diameter of the eggs from the very moment of placing them in the incubating apparatus. For better orientation, the volumes and numbers of eggs corresponding to their diameters are given in Table 39. One apparatus of the Rückel–Vacek type of normal dimensions, with a tray 47 × 48 cm in dimensions and 15 cm deep will hold 10 to 15 thousand eggs.

Although data on the fecundity of the huchen have been discussed in Chapter 9, it should be briefly repeated that in practice, one may expect an "obtained fecundity" from 7400 to 27,000 eggs per female; on the basis of long-term experience, an average may be set up between 2 and 12 thousand eggs per female 80–84 cm in average length and 6–7 kg in average weight, which corresponds to the dimensions of huchen females most frequently occurring on our farms. Thus there are 1800 to 2000 eggs per kilogram of female body weight. Heavier females produce higher absolute but lower relative numbers of eggs. For example, on the huchen farm at Dubná Skala, 330,000 eggs were obtained from 23 females 267 kg in total weight in 1965, i.e. 14,348 eggs per female 11.6 kg in average weight, but only 1236 eggs per kilogram of female body weight, which is scarcely half that of females 6–7 kg in body weight.

Thus eggs from one female can roughly be incubated in one apparatus of the Rückel–Vacek type (Pl. 21).

The water supply to the apparatus should be adjusted so as to vary between 50 and 100 ml s^{-1}, i.e. about 3–6 litres per minute. Each apparatus should have its own supply of fresh water, although in an extreme case one and the same amount of water can be used to supply up to four apparatus of the Rückel–Vacek type, or 6 hatching trays in succession. During the hatching of embryos, the water inflow must be temporarily decreased so that the water current does not carry away egg shells or even the hatched young, or that it does not press them against the outflow screen.

The incubation period between insemination and the "eyed egg" stage (when the eggs can be handled and even transported to greater distances) lasts 230–270 temperature units on average; 290–300 temperature units up to the hatching phase.

TABLE 39

Number of huchen eggs per 1 l volume in dependence on egg diameter and volume

Egg diameter (mm)	Egg volume (mm^3)	No. of eggs per 1 l
4	33.51	29841
4.1	36.09	27711
4.2	38.79	25778
4.3	41.63	24021
4.4	44.60	22420
4.5	47.71	20959
4.6	50.97	19621
4.7	54.36	18395
4.8	57.91	17269
4.9	61.60	16233
5.0	65.45	15279
5.1	69.46	14398
5.2	73.62	13583
5.3	77.95	12828
5.4	82.45	12129
5.5	87.11	11479
5.6	91.95	10875
5.7	96.97	10313

The duration of the incubation period is most influenced by water temperature which may vary considerably during that period (Fig. 32), and by oxygen content (Chapter 17).

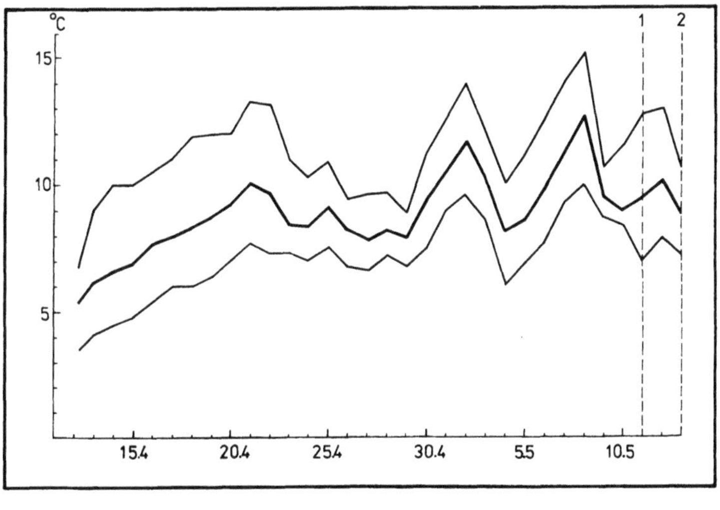

Fig. 32. Water temperatures during incubation of huchen eggs in the hatchery of the Slovak Anglers' Union at Dubná Skala in 1969. Thinner curves denote temperature ranges, thicker curve mean temperatures. 1 – eyed egg stage; 2 – hatching. Abscissa, dates (orig. L. Skácel).

During incubation it is important to treat the eggs regularly, emphasis being put on prevention of fungal infection in a manner described below (Chapter 29). Losses among properly treated eggs are low, not exceeding 5–8% as a rule.

Like incubation, hatching also is influenced by water temperature. Usually, hatching is most intensive during the first three days, the remaining eggs hatching within another three days. At that time, it is necessary to remove not only sterile eggs (as long as these had not been removed previously) but also empty egg shells, to prevent stoppage of outflow screens. In the Rückel–Vacek apparatus, the upper inflow is changed into the lower by shifting the tray. This serves not only to decrease stoppage of the outflow screen with egg shells, but above all to prevent the soft yolk sacs of the free embryo from being pressed against, or sucked into, the openings of the tray. A temporary decrease of the water inflow can also be recommended.

Throughout the incubation period, the eggs must be protected against direct sunshine, the same as hatched young until about 2/3 of their yolk sac has been absorbed.

After hatching has ended, the apparatus must be thoroughly cleaned; all impurities and dead young must be removed.

A sufficient supply of water of good quality during incubation and hatching is a vital factor. The water entering the apparatus must be clean, without any admixtures whatever. In the contrary case there is a risk of the eggs becoming infected with fungus or suffocated by suspended solids. For this reason, spring water combined with surface water if need be is most advantageous. If the apparatus is supplied solely with surface water it is essential that it is filtered. Even though the eggs do not require a high dissolved oxygen content of water, the water supply is adjusted for maximum input that does not move the eggs. The water temperature should not drop below 4° C and it should not show too large oscillations. Therefore it is important not only to check the water temperature regularly but also to provide the possibility of pre-heating. There should be two boilers independent of each other (e.g. one electric and the other connected with central heating). The oxygen content should not drop below 7 mg l^{-1}.

27. *Rearing Young Huchen*

After the rearing apparatus has been cleaned, the huchen young are divided so that there are 5 to 7 thousand per apparatus. The water supply is adjusted so that it is higher than during hatching, but it must not drag down the young as also concentrate on one spot. Since the alevins are still photophobic at this stage, direct illumination of the apparatus should be avoided. The motility of the alevins increases parallelly with the absorption of the yolk sac. When about 2/3 of the yolk sac have been absorbed the alevins begin to change to exogenous food and therefore supplementary food should be provided from this period onwards. It is to be emphasized that correct supplementary feeding of huchen alevins is among the most important point of the whole huchen farming. In fact, this is what decides the extent of losses, which are usually highest in this very period of change from endogenous to exogenous nutrition.

Live food, i.e. zooplankton, is considered most suitable. In this developmental stage, huchen alevins are well adapted to preying upon zooplankton: they are quite motile and have well shaped head and a big mouth with a wide gape. At first, cladocerans are most suitable, being abundant in the ponds at this time; later they should be replaced by bigger crustaceans, particularly freshwater shrimps. Supplementary food is first offered to the alevins in the highest situated apparatus. The food should be offered four to six times a day and the water inflow should be slightly decreased so that the zooplankton is not washed away. Towards the end of this period, usually lasting 2–3 weeks, live food should be partly and gradually replaced by finely minced liver, spleen or a fodder mixture consisting essentially of finely minced fresh fish, spleen or liver, curds, and frozen zooplankton.

Meticulous care should permanently be paid to keeping the water clean by removing all remains of unused food, faeces, dead alevins, etc. As soon as the alevins have attained around 16–20 mm in length it is possible to pass to a further stage of rearing, which can be in apparatus, troughs or various vessels, ground pools, ponds, or the young can be stocked in natural waters.

REARING ALEVINS IN HATCHING APPARATUS

Prolonged rearing of alevins in hatching apparatus should be considered a more or less temporary measure. As a rule, it is carried out only when the alevins are expected to be sent to fishery organizations for stocking in free waters, or to be transferred into ponds of another huchen farm. In such cases, however, the alevins cannot be kept in the apparatus for more than another three weeks, since their uneven growth results in increasing cannibalism, and the high density in the jars increases the risk of an outbreak of infection. All principles stated in the preceding paragraphs are valid for rearing the alevins in hatching apparatus. Since at that time the alevins have completely lost photophobia and orientate themselves visually when foraging, sufficient illumination must be provided at the time of feeding.

REARING ALEVINS IN TROUGHS

The purpose of transferring the alevins into troughs is to rear huchen fingerlings 3–5 cm in length. For this purpose, troughs of various sizes and made of various materials are used; those made of glass fibre have been found most satisfactory, as they are light, easy to handle, and hygienic. However, concrete troughs with a natural bottom

TABLE 40

Growth of young huchen in troughs from the initial length of 14 mm

No. of days after stocking	Average *Tl* (mm)
5	16.5
10	18
20	20
25	25
35	28
60	40
75	50
90	65

can be used as well. The standard troughs used in Czechoslovakia are 400 × 40 cm and 20–40 cm deep. In Yugoslavia, good results have been obtained even with smaller troughs, 120 × 40 × 25 cm. Large troughs, 700 × 70 × 70 cm, are usually placed outside the hatchery, and should be roofed. Instead of troughs, square containers are sometimes used, made of glass fibre, 170 × 170 × 30 cm, with rounded edges, a central outlet and a peripheral inlet, in which the water moves circularly. The water supply to these containers is regulated so that the water is exchanged six to eight times an hour. The current velocity should be 10–20 cm s^{-1}, which equals 0.25–0.30 l s^{-1} (Harsányi 1982).

In troughs the huchen may be reared even longer, until they have attained 9 or at most 10 cm in length. Rearing huchen is more exacting than rearing other salmonids, as it requires not only systematic watching, but a high level of hygiene and frequent feeding with almost exclusively live food.

In spite of numerous experiments, feeding the huchen with granulated fodder has not proved satisfactory (Jungwirth 1977). Therefore in trough cultures one is entirely dependent on using classical food: zooplankton, tubifex worms, fly larvae, fish fry or other live foods are preferred, but unfortunately not always available. For this reason they are replaced, in practical farming, by finely minced spleen, liver, fish, hard-boiled eggs and curds, offered either separately or in the form of a mixture with vitamins and mineral substances added, which is more advantageous from the nutritive point of view. So far, there is no unified recipe for such a mixture; its composition depends on the breeder's experience or, most frequently, on the availability of the individual ingredients.

A trough culture is determining as to hygiene, which is why the troughs must be continuously inspected and all contamination removed; the food must be offered several times (2–15 times) a day and in such quantities as can be consumed by the fingerlings.

Because of very high population densities, small space and especially the kind of food offered, huchen fingerlings reared in troughs have a rather slow growth rate (Table 40).

Losses in trough cultures average around 30%, i.e. they are smaller than in ground pool cultures, but the costs are higher, the growth is slower and, because of the high population density, the risk of an outbreak of dangerous epidemics is considerable.

REARING YOUNG IN GROUND POOLS

The original, but later abandoned way of rearing huchen young in ground pools has recently been used with increased frequency. This way has an advantage against other methods in a smaller risk of failure, lower costs and stronger fingerlings produced, which compensate for the comparatively higher losses due to uneven growth and high cannibalism of the huchen. The pools used for this purpose are of different sizes; most satisfactory are pools 125 – 300 m^2 in area, rectangular in shape, with a 1:3 to 1:4 side ratio. In pool cultures, success depends primarily on the quality and amount of natural food as well as on its availability, which depends on water transparency and temperature, particularly during the first few days after the young have been transferred to the pool. These facts must be taken into account when deciding on the density of the stock, the latter usually varying between 20 and 100 individuals per square metre of water surface. Later on, food is supplied at least five times a day and, if possible, live fish (minnow, nase carp or other cyprinid young) are offered to the young huchen as soon as they attain the length of 40 mm. The daily dosage of food increases with increasing size of the young huchen, which minimizes the risk of cannibalism.

Rearing huchen young in ground pools resembles to a considerable extent their development in natural conditions and therefore the differences in individual growth are greater than in trough cultures. The fish attain 70 – 80 mm in length within three months and 90 – 120 mm within five months. Occasionally even individuals as big as 16 – 18 cm may occur. Naturally, the great differences in size increase the risk of cannibalism even with a sufficient food supply. For this reason, when conditions permit, the stock should be sorted by size or released in natural conditions in time.

The rearing results depend primarily on stock density. According to experience so far, survival is up to 65% at an initial density of 20 alevins per m^2. When 50 alevins per m^2 are stocked survival is only 40%, hence the losses amount to 60% as against 35% in the former case. Naturally, even growth is slower, by about 20%.

Perfect hygienic conditions must be maintained even in pool cultures. To minimize the risk of an outbreak of infectious diseases it is an advantage to apply preventive baths in a solution of malachite green (0.2 mg l^{-1}) for a whole day every two weeks. If ichthyophthiriosis is detected, a long-term bath (five days) in malachite green at the above concentration is applied.

REARING YOUNG IN NATURAL CONDITIONS

This means releasing and rearing huchen young in specially selected or even adapted small streams from which the grown-up fish can migrate into the main stream or are fished out and transferred elsewhere. This manner of rearing has its shortcomings as well as advantages. The costs are low and are only connected with the adaptation of the selected stream or section before the young are released, or with the autumn fishing. The growth of young in such a stream is usually faster than in pools or troughs. On the other hand, uneven growth results in too big differences in individual sizes and in higher losses than with other rearing methods; they attain around 50% on average. However, they may be even higher, 80–95%, particularly if predators were not removed from the selected stream, if there is a lack of natural food, if the water discharge is unbalanced, etc.

Tributaries of streams in which huchen occur naturally are most suitable for this manner of rearing young (Pl. 22.2). The young thrive in streams with a natural occurrence of grayling and particularly where minnow are abundant. Streams in which barbel, nase carp, chub and spirlin occur, and meadow brooks are also suitable. In principle, the selected stream should not be too big, its slope should be gentle, its bottom and banks sufficiently diverse, its water not too deep, and pools alternating with rapids. Besides, the stream should be sunlit and contain abundant zooplankton, zoobenthos, or young of non-predatory fish. It is an advantage if the stream is separated from the major stream by a natural or artificial barrier preventing predatory fishes from moving into it. Streams passing woodlands, with steeper slopes and flowing in beds resembling a channel, as well as those with very cold or even spring water, are unsuitable for this purpose.

Before releasing young huchen the stream or section selected is fished out by means of an electric fishing-gear, removing all predatory fish, including large brown trout, rainbow trout, chub, barbel and, if small young are released, also all large sculpins.

The stock density is determined by the kind (age) of young to be released. In the case of free embryos the density should be 2.5–3.5 individuals per m^2; if alevins with yolk sac already absorbed and adapted to exogenous food are released the rate is 1.5–2.5 individuals per m^2 of water surface.

Although huchen will leave such streams when grown up it is better to fish them out from the rearing stream by October at the latest and to transfer the fingerlings to the main streams. In suitable conditions the fish grow very fast, attaining 10–14 cm in size by October.

GROWTH OF YOUNG

Huchen eleutheroembryos (free embryos) have after hatching distinct differences in size, undoubtedly caused by differences in egg size, a fact known to occur in other fish species as well (Nikol'skii 1974). According to the data published by Balon

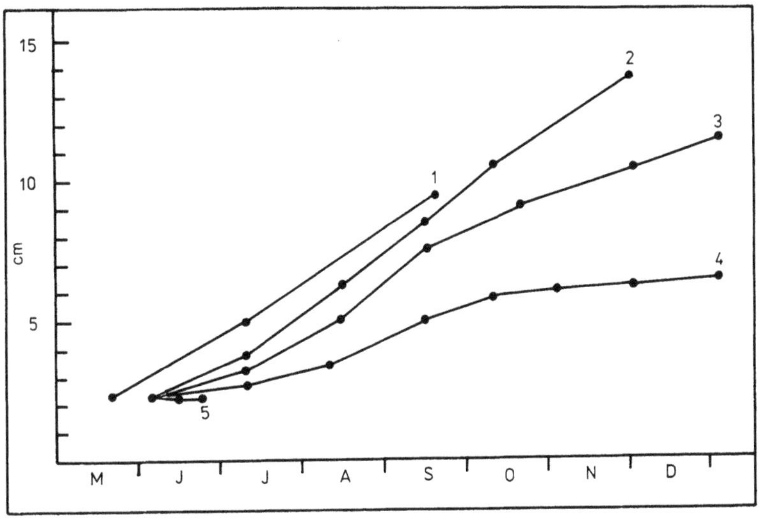

Fig. 33. Comparison of linear growth during the first year of life of young huchen fed with different food and reared in different environments. 1 – a pond with natural food (zooplankton, chironomid larvae, oligochaetes, fry of cyprinid fishes); 2 – rearing pond, on natural food; 3 – ditto, fed with minced fish; 4 – ditto, fed with minced liver; 5 – ditto, fed with protein granules (adapted from Jungwirth 1978).

(1956), although the author does not mention egg size, the eleutheroembryos studied were around 14.4 mm in average *Tl* on the first day after hatching, which agrees with our own observations. However, Nieslanik measured the lengths of eleutheroembryos hatched from eggs that were 5 mm in diameter and came from a female huchen weighing 20.8 kg stripped in the hatchery at Martin; he found an average *Tl* of 22 mm and an average weight of 84 mg. According to Jungwirth (1977), the young are 13–15 mm long when they have absorbed around 3/4 to 4/5 of the yolk sac. Naturally, such differences in size are manifest, even during subsequent periods, as considerable differences between the sizes of huchen of identical age. Further differences are due to different environmental conditions, especially the kind, quality and availability of food. This is clearly seen in Fig. 27, showing the growth of young-of-the-year huchen in a natural environment and under various rearing conditions, compiled from the data by Bastl and Kirka (1959). Growth differences among reared huchen young due to different feeding and environmental conditions can be inferred from the foregoing paragraphs; however, they are most clearly illustrated in Fig. 33, showing the experimental results obtained by Jungwirth (1977 b). It is seen that huchen fed with live food were twice as long on an average as those of the same age, but fed with liver and, as emphasized by the author, the former resembled in their growth huchen from natural conditions. A marked influence of food not only on growth but also on the length–weight relationship in the huchen was pointed out by Bastl and Kirka (l. c.); the authors found that huchen from a pond culture exhibited a lower value of the regression coefficient than those from running waters, and that huchen young from a pond culture were at first heavier than those of the same size from running waters, but having attained over 61 mm in body length the weight growth of the former was much slower, indicating that the food supplied was unsatisfactory in terms of both composition and quantity. It is of benefit for a breeder to utilize this experience to the full. First of all, he must take into account the individual differences in growth and divide the huchen stock accordingly; huchen of equal size should be kept together. Since the huchen is a predator, the breeder must forego its cannibalism by supplying sufficient amounts of live food in due time.

FEEDING

Provided a fish farm designed for huchen breeding fulfils all the requirements mentioned above, then food and feeding are the major factors deciding the success and quality of the huchen culture. In part, these problems have been alluded to in the preceding chapters, and this section based on our own and foreign experience, is a brief summary of the requirements of the huchen as to food and feeding.

The most suitable and the most natural food for young huchen includes animals ranging from cladocerans through freshwater shrimps and brine shrimps up to fish fry of various sizes. If possible, this kind of food should be provided; however, circumstances often do not allow it. Thus, in rearing young in troughs or ground pools, with increased stock densities, supplementary food must be provided in most cases. This consists mainly of finely minced spleen, later also liver, hard-boiled eggs, curds and minced fish. Although the above food items may be offered singly it is better to use them in a mixture, adding a small amount of granulated trout fry food (EWOS, Clark). The possibility of enriching the diet with vitamins or incidentally antibiotics may be considered. In practice, good experience has been obtained with frozen zooplankton, which can be prepared in advance and stored for almost unlimited time. When dead fodder is offered it is essential slightly to increase the water flow so that the fodder particles may float on the water surface or in the water column as long as possible and stimulate, by their apparently active movements, the young huchen to seize them.

Feeding with granulated fodder mixtures usually used in rearing brown trout or rainbow trout has not proved satisfactory either in Czechoslovakia or abroad (Jungwirth 1977a, b). Huchen young practically ignore this diet, their growth slows down, and the mortality rate increases rapidly. Granulated fodder mixtures may only be used as a minor supplement to the preferred food. Thus Skalin (1983), for example, fed huchen young and later also huchen fingerlings with a special fodder mixture in which the content of essential amino acids was qualitatively and quantitatively the same as in cladocerans or in the young of *Chondrostoma nasus*. Before feeding, the mixture was blended with beef liver in a 1:2.85 ratio. While the huchen accepted this diet their growth was considerably delayed: after 83 days, they attained 74 mm and 2.53 g on average, and only 219.2 mm and 89.2 g after 769 days.

Plate 1. The huchen, *Hucho hucho hucho* (1). The relatively large adipose fin is typical of the whole genus *Hucho* (photo J. Dubeň). A huchen in dorsal view, (2) showing its strong, torpedo-shaped body and wide skull, and suggesting its predatory habits (photo J. Dubeň).

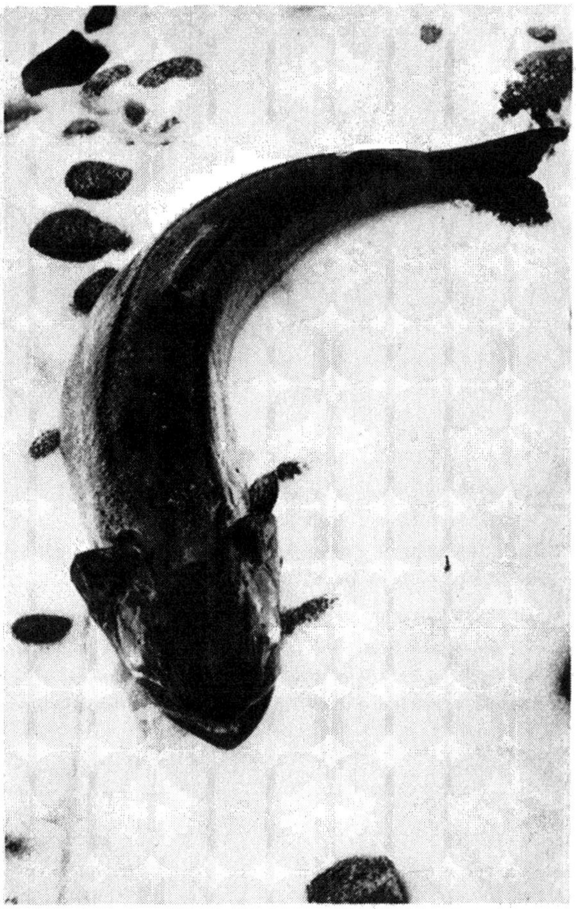

Plate 2. Skiagraph of a huchen 315 mm in *Tl* (1) (photo J. Nieslanik). Detailed view of the head of a huchen 12.5 kg in weight (2) (photo J. Dubeň).

Plate 3. Detailed view of the lateral line region in the middle of a huchen's body, showing the typical markings and scales inserted in the skin (1) (photo J. Dubeň). Key scales of *Brachymystax lenok* (2), *H. h. taimen* (3), *H. h. hucho* (4, 5) and *Salmo trutta* m. *lacustris* (6). Scales of Huchonini are longitudinally protracted, ellipsoidal and widened at their posterior part (from Holčík 1982).

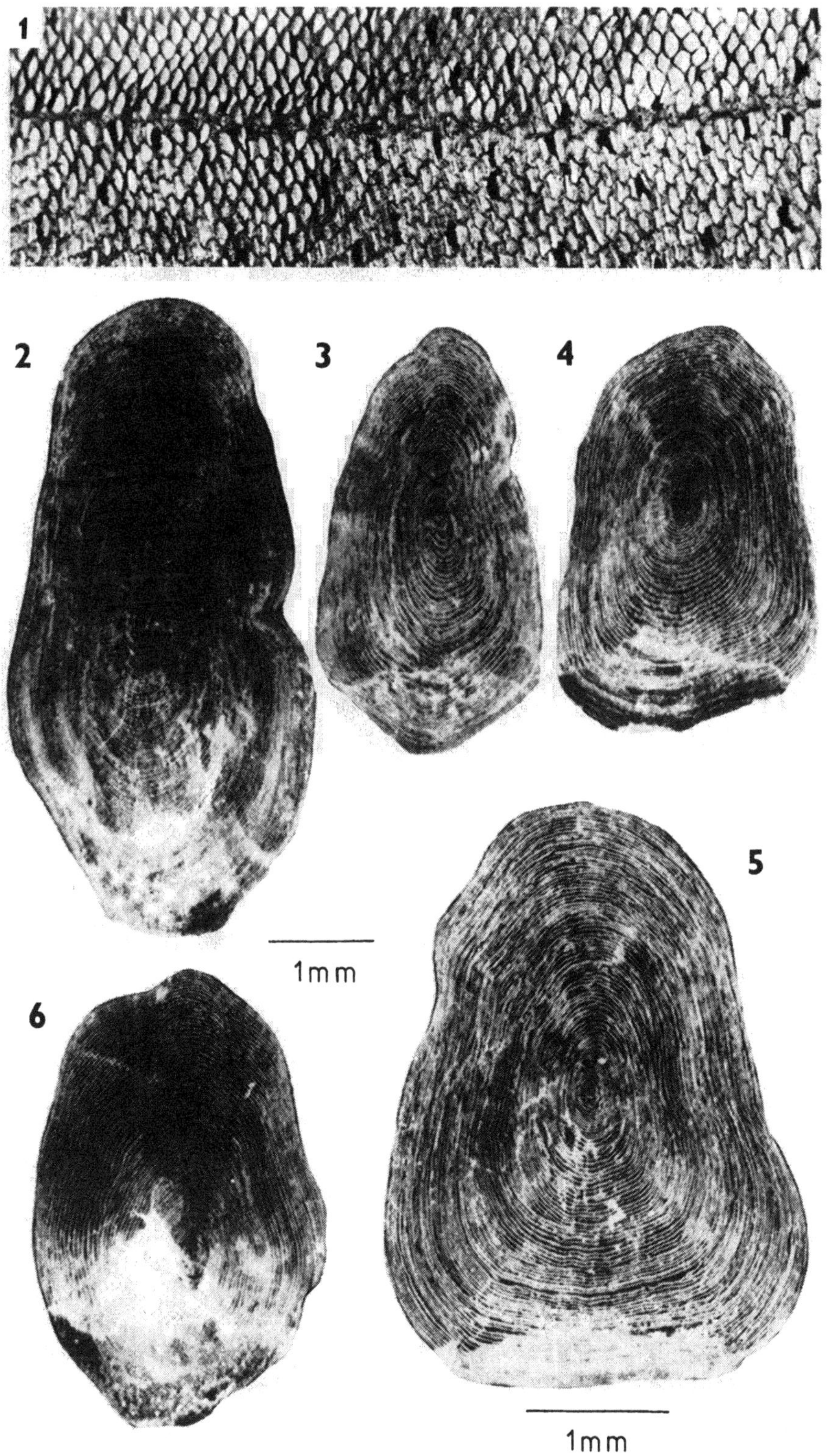
1
2
3
4
5
6
1mm
1mm

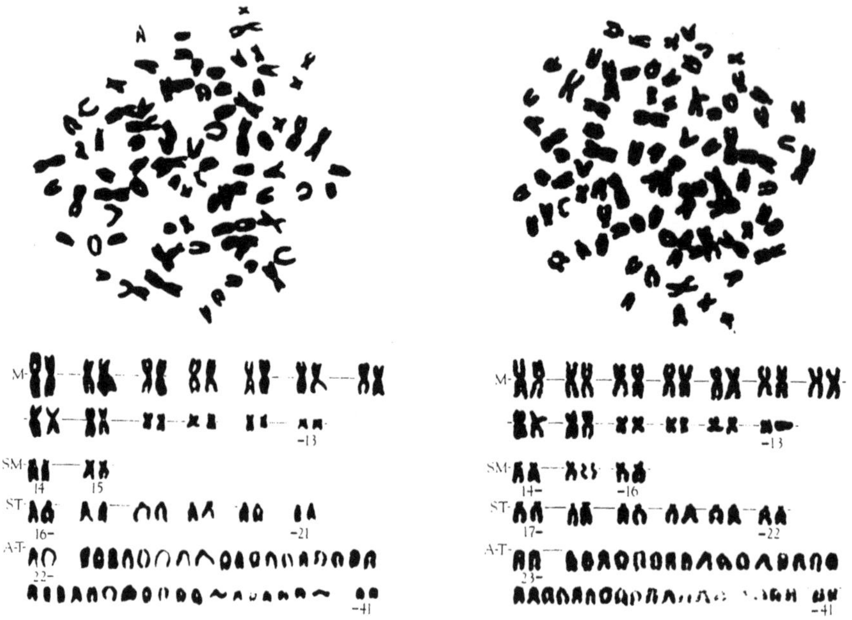
M-
-13
SM-
14
15
ST-
16-
-21
A-T-
22-
-41
M-
-13
SM-
14-
-16
ST-
17-
-22
A-T-
23-
-41

Plate 5. The Váh river near Ivachnová (1). At present this is the most important huchen locality in the Váh river at all (photo J. Dubeň). The confluence of the Orava and Váh rivers (2) creates favourable conditions for the occurrence of huchen (photo J. Dubeň). The Orava river below Oravský Podzámok.

◄ Plate 4. A huchen, weighing 13 kg, from the Orava river (1) (the Danube river system) shows more numerous markings which often occur on the adipose fin as well (courtesy of the magazine Poľovníctvo a rybárstvo). Karyotype of the huchen from the Poprad river (2) (the Vistula river system) and that of a huchen from the Sava river (3) (the Danube river system) (photo P. Ráb).

Plate 6. The Orava river below Oravský Podzámok (1). This section is known to house big huchen (photo J. Dubeň). The Turiec river near Socovce (2). The rather narrow but deep stream of the Turiec contains even at present, huchen up to 30 kg in weight and it is the only river in Slovakia which is hardly affected by anthropic influence, especially pollution and canalization. A section of the stream 21 km long is a protected huchen locality – the only huchen reserve in the world (photo J. Dubeň).

Plate 7. The Hron river in the environs of Brezno (1). The narrow and shallow stream has only a small huchen population (photo J. Dubeň). The Liptovská Mara reservoir (2), a man-made lake, was erected by damming the Váh river. Here, as in most other submountane reservoirs and lakes, the huchen populate the littoral parts and those in front of the mouths of tributaries (photo J. Dubeň).

Plate 8. The Dunajec river between the villages Majere and Lysá nad Dunajcom (the Vistula river system) (1). Introduced in the past, huchen found here good living conditions and created a self-sustaining population. Before the erection of waterworks on the Vistula river, salmon *(Salmo salar)* and sea trout *(Salmo trutta* m. *trutta)* also migrated up to this stretch (photo J. Dubeň). The Poprad river (2) (the Vistula river system) is one of the rivers where the huchen has been successfully acclimatized (photo J. Dubeň). The Hornád river (the Danube river system) in the Slovenský raj Mts. (3). Introduction of huchen was followed by a decline of populations of brown trout *(Salmo trutta* m. *fario)* and grayling *(Thymallus thymallus),* so the huchen population had to be removed later (photo J. Dubeň).

Plate 9. The Mutnianka stream (shown here in the environs of Beňadov) (1) used to be one of the well-known spawning streams for a part of the huchen population in the Biela Orava river. The photograph documents the summer status, when the stream suffers from lack of water (photo J. Dubeň). A pair of huchen in a spawning site in the Revúca stream (a tributary of the Váh river) (2). Right, the female, left, the male (total length about 850 mm, weight about 6 kg) (photo J. Dubeň).

Plate 10. A newly cleared huchen redd in the Váh river near Ivachnová (1) (photo J. Dubeň). A male huchen (2) with typical "scratches" on the top of its head, inflicted in duels during the spawning period (photo L. Skácel).

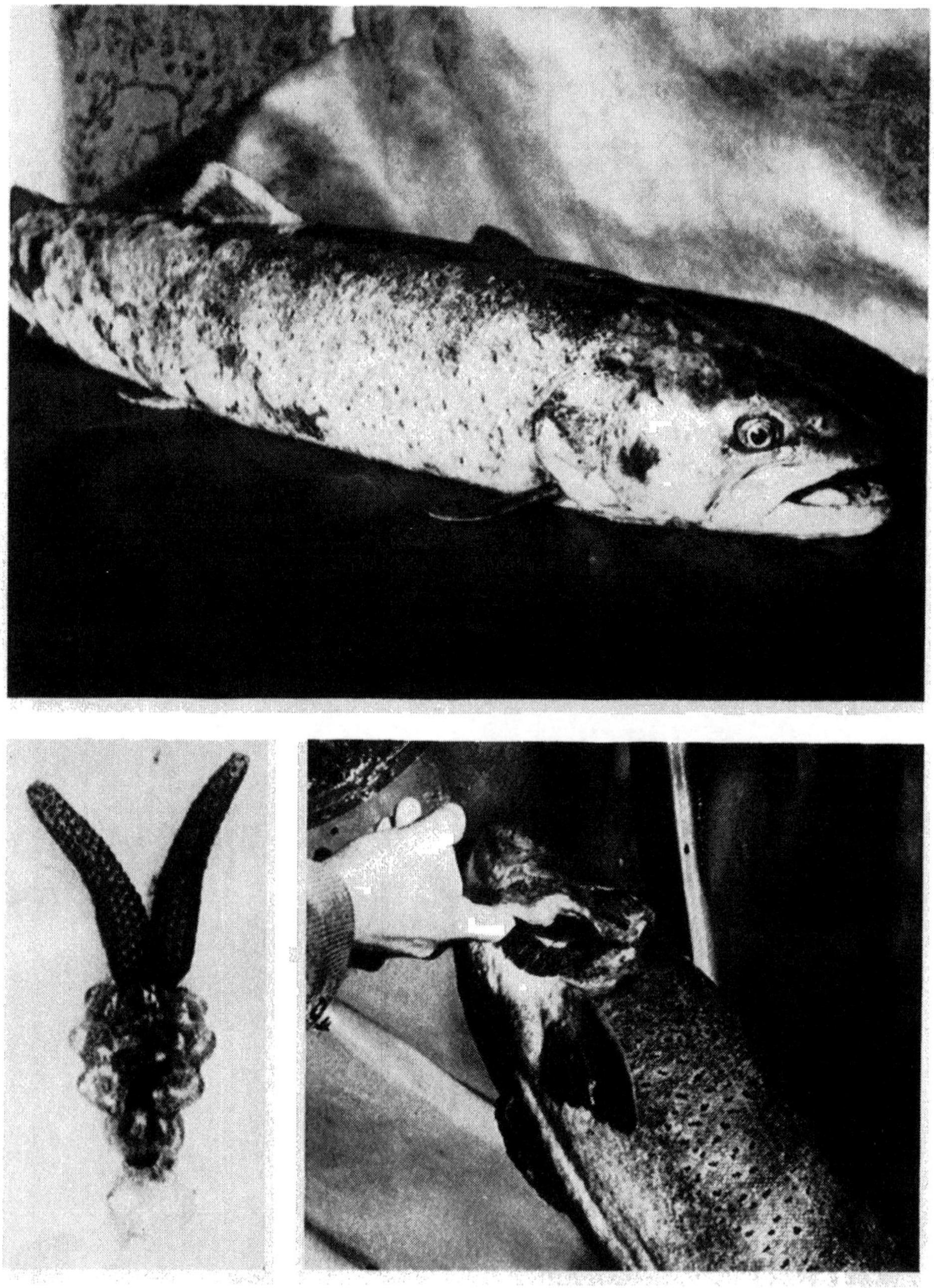

Plate 11. Huchen are sensitive to infestation with the parasitic fungus *Saprolegnia* sp. which usually has lethal consequences (1) (photo J. Nieslanik). *Basanistes huchonis* (2), a typical and specific parasite of the huchen (photo J. Nieslanik). Duels during the spawning period may result in serious injuries: in the huchen shown here (3), part of its opercle is missing (photo L. Skácel).

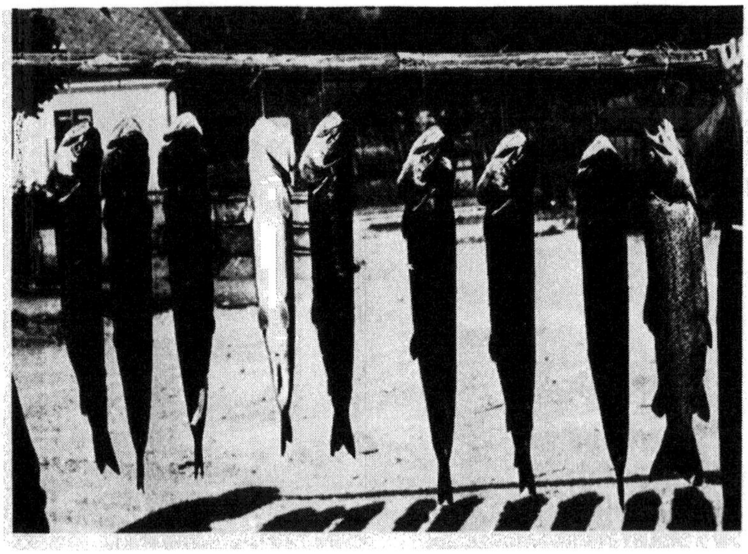

Plate 12. (1) Atypical shape of head, probably due to a shortening of neurocranium. Total length of fish 840 mm, weight 7 kg (photo J. Dubeň). (2) Mass mortality of huchen, due to high water temperatures, may occur in summer, at low levels and in hot weather, like that in the Orava river in August 1952 (photo J. Nieslanik).

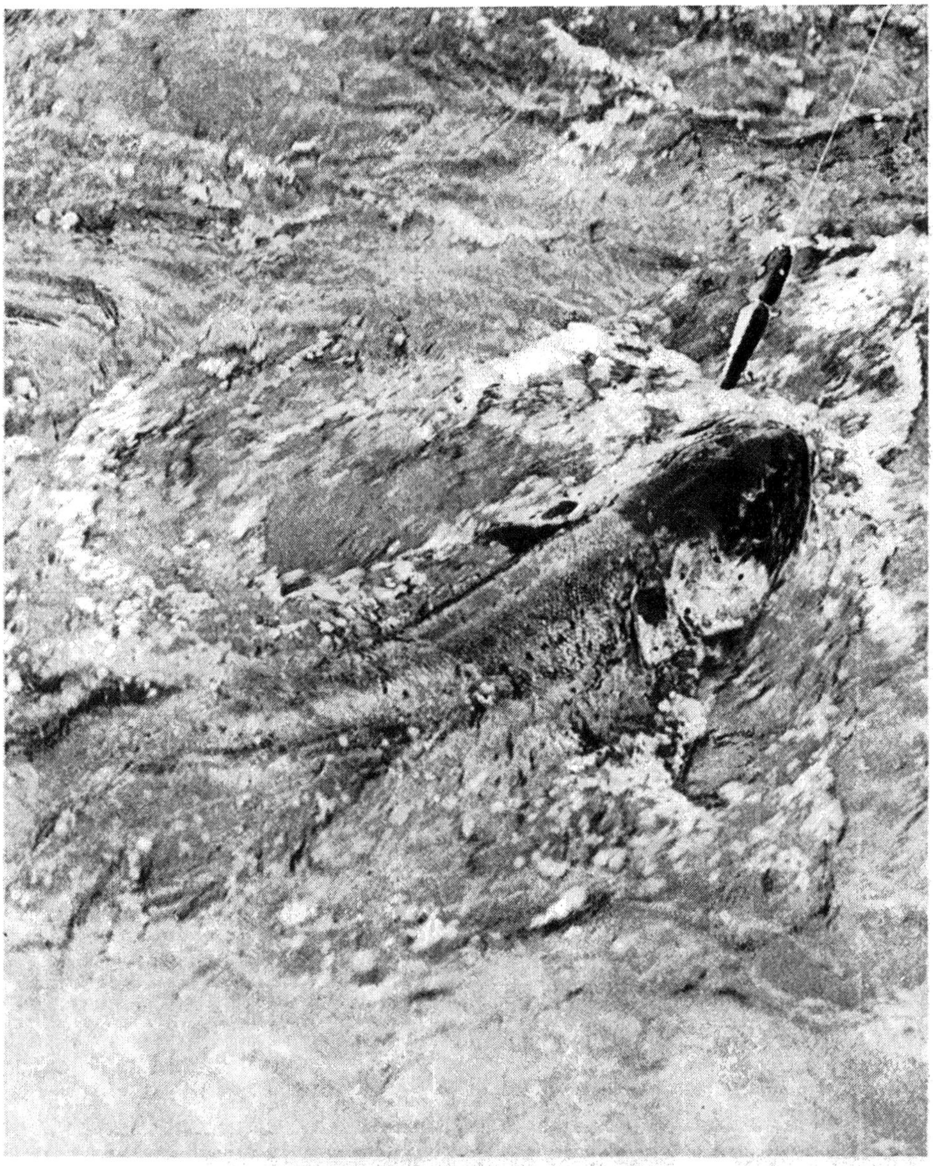

Plate 14.1. Overcoming a huchen over 10 kg in weight (the one shown here (1) was a female 1050 mm in *Tl* and 12.5 kg in weight) gives a never-to-be-forgotten thrill (photo J. Dubeň).

◀ Plate 13. Trolling a dead fish is the most effective way of huchen angling (1) (photo J. Dubeň). Some of the most commonly used artificial bait (2) for huchen fishing at present (photo J. Dubeň).

Plate 15. One of the ponds in the historical hatchery of Samo Ivaška, the pioneer of breeding huchen in captivity, at Martin-Podstráne (photo J. Dubeň).

◀ Plate 14.2. Surrender... (2). The huchen is so exhausted by frantic attempts to escape that it puts up almost no resistance in the phase of the fighting. The huchen shown here was 9.5 kg in weight (photo J. Dubeň).

Plate 16. Fishing huchen for breeding purposes in Slovakia (1). For this fishing special sleeve nets are used, fastened to a wooden semicircular frame, into which the huchen are driven by means of an electrical shocker (photo L. Skácel). However, angling (2) is the best method of obtaining parent fish in the wild (photo J. Duben).

Plate 17. Rearing a good shoal of parent fish is the primary condition of successful closed-cycle huchen farming. Shown here (1) is a part of the shoal of young parent huchen in the hatchery of the Slovak Anglers' Union at Podsuchá (photo L. Skácel). A parent huchen is fished out from a pond with lowered water level with a special net (2) (photo J. Dubeň).

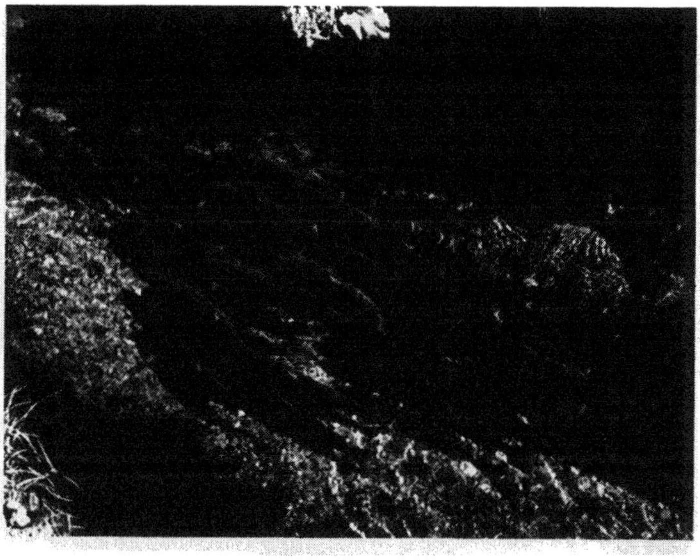

Plate 18. (1) In a stretcher (in the rear) a huchen is transported to an anaesthesia vessel or directly into a handling cradle (in the foreground) (photo L. Skácel). (2) A bigger huchen must be anaesthetized before hand-stripping (photo L. Skácel).

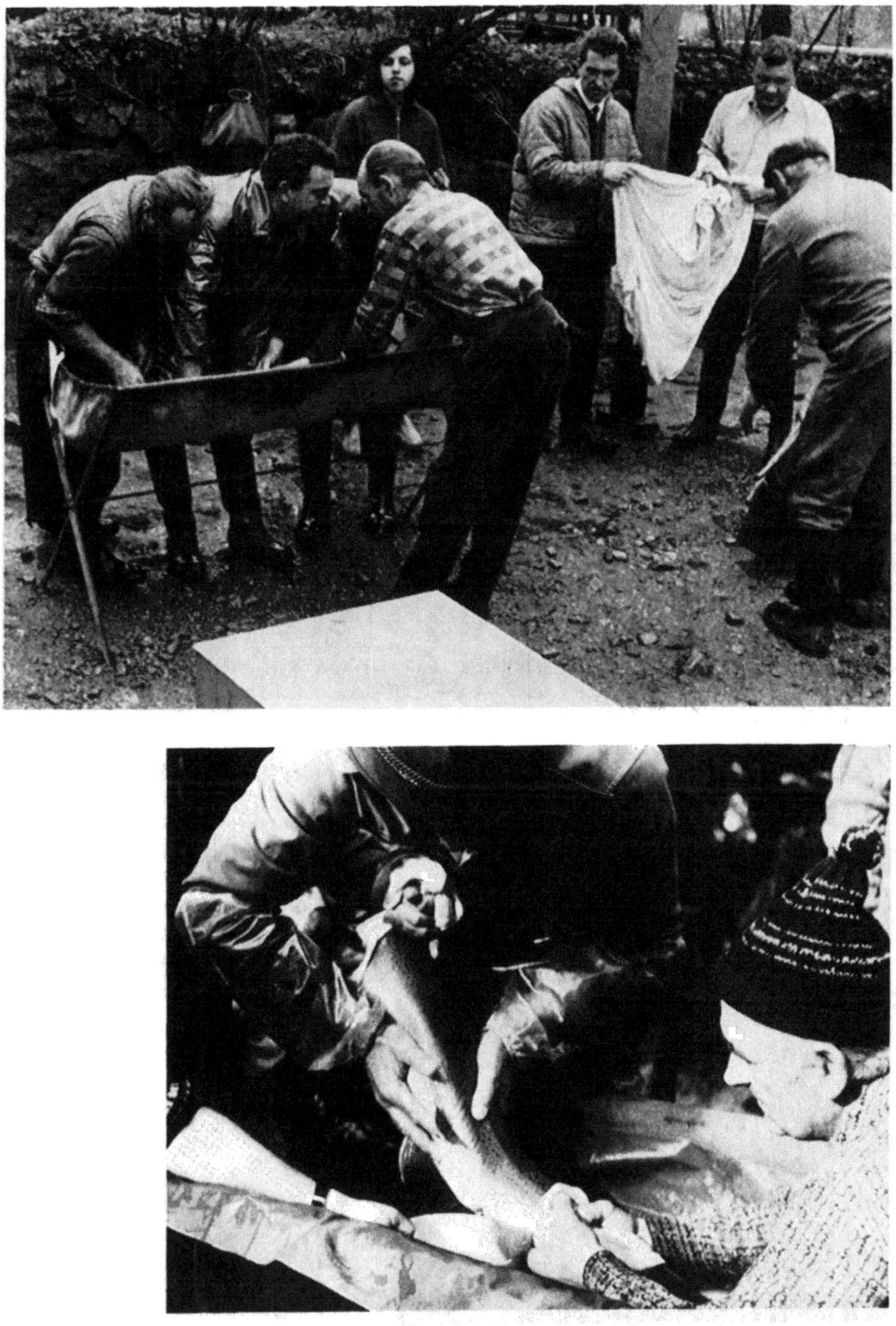

Plate 19. (1) A working group hand-stripping a huchen (photo L. Skácel). (2) Hand-stripping: two workers hold the huchen, a third one carefully presses the abdomen of the fish to strip it of its sexual products, which a fourth worked catches up in a dish (photo J. Dubeň).

Plate 20. (1) A female huchen, stripped of her eggs, is released in a special pond to finish spawning in a natural way. Handling such a big huchen requires using cautions and taking care (photo J. Dubeň). (2) Between two successive hand-strippings, a male huchen (to be used several times) recovers in a special container in which water is continuously exchanged (photo L. Skácel).

Plate 21. Fertilized huchen eggs are incubated in apparatus of the Rückel –Vacek type (1) or of a circular type (2) (photo J. Dubeň).

Plate 22. (1) Hatched huchen eleutheroembryos (photo J. Dubeň). (2) Huchen juveniles are released in small and shallow streams with a good fodder basis (photo J. Dubeň).

Plate 23. Huchen from the Váh river (the Danube river system) near Bešeňová (Slovakia). Total length 1050 mm, weight 12.5 kg (1) (photo J. Dubeň). The taimen, from the lake Dood Cagaan, Darkhat valley (the Enisei river system), Mongolia, July 1969. Total length about 1100 mm, weight about 17 kg (2) (photo K. Hensel).

Plate 24.1. Although the coloration of a taimen will change after its death the red pigmentation of its abdominal part and of its caudal and anal fins will persist for a long time (photo K. Hensel).

Plate 24.2. Huchen are phylogenetically closest to the genus *Brachymystax* Günther, 1866. *Brachymystax lenok,* shown here, is the only species of this genus (about 400 mm *Tl;* the Shishkhid river, some 5 km below its outflow from Lake Dood Cagaan, Mongolia, in the Enisei river system) (photo K. Hensel).

Plate 25. Close-up of the head of a female huchen, about 13 kg in weight (1). Note the big eye, the long and wide upper jaw and the large mouth (photo P. Áč). The white gape of the male huchen (2) contrasts with the copper red of his back (photo P. Áč).

Plate 26. The large eggs of a huchen (1) are amber-yellow, orange-red, or even red. They are slightly sticky, but the stickiness disappears after rinsing (photo P. Áč). The sexual dichromatism of the huchen is discernible for a long time and even at a distance (2). Upper, the male; lower, the female (photo J. Dubeň).

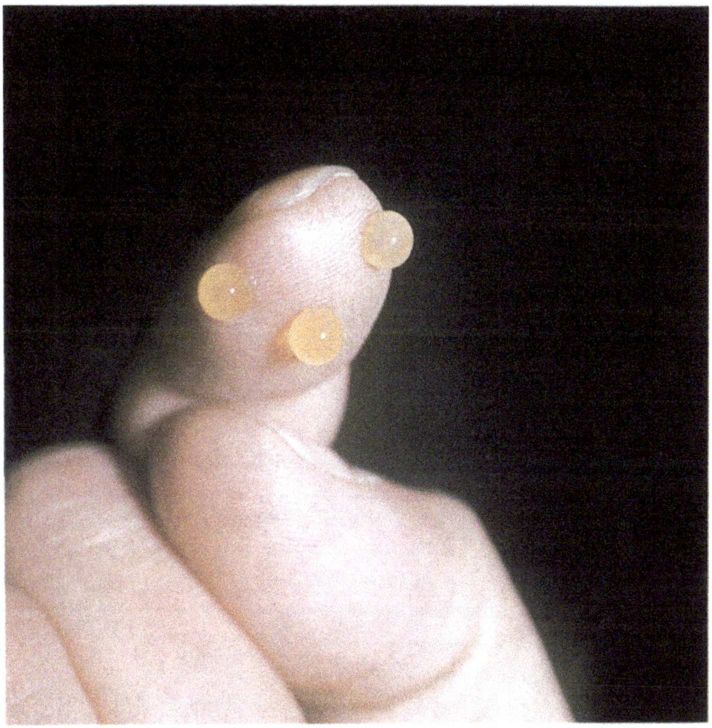

Plate 27. Young, sexually immature salmonines show typical juvenile coloration, characterized by the presence of parr marks (1). The photograph shows a young taimen (below), compared to a young Arctic grayling *(Thymallus thymallus arcticus)* from the Enisei river system in Mongolia (photo K. Hensel). The huchen retains its parr marks for a long time (2). The illustrated specimen had 350 mm in *Tl* (photo P. Áč).

Plate 28. One of the biggest huchen caught in recent years (1): total length 1360 mm, weight 26 kg, caught in the Austrian stretch of the Drava river on 28 Feb. 1965 (courtesy of the magazine Poľovníctvo a rybárstvo). A capital huchen caught in the Turiec river on Christmas 1977 (2). Total length 1280 mm, weight 23.8 kg (courtesy of P. Mamatej).

Plate 29. The differences in coloration between a male (1) and a female (2) huchen are particularly marked during spawning time. At the same time, the photographs demonstrate the manner of holding the huchen during hand-stripping (photo J. Dubeň).

Their coefficient of condition, calculated from Skalin's data, was only 0.62–0.84. All the above values are well below those for huchen growing in natural conditions or those fed with live food (see data in Chapter 10).

The feeding frequency should be such as to supply the huchen young with the maximum possible amount of food which they are able to consume. It is better to supply smaller amounts of food several times a day. Besides feeding by hand, automatic feeders can also be used. Jungwirth (l.c.) used an automatic device consisting of a meat mincing machine combined with a timer with a minute and a second relay. The automatic device was set so as to supply huchen young with food at ten-minute intervals during the light phase of the day, and huchen fingerlings at twenty-minute intervals. However, due to this high feeding frequency the amount of food used was about twenty times that actually consumed by the fish.

It is important to control water temperature during feeding. Our own as well as Austrian experience indicates that young huchen growth best at water temperatures above 10° C. According to Jungwirth (l.c.), the optimum water temperature is around 16° C, while at 6–7° C the growth drops to zero values.

In offering live food, good experience has been obtained with the method used in the State Fishery Farm at Kláštor pod Znievom. Previously prepared ponds with abundant zooplankton are stocked with a smaller number of huchen and a larger number of young rainbow trout, which are then provided with abundant supplementary food. It has been found that in such a culture the huchen act as "sanitary police", since they prefer weaker, stunted and diseased rainbow trout. Using cyprinid fishes for the same purpose one should expect about 350–400,000 young cyprinids per 10,000 huchen fingerlings 4–6 cm in initial total length for a period of 90 days, i.e. between early August and the end of October (Pavlík 1955).

LOSSES

Even if the utmost care is taken during rearing, losses among huchen young cannot be avoided. During incubation, even in spite of good egg fertilization, losses average about 5%. To minimize losses in this period it is important that the water temperatures in the incubation apparatus vary at most within the range 6–12° C (Jungwirth and Winkler 1984). Further mortality, varying between 5 and 10% on average should be expected to occur between hatching and the change to exogenous nutrition of the young. In this period, diseases and water quality are the decisive factors.

During the subsequent period, the losses depend primarily on the kind, amount, availability and quality of food supplied. At the same time, the danger of pests, cannibalism or escape of alevins and fingerlings through outlets should not be overlooked. Since data on the survival rates of huchen young under different conditions are scanty and differ considerably, we shall summarize them all here.

During the early history of huchen farming, losses were incomparably higher than at present. Thus an experiment carried out by Köttl at Kláštor pod Znievom in 1913 resulted in 74% losses between stocking huchen fingerlings in earthen ponds in May and fishing them out in October (Ivaška 1951). The kind of food offered to the huchen in that experiment is not known. Köttl estimated that 50,000 huchen alevins were required to obtain one thousand yearlings, that is, the losses amounted to about 98%. Ivaška's first experiments were fairly promising – with losses between stocking young in June and fishing out fingerlings in September of the same year or in May of the next year amounting to 55–57%. In those experiments the huchen young were not supplied with supplementary food but the experimental pools had abundant plankton and freshwater shrimps *(Gammarus* sp.). Ivaška's next experiments were not as successful, as the losses in all four pools varied between 92 and 97% from stocking in June to fishing in mid-November. The huchen were not fed and subsisted on natural food available in the pools. Ivaška believed that cannibalism was the major cause of the failure. He recommended the principle which, in essence, has been observed until present: the huchen should be divided by their size at least twice during the first year and, at the same time, they should be fished by the end of October at the latest.

On the State Fishery Farm at Kláštor pod Znievom, a pond 1.8 ha in area, supplied with spring water, with abundant summer zooplankton, freshwater shrimps and caddis fly larvae (Trichoptera), was stocked with 20,000 alevins (i.e. about 11,000 ind. ha^{-1}) in spring. The losses were only 20% by autumn, but two years later, with the same stock, the losses amounted to 97.5%. The farmer believed, however, that most of the loss was due to young having escaped through the outlet, which was later found to have been damaged. In 1978, in the same farm, alevins were stocked in five pools 120 m^2 in area at a rate of 10,000 per pool (i.e. around 833,000 ind. ha^{-1}), and another pool 144 m^2 in area was stocked with 15,000 individuals of the same fry (i.e. 1.04 million ind. ha^{-1}). The young were regularly supplied with food, especially with spleen. The pools were wintered before being stocked with huchen alevins. Their channels were limed before filling and the grass was mown and removed. In spite of the high stock density the losses were only 57%; in all, 28,000 huchen fingerlings were fished out at the end of September.

Further data come from the huchen farm of the Slovak Anglers' Union (SAU) at Považská Bystrica. In June 1978, three oblong pools 300 m^2 in area were stocked with a total of 75,000 young alevins with remains of yolk sacs. Thus the stocking rate was around 833,000 ind. ha^{-1}. Immediately after stocking, the alevins were fed with a mixture of minced spleen and liver, applied uniformly to the whole water surface of the pools and, at the same time, zooplankton and *Tubifex* worms were regularly imported into the pools. At the end of September, the stock was fished, yielding a total of 52,000 huchen fingerlings 8–9 cm in total length, of quite uniform size. Thus the total loss was only about 31%, in spite of the high stock density.

Rather high mortality rates were recorded on the huchen farm of the SAU at Dubná Skala in 1973–1977, where they reared huchen young from the hatchery of the local organization of the SAU in Martin. The average survival rate was only 29.7% ($s = 18.18$, $c.v. = 61.09$, $n = 13$), ranging from 7 to 60%. There the young were reared in earthen ponds of various sizes (26 × 6, 27 × 4, 35 × 20 m; circular pond 12 m in diameter) and fed with minced spleen and liver, young minnow and nase carp and, occasionally, ant pupae when alevins had attained a length of 40 mm. In all pools, however, ichthyophthiriasis and myxosomosis occurred permanently.

The direct influence of the kind of food on the extent of losses is illustrated by data of Jungwirth (1977), a diagram of which is shown in Fig. 34. It is evident that the least losses occurred in intensive tank cultures where the mortality rate among young fed with zooplankton, chironomid larvae and *Tubifex* worms was about 28% within five and a half months; only 18% within six and a half months when the young were fed with live cyprinid young; and 37% within the same time span when the young were fed with minced liver. A total failure resulted from feeding the young with granulated fodder, when all young died within three weeks after stocking. As already mentioned above, the kind of food supplied became also manifest in the body lengths attained. Unfortunately, the author does not give the stocking densities.

The same author studied the survival of huchen young reared under various environmental conditions. His data (Fig. 34) indicate that the least losses occur in waters from which other fishes had been removed.

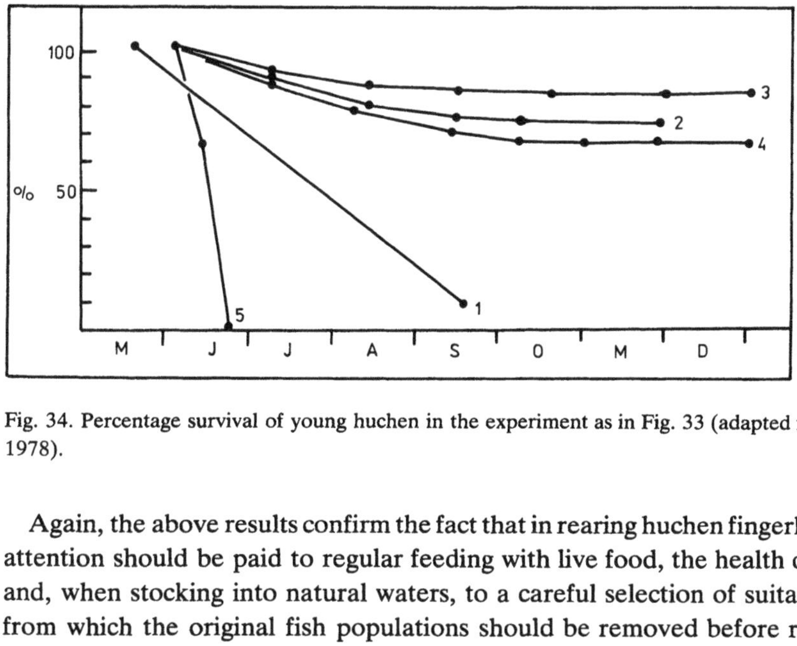

Fig. 34. Percentage survival of young huchen in the experiment as in Fig. 33 (adapted from Jungwirth 1978).

Again, the above results confirm the fact that in rearing huchen fingerlings, utmost attention should be paid to regular feeding with live food, the health of the young and, when stocking into natural waters, to a careful selection of suitable streams, from which the original fish populations should be removed before releasing the huchen young. It appears that for intensive cultures, stock density exerts almost no influence on the losses.

28. *Rearing Recruits*

SELECTION AND SORTING

Maximum possible care should be given to rearing recruits, as it is this group which forms the basis of the future stock of parent fish and is thus the basic condition of successful huchen farming. By carefully selecting from a sufficiently numerous stock, the breeder replaces natural selection to a considerable extent. In practice, one should concentrate particularly on the first and second age groups. Selecting from the first age group, one concentrates on the so-called "bolters", i.e. individuals which differ from the remaining stock in having fastest growth and which attain the greatest dimensions at the end of the growing season. Selection and sorting are carried out throughout the season and the selected individuals are kept separately. A similar procedure is maintained when selecting among individuals in the second age group, concentrating on the fastest growing fish, which are separated again, but here it is sufficient to select only once, at the end of the growing season. Naturally, one-year-old fish are kept separate from two-year-old ones to prevent cannibalism. Besides growth, note is also taken of the general appearance and body form of the fish. Individuals showing abnormalities are excluded from breeding. As to feeding, supplementary fodder (especially minced fish) may be used during the first and second years but live fish must be provided from the end of the second growing season onwards.

Individuals selected from the third age group may already be kept together with those of the fourth age group, since the risk of cannibalim is now lower. Provided sufficient space, however, separate rearing of the two groups is preferable until the first hand-stripping. The fish are given abundant food, often live fish, and further selections are made primarily for good appearance and condition. This is also valid for older age groups. Again, utmost care is paid to feeding with live fish in sufficient quantities. Insufficient or unsatisfactory food results in delayed sexual maturation, even though the fish apparently exhibit good growth. It has appeared that sexual maturation may be delayed up to 9 years of age, or that the fish may become sterile. In culture, male huchen usually attain sexual maturity and are capable of spawning in the fifth year of life, females one year later, but exceptionally not until the seventh year of life.

Recruits of age groups 5–7 are kept together, males and females, and when they have attained sexual maturity they are used for hand-stripping; thereafter they are included in the stock of parent fish. Those fish that do not attain sexual maturity by the seventh year of life are excluded from breeding even if they are healthy, and in good condition. On the first hand-stripping, particular attention is paid to the quality of sexual products obtained from the recruits. In order to be able to check it in subsequent year it is necessary to mark the individuals (particularly the females) with suitable tags.

GROWTH

Since growth and size or weight attained by the fish are still the most important criteria for selecting and gathering a cohort of recruits, particular attention is paid to these criteria. Although huchen have not yet been specially selected for rapid growth and therefore show considerable individual variation in this point, selection will gradually narrow the variation range and uniformity will follow. In our experience, the growth of huchen during the first five years of life is well illustrated by the data in Table 41.

TABLE 41

Growth of huchen in a pond culture

Age (years)	*Tl* (cm)	*w* (kg)
1	15—25	100—200
2	25—45	250—800
3	40—55	500—2000
4	60—65	1500—3000
5	60—70	2500—4500

FEEDING

While a mixture of minced fish, spleen, liver and fish meal with the addition of a small amount of granulated fish fodder may be fed to the huchen even during the second year of life; such food should be considered as a substitute and ideally only live fish should be provided. For this purpose, one can utilize any available, less valued or coarse fish coming from streams, reservoirs or ponds, but their health must be verified. Fish coming from drainage areas known for, or suspected of, the presence of diseases or particularly abnormal occurrence of parasites must not be used before they have been subjected to a disinfectant bath.

Being a predator, the huchen will accept practically any fish species. However, it prefers certain species, taking others reluctantly. The fishes to be fed to the huchen should be selected accordingly. The most suitable fish species are those occurring in the environment typical of the huchen, i.e. in submontane streams, such as minnow, gudgeon, spirlin, chub, dace, nase carp and barbel. Bleak and roach are also taken quite willingly, less so fish species from lowland streams, such as crucian carp, bream, tench and perch. In practice, however, it is the availability of live fish that decides, and thus the huchen are offered those fish species which they take less willingly but to which they get used after some time. Such fish are first screened by size and the smallest offered first, particularly when fish are offered for the first time. In a specialized huchen farm it is an advantage to provide for rearing forage fish, especially the nase carp. Rearing of this species in captivity has been improved and the method offers no difficulties.

The passing from one kind of food to another, particularly a temporary replacement of live fish by dead ones or by a substitute, may cause the huchen to refuse the food offered even for several days. This, however, need not cause worries, for a hungry huchen will sooner or later accept even such food. It is essential in such cases to offer food slowly and in small amounts, watching whether the huchen take the food when it hits the water surface. Even fish that died during transportation may be used as food, but care must be taken that they are fresh. If they cannot be fed at once they must be preserved by freezing. Later they are thawed and fed to the huchen.

The huchen are fed 2–3 times a day and food is offered as long as it is taken by them. The food must be offered so as to prevent the huchen from being injured by hitting the banks of the pool. In rainy weather, when the water becomes turbid, feeding may be interrupted, since in such situations the huchen – a predator relying on eyesight – decreases or interrupts food intake even in natural conditions.

LOSSES

Minimum and negligible losses occur during rearing recruits and later parent fish. The losses are mainly due to the exclusion of individuals unsuitable for further breeding or to rather serious injuries caused by careless handling during translocations, hand-stripping and feeding. However, the health of the huchen must receive due care, especially introduction of any infection must be prevented. The losses among the huchen in age groups 2 to 5 usually do not exceed 5–10%, being the highest in age group 2.

29. Sanitary Problems

Although sanitary problems have already been mentioned in the preceding chapters, we consider it necessary to return to them specifically; as a matter of fact, this is a serious and continuous process that forms an important link in the chain of measures conditioning the success or failure of huchen farming. Basically, the major problem is prevention of both introduction and outbreak of a fatal infection, and of careless handling. At the same time, successive developmental stages have their particular problems, too.

DURING INCUBATION AND HATCHING

Prevention consists in the regular treatment of eggs, i.e. in removing non-fertilized and dead eggs and all contaminants, and in applying regular disinfectant baths. The eggs are bathed in such baths every third day, beginning on the first day on which they were placed in the incubator. The purpose of these baths is mainly to prevent fungous infection. The most commonly used bath is a 40% solution of formaldehyde, diluted with water in the ratio 1:2000 to 1:5000. The action of the solution lasts 30 minutes. A solution of potassium permanganate at a concentration of 1:10,000 can also be used, the exposition time being 10 minutes. The baths must be left out in the hatching period, but only if the health condition of the hatching eggs is good and if no previous infection has occurred.

After having reached the "eyed egg" stage, the developing embryo is quite resistant to shocks. This fact can be utilized to remove non-fertilized eggs, particularly if they are to be expedited. On carefully pouring the eggs several times from one vessel to another, the non-fertilized eggs turn white and can be readily recognized and sorted out.

In any handling of the eggs, strict cleanliness should be watched meticulously. For handling and in the hatchery in general, only utensils for this purpose and reserved exclusively for the hatchery are used.

IN REARING ALEVINS AND FINGERLINGS

Here, too, maintaining utmost cleanliness in all rooms of the rearing plant and removing all contaminants from the apparatus, troughs and pools is the most important requirement. Disinfectant baths must be applied from time to time to the condensed stocks. These baths consist in a solution of malachite green in a concentration of 0.2 mg l^{-1}. The bath is applied approximately every two weeks, the exposure time being 24 hours in troughs and up to five days in ground ponds. This measure prevents an outbreak of ichthyophthiriasis. In case of an outbreak of this disease, or if an infectious disease of gills is detected, baths in a solution of copper sulphate at a 1:2000 concentration (5 g copper sulphate dissolved in 10 l water) for 60 seconds were found satisfactory.

Methylene blue is also used, either as a bath or applied with a tampon directly to the diseased spot. The first method is applied to eggs, similarly as with other salmonid species. The concentration, applied to control bacterial infections and ectoparasites, is 0.2 g l^{-1} water. The bath is applied for seven days at water temperatures below 10° C. A concentrated solution is used for local treatment. Besides, this preparation may also be added to the fodder in a concentration of 1 g per 1–2.5 kg fodder.

In rearing fingerlings in ground pools, moreover, one must remove predators, such as amphibians, water beetles and dragon-fly larvae, and deny access to fish-eating birds. The fodder supplied must be of high quality; substituting fodder must be fresh.

IN REARING RECRUITS AND PARENT FISH

Preventing outbreaks of diseases among recruits and parent fish consists mainly in offering sanitarily faultless food and in preventing injuries that may be caused in translocation and distribution of the fish in ponds, or during hand-stripping. One should bear in mind that such big fish as huchen require careful handling: their considerable weight and strength increase the probability of suffering an injury (of skin and especially eyes), which then provides access to diseases. Therefore, the fish must be carefully treated with an antifungal bath after each handling. In the case of huchen used for hand-stripping, the bath is applied as soon as the fish recover from anaesthesia. The best experience has been obtained with baths in a solution of malachite green at a concentration of 1:150,000 in a solution of potassium permanganate at a concentration of 1:10,000 or in a 1.5 per cent solution of common salt. All baths are applied for 5–10 minutes. When using common salt bath which, while being the least used, has been found the most satisfactory, the water is not aerated. The malachite green bath must not be used at higher water temperatures. It is of advantage if a special tank is provided for the purpose of prophylactic baths on a huchen farm.

According to a personal communication from P. Mamatej, even serious fungous infestation of parent fish can sometimes be coped with by repeated application of 40% formaldehyde in a cotton wad to the infested skin areas.

The water quality in pools containing recruits and parent fish can be unfavourably affected by remains of unused fodder, particularly if dead fish are offered. Very good results have been obtained by stocking the pools with burbot which act as a "sanitary police" and reliably remove both unused food remains and huchen excreta.

30. Dispatch and Transport

As in other salmonids, certain fundamental principles must be observed in transporting huchen of any size or age group. Transport free of losses is conditioned especially by the following factors: health of the fish, their nutritional status, filling of their digestive tracts, distance to which they are to be transported, size and kind of transporting containers, quality and purity of water, its temperature and oxygen content, the manner of its regeneration, season of the year, economic aspects, etc.

The huchen belongs to the fish species which require a high oxygen content of water. Thus the water temperature during the transport should also be kept low, with the values of oxygen saturation approaching 100%. Lower water temperatures slow down metabolism, and hence also the oxygen demand. The oxygen consumption of fishes requiring good aeration of water, relative to its temperature (in terms of mg O_2 per kg of fish) has been tentatively determined as follows: 130 mg at 5° C, 160 mg at

10° C, and 190 mg at 15° C. This is net oxygen consumption, not considering other factors. The water temperature of 15° C should be the upper limit at which one can still transport those fish which are not destined to be consumed but to be used for further breeding. The above data are tentative, as already mentioned. They may be affected especially by such factors as water purity, organic matter content, and size of the fish transported (the smaller their size the greater their oxygen consumption), etc.

Today, various kinds of containers are used to transport fish, from the smallest ones, 25–50 l in volume, up to the biggest ones, holding 2000 l (boxes or tanks), or specially adapted vehicles. In classical transport, wooden barrels 25 to 250 l in volume were used, in which the water was mechanically aerated by splashing. For the transport, clear water witout admixtures and of low temperature should be used; if aerators or oxygen are not used, the huchen should not be transported in well or spring water.

TRANSPORT OF EGGS

Eggs can be transported immediately after insemination, rinsing and swelling, using vacuum flasks or transporting trays. If they have to be transported to great distances it is of advantage to transport them only in the "eyed egg" stage. Non-fertilized eggs without ovarial fluid may be transported shorter distances in vacuum flasks; similarly also milt can be transported in vacuum flasks at a temperature slightly above freezing. In the "eyed egg" stage, the eggs are transported dry in standard polystyrene transporting boxes; they are placed in these boxes evenly distributed over trays with a bottom of netting, placed one over another. The top tray is provided with a layer of crushed ice, the bottom tray with a draining inlay of foam rubber, sawdust, moss, etc. A tray 25 × 25 or 30 × 30 cm holds about 5000 eggs. Care should be taken that the eggs be approximately equal in size. In this way the eggs can be transported for over 70 hours without suffering any losses.

TRANSPORT OF ALEVINS

Huchen alevins can only be transported after having absorbed most of the yolk sac contents and are capable of rising from the bottom and of swimming. The most satisfactory manner of transporting alevins is in polyethylene bags 75–100 kg in volume. Two such bags are used, one inserted in the other, to decrease the risk of damage. The bag is filled to 1/3 of its volume (i.e. to 20–30 l) with clean and cold water. Then about 5000 alevins (for transport lasting 6–10 hours) or up to 10,000 alevins (for less than 6 hours) are placed in the bag, air is pressed out and the space over the water is filled with oxygen at a pressure of about 0.02 MPa (= 0.2 atm). The end of the bag is folded and tied up thoroughly and placed in a suitable container to protect it from being damaged and, at the same time, to provide thermal insulation.

Care should be taken that the water temperature in the bag does not rise above 20° C during transport.

If need be, common barrels, boxes or tubs used for transporting stocking fish can be used to transport alevins. A vessel 150–200 l in volume can be used to transport at most 50,000 alevins; the vessel is filled with water to only 2/3 or at most 3/4 of its volume. In the past, good experience has also been obtained using vertical containers made of varnished galvanized iron sheets, oval in cross section, with a lid fitted with a valve serving to fill a smaller vessel inside the container with compressed air, which oozed into the water at the bottom.

TRANSPORT OF FINGERLINGS

Smaller amounts of fingerlings (100 ind. per 20 l water for a transport not longer than 5 hours) can be transported in the same way as the young (see above).

For transporting larger amounts of fingerlings, bigger vessels fitted with means of aeration or oxygenation of water must be used. The aeration apparatus must work reliably, and therefore must be thoroughly checked before the transport. Besides, the fingerlings must be transported in darkness to prevent possible cannibalism and, at the same time, limit the motion of the transported fish.

For transporting huchen summerlings and yearlings, the breeders in Czechoslovakia mainly use laminated fiberglass containers of two kinds. Smaller containers are 386 l in volume, slightly tapering towards the bottom and rounded at the edges. They are fitted with a double, perforated lid. The container is provided with a valve and a device permitting dispersion of the air or oxygen supplied to the water, and with an outlet at the bottom. For transport, the container is filled with water to only 2/3 of its volume, i.e. with 200–250 l clean, well aerated and cold water. One such container holds up to 1000 huchen fingerlings 10–15 cm long, or 10–15 kg, if transport takes place in early spring or in autumn. In summer or in warm weather the stock is decreased to about a half the amount.

Large transporting containers are similar in construction, but their volume is about 1300 l. They are filled with about 600–800 l water and can be used to transport up to 500 fingerlings of the above size, i.e. as much as 50 kg. When the transport is to last longer than 10 hours or at higher temperatures the stock is decreased to a half.

However, ordinary fish barrels 250 l in volume, with a close-fitting lid, have still been satisfactorily used for transporting huchen fingerlings. Again, they are filled with water to only 2/3 of their volume and can be used to transport 500–1000 fingerlings. They are mainly used for transport shorter distances (for less than 5 hours) by means of lorries; the water in the barrels is aerated by continuous movement during transport, and thus it is recommended not to stop during transport.

If a smaller amount of fingerlings is to be transported in cold weather and at low water temperature it is advisable to feed the fish before the transport to prevent cannibalism. Normally, however, the fish should not be fed for at least two days before the transport, and if possible they should be transported early in the morning. When the transport proceeds normally, the huchen are lively, dark in colour and mostly stay near the bottom. As soon as the fish are observed to rise to the water surface, turn paler in colour or gasp for breath, more air or oxygen or some fresh water must be let in.

TRANSPORT OF RECRUITS AND PARENT FISH

Huchen recruits are transported similarly as brown trout. Larger containers with an aeration device are used for this purpose. In 800 l water up to 100 kg fish can be placed.

Particular attention should be paid to transporting parent fish, which occurs only rarely, however. It must be remembered that the transporting container must have a sufficient entry port which allows putting in or taking out of the fish without any risk of injuring them. In principle, the parent fish must not be transported during heavy frosts or at high temperatures, the water must be thoroughly aerated and the transport should not last too long. Fish caught in the wild can be transported singly in polyethylene bags, filled with water so that only half of the fish's body is submerged, and the remaining volume is filled with oxygen.

Epilogue

It is not usual for a factual publication like this to be concluded by a special epilogue. We have decided, however, to make an exception to this rule. The reason is that we wish to take advantage of this opportunity to underline the fact that, regardless of the vast range inhabited, Eurasian huchen is among species acutely endangered with extinction. In this connection we must re-affirm the statement of Kasecker (1970) that no other fish species has been decimated by man to such extent as the huchen. The situation of the nominate form, the huchen, is especially critical. Its present range is broken into short sections of individual streams and it is suffering from ever increasing civilization pressure. The fact that the huchen has not disappeared altogether should be attributed to its admirable ecological adaptability rather than to deliberate endeavours to save it, although the latter should in no way be underestimated.

It is a well-known fact that in modern concepts of nature conservation the protection of a species is inseparable from that of its environment. This postulate is doubly true as regards the protection of aquatic fauna: the hydrosphere has suffered from man's interference much more than the pedosphere or the atmosphere. This is particularly true of inland waters, which are affected not only by direct interference but, in one way or another, by interference with the surrounding landscape or, not infrequently, even with distant regions. In the first place, this relates to water pollution. Rivers and lakes are affected by pollution both directly (by industrial waste waters from factories and municipal sewage from towns on their shores and by acid rains) and indirectly (by polluted surface waters, seepage of contaminated ground water, and wash-off of chemicals from surrounding arable land and woodland). Thus the problem of protection of aquatic organisms is primarily that of protecting their environment. This postulate is of cardinal importance for the protection of the huchen, a species inhabiting a sharply defined and rather narrow range of aquatic habitats. Moreover, the huchen, like all species attaining large body dimensions within a limited time is particularly vulnerable.

In our opinion, there is no need to discuss whether or not to protect the huchen. Leaving aside the obvious duty resulting from the necessity of conserving natural resources which, nota bene, has been codified in every developed state, the following concrete arguments support the saving of the huchen:

(1) It is the biggest salmonid species and, at the same time, one of the biggest freshwater fishes at all;

(2) In growth rate it ranks among the fastest growing freshwater species;

(3) In Europe it is the only specialized piscivorous predatory fish of the submontane zone;

(4) It is a species of considerable ecological tolerance, including a fairly high degree (compared to other salmonids) of resistance to water pollution;

(5) It is a potentially valuable object of genetical investigations and breeding experiments;

(6) It is a phylogenetically very ancient relict form, close to the ancestral form of salmonid fishes;

(7) It is a highly prized game fish.

Clearly, the huchen is an important species in both theoretical and practical respects, and the inevitability of conserving it is dictated not only by scientific, historical and cultural, but also economic points of view.

The following measures for conservation and improvement of the species can be recommended on the basis of present knowledge and experience:

(a) The protection of streams populated by the huchen should be increased by consistent improvement of their purity, and by preserving the continuity of sections populated by the huchen;

(b) Catchment areas of streams populated by the huchen should be protected against excessive application of pesticides and inorganic fertilizers, particularly in areas comprising the spawning grounds of the species;

(c) In suitable streams or stream sections least affected by landscape cultivation, huchen reserves should be established, which, at the same time, could serve as model objects for investigations of the ecology and rational exploitation of the huchen;

(d) Huchen fishing should be regulated by legislation and on the basis of knowledge of the bionomics of the species;

(e) Specialized, closed-cycle huchen farms should be established, aimed at producing advanced stocking fish. Within one country there should be several rather than one big farm, to prevent disastrous consequences of any breakdowns;

(f) Acclimatization and re-acclimatization experiments should be conducted on a scientific basis, their results should be systematically followed up and evaluated.

In endeavours to conserve the huchen, the primary prerequisite and a *conditio sine qua non* is a thorough knowledge of the species. As emphasized in many places in this book, knowledge of the huchen is still fragmentary and incomplete. It is a sad reality that our current knowledge has been obtained by unsystematic and often only occasional studies or observations by various authors. Data on the nominate form date mostly from the first half of the 20th century and are often contradictory and not comparable. Subsequent additions have been scant, even in spite of the fact that the need to protect and conserve the huchen has become increasingly urgent.

However, research alone would be impotent, should it not meet with the understanding and support of both practical fisheries and authorities of the state administration. Conservation of the huchen and taimen is a problem for the whole society and its soluting requires a joint and coordinated effort of all parties. Naturally, such attitudes should be applied not only within one country, but also on an international basis, as the conservation of this species concerns several countries, mainly those along the Danube. In conclusion, therefore, we appeal to all responsible authorities to change their attitude towards the problem, to put away narrow, often local and sometimes even personal and temporary interests, and to join in a common endeavour to save the huchen, so that not only the present but also future generations may rejoice and benefit from this unique and splendid jewel of nature.

References

ADDLASSNIG-JASER, H., 1959: Ein Riesen-Huchen. Wochenschau **9:** 32.

AMBROS, C., 1959: Zvieracie zvyšky z doby bronzovej z Gánoviec, okr. Poprad. Slov. Archeol. **7:** 47–70.

ANBINDER, E. M., GLUBOKOVSKII, K. M. and POKOZII, N. V., 1982: Kariotip sakhalinskogo taimenya. Biol. Morya (2): 59–60.

ANDRAŠČÍK, M., 1980: Štatistické metódy v ichtyológii. M. Sc. Thesis. Comenius University, Bratislava.

ANDRESKA, J., 1974: Rybářství slovenských hor. Rybářství **8:** 190.

ANONYMUS, 1897: Große Huchen. Allg. Fisch. Ztg. **22** (N.F. 12): 87.

ANONYMUS, 1905: Versuche zur Einbürgerung des Huchens in der Themse. Allg. Fisch. Ztg. **30** (N.F. 20): 468.

ANONYMUS, 1910: Ein schwerer Huch. Oesterr. Fisch. Ztg. **7:** 13.

ANONYMUS, 1928: Ein guter Huchenfang. Allg. Fisch. Ztg. **53** (N.F. 43): 140.

ANONYMUS, 1933: Zwei schwere Huchen. Oesterr. Fisch. Ztg. **30:** 29.

ANONYMUS, 1969: Recent introductions of fish, shrimps and oysters. FAO Fish Cult. Bull. **2:** 15.

ANONYMUS, 1973: Sulec v Španiji. Ribič **32:** 255.

ANONYMUS, 1976a: Rekordni sulec iz Drave. Ribič **35:** 97.

ANONYMUS, 1976b: Information. Blinker **3:** 9.

ANONYMUS, 1986: Lov na hlavatky. Mladá Fronta **42** (10): 1.

ANTIPA, G., 1909: Fauna ihtiologiă a Romăniei. Acad. Rom., Publ. Fond. Adamachi, Bucureşti.

BAJKOV, A., 1924: Původ lososů rodu *Hucho.* Příroda **17:** 217–220.

BAKO, J., 1930: Hlavátka. Rybářský Věst. **10:** 83–84.

BAKO, J., 1946: Rybnatosť slovenských vôd po vojne. Čs. Rybář **1:** 182–184.

BALON, E. K., 1956: Vývoj hlavátky (*Hucho hucho* L.) počas endogénneho spôsobu výživy po vyliahnutí. Poľnohospodárstvo **3:** 433–455.

BALON, E. K., 1967: Vývoj ichtyofauny Dunaja, jej súčasný stav a pokus o prognózu ďalších zmien po výstavbe vodných diel. Biol. Práce **13:** 1–121.

BALON, E. K., 1968: Opis dokladového exemplára hlavátky *Hucho hucho* Linnaeus, 1758 v nížinnej zóne československého úseku Dunaja. Acta Rer. Natur. Mus. Nat. Slov. **14:** 91–94.

BALON, E. K., 1975a: Reproductive guilds of fishes: A proposal and definition. J. Fish. Res. Board Can. **32:** 821–864.

BALON, E. K., 1975b: Terminology of intervals in fish development. J. Fish. Res. Board Can. **32:** 1663–1670.

BALON, E. K. (Ed.), 1980a: Charrs. Dr. W. Junk, The Hague.

BALON, E. K., 1980b: Comparative ontogeny of charrs. *In:* Charrs. Ed. E. K. Balon. Dr. W. Junk, The Hague, pp. 563–606.

BALON, E. K., 1985: The theory of saltatory ontogeny and life history models revised. *In:* Early Life Histories of Fishes: New Developmental, Ecological and Evolutionary Perspectives. Ed. E. K. Balon. Developments in Environmental Biology of Fishes **5:** 13–30.

BĂNĂRESCU, P., 1960: Einige Fragen zur Herkunft und Verbreitung der Süßwasserfischfauna der europäisch-mediterranen Unterregion. Arch. Hydrobiol. **57:** 16–134.

BĂNĂRESCU, P., 1964: Fauna Republicii Populare Romine. Pisces – Osteichthyes. Fauna Rep. Soc. Romania 13. Editura Academiei Republicii Populare Romine, Bucureşti.

BARDOUN, A., 1946: Přednáška a film o chovu hlavatek. Čs. Rybář **1:** 47.

BASKAKOV, N. A. (Ed.), 1967: Russko-karakalpakskii slovar'. Izd. Sov. entsyklopediya, Moskva.

BASTL, I.: 1958: Hospodársky význam hlavátky. Čs. Rybářství **10:** 113–114.

BASTL, I. and HOLČÍK, J., 1968: K otázke plodnosti a potravy hlavátky v Oravskej priehrade. Poľov. a Rybár. **20:** 18.

BASTL, I., HOLČÍK, J. and KIRKA, A., 1975: Ichtyologický výskum Karpatského oblúka. 6. Ichtyofauna chráneného náleziska hlavátky na rieke Turiec. Acta Rer. Natur. Mus. Nat. Slov. **21:** 191–224.

BASTL, I., HOLČÍK, J. and KIRKA, A., 1976: Niekoľko poznámok k príčinám úbytku hlavátky v Turci a k problémom jej ochrany všeobecne. *In:* Hlavátka podunajská. Compiled papers. Príroda, Bratislava, pp. 65–74.

BASTL, I. and KIRKA, A., 1959: Príspevok k biometrike, výskytu a rastu mladi hlavátok (*Hucho hucho* Linné 1758) z tečúcich vôd a umelého chovu v prvom roku života. Biol. Práce **5** (4): 41–109.

BEHNKE, R. J., 1968: A new subgenus and species of trout, *Salmo (Platysalmo) platycephalus* from south-central Turkey with comments on the classification of the subfamily Salmoninae. Mitt. Hamburg. Zool. Mus. Inst. **66:** 1–15.

BEHNKE, R. J., 1980: A systematic review of the genus *Salvelinus. In:* Charrs. Ed. E. K. Balon. Dr. W. Junk, The Hague, pp. 441–481.

BENAZZI, M., 1973: Cytotaxonomy and evolution: General remarks. *In:* Cytotaxonomy and Vertebrate Evolution. Eds. A. B. Chiarelli and E. Capanna. Academic Press, London–New York, pp. 3–16.

BEREZOVSKII, A. I., 1924: Ikhtiofauna ozer Minusinskogo i Achinskogo okrugov Eniseiskoi gubernii. Trudy Sib. Ikhtiol. Lab. **2:** 13–75.

BERG, L. S., 1909: Ryby basseina Amura. Zap. Akad. Nauk (8), Fiz.-mat. Otd. **24:** 1–270.

BERG, L. S., 1940: Sistema ryboobraznykh i ryb, nyne zhivushchikh i iskopaemykh. Trudy Zool. Inst. Akad. Nauk SSSR **5:** 87–517 (2nd ed. 1955 – ibid. **20:** 3–286).

BERG, L. S., 1948: O proiskhozhdenii forelei i drugikh presnovodnykh lososevykh. *In:* Pamyati akademika S. A. Zernova. Izd. Akad. Nauk SSSR, Moskva–Leningrad, pp. 159–172.

BERG, L. S., 1948–1949: Ryby presnykh vod SSSR i sopredeľnykh stran. Vol. 1, 3. Izd. Akad. Nauk SSSR, Moskva–Leningrad.

BERG, L. S., 1955: Sistema ryboobraznykh i ryb, nyne zhivushchikh i iskopaemykh. 2nd ed. Trudy Zool. Inst. AN SSSR **20:** 1–286.

BERG, L. S., BOGDANOV, A. S., KOZHIN, N. I. and RASS, T. S. (Eds.), 1949: Promyslovye ryby SSSR. Opisaniya ryb. Pishchepromizdat, Moskva.

BERGERON, J. F. and BROUSSEAU, J., 1981: Guide des poissons d'eau douce du Québec. Gouvernement du Québec, Ministère du Loisir de la Chasse et de la Pêche, Direction Générale de la Faune, Québec.

BLAHÁK, P., 1972: Další doklad výskytu hlavatky podunajské (*Hucho hucho* Linné 1758) v československém úseku Dunaje. Acta Rer. Natur. Mus. Nat. Slov. **18:** 75–78.

BLANC, M., BANARESCU, P., GAUDET, J.-L. and HUREAU, J.–C., 1971: European Inland Water Fish. A Multilingual Catalogue. Fishing News (Books) Ltd., London.

BLOCH, M., 1782: Ökonomische Naturgeschichte der Fische Deutschlands. III. Berlin.

BOGYI, J., 1962: O hlavátke v slovenských vodách. Čs. Rybářství **17:** 150–151.

BOHL, M., 1977: Erhalt umweltbedrohter Wasserorganismen. Ein Beitrag zur Aufzucht und Haltung von Huchen. Fischer u. Teichwirt **28:** 112–114.

BOHL, M., 1979: Untersuchungen zur Bestandserhaltung umweltbedrohter Huchen (*Hucho hucho* L.). Fischer u. Teichwirt **30:** 122–125.

BOHL, M., 1982: Zu Formen der Huchenschwimmblase. Fischer u. Teichwirt **33:** 362–363.

BORISOV, P.G., 1928: Ryby reki Leny. Trudy Yakut. Kom. Akad. Nauk SSSR **9:** 1–181.

BORNE, M., 1942: Die Anglerfischerei. 8th ed. Paul Parey, Berlin.

BRGLEZ, J., 1966: Metljaji v prebavilih sulca iz Ljubljanici. Ribič **25:** 114–116.

BRODBECK, W., 1953: Woher auch "Donaulachse" im Oberrhein? Schweiz. Ztg. Sportfischer **4/5:** 152–156.

BUDAGYAN, F. E. (Ed.), 1961: Tablitsy khimicheskogo sostava i pitatel'noi tsennosti pishchevykh produktov. Medgiz, Moskva.

BUDAJ, O., 1976: Hospodárske aspekty a perspektíva chovu hlavátky podunajskej na Slovensku. *In:* Hlavátka podunajská. Compiled papers. Príroda, Bratislava, pp. 22–31.

BUKIREV, A. I., 1967: Kamskii losos – *Hucho taimen* (Pallas). Izv. GosNIORKH **62:** 39–56.

BUSHUEV, V. P., 1983: Biologiya taimenya *Hucho perryi* (Brevoort) iz reki Kievka (Yuzhnoe Primor'e). *In:* Ekologiya i sistematika presnovodnykh organizmov Dal'nego Vostoka. Sbornik nauchnykh trudov: 61–72.

BUŞNIŢA, TH. and ALEXANDRESCU, I., 1963: Atlasul peştilor din apele R.P.R. Editura Ştiinţifică, Bucureşti.

BUTKOVSKÝ, S., 1930: Zkušenosti z umělého chovu hlavatek. Rybařský Věst. **10:** 103–105.

BYKHOVSKII, B. E. (Ed.), 1962: Opredelitel' parazitov presnovodnykh ryb SSSR. Izd. Akad. Nauk SSSR, Moskva–Leningrad.

CACUTT, L., 1979: British Freshwater Fishes. Croom Helm, London.

CARLANDER, K. D., 1970: Handbook of Freshwater Fishery Biology. 3rd ed. The Iowa State Univ. Press, Ames.

CAVENDER, T. M., 1980: Systematics of *Salvelinus* from the North Pacific Basin. *In:* Charrs. Ed. E. K. Balon. Dr. W. Junk, The Hague, pp. 295–322.

CAVENDER, T. M. and MILLER, R. R., 1972: *Smilodonichthys rastrosus,* a new pliocene salmonid fish from western United States. Bull. Mus. Natur. Hist. Univ. Oregon **18:** 1–44.

CHIDRAVADIVELU, K., 1972: Growth of *Hucho taimen* (Pallas, 1773) in the upper Enisei river of Mongolia. Věst. Čs. Spol. Zool. **36:** 172–178.

CLEMENS, W. A. and WILBY, G. W., 1961: Fishes of the Pacific Coast of Canada. Fish. Res. Board Can. Bull. **68:** 1–443.

COLLETTE, B. B., 1977: Epidermal breeding tubercles and bony contact organs in fishes. Symp. Zool. Soc. London **39:** 225–268.

CORREDERA-MARTIN, J. M., 1981: Los salmones del coto de Villagonzalo. Gac. provincial, 9 July 1981: 11.

CROSSMAN, E. J., 1968: Changes in the Canadian freshwater fish fauna. Symposium on Introduction of Exotic Species, Dept. of Lands and Forests, pp. 1–20.

CZECZUGA, B., 1975: Carotenoids in fish. IV. Salmonidae and Thymallidae from Polish Waters. Hydrobiologia **46:** 223–239.

D., 1970: Zum Huchenvorkommen in Oberbayern. Allg. Fisch. Ztg. **95:** 791.

DOGEL', V. D., 1954. Oligomerizatsiya gomologichnykh organov. Izd. Leningradskogo universiteta, Leningrad.

DOROFEEVA, E. A., 1977: Izpol'zovanie dannykh kariologii dlya resheniya voprosov sistematiki i filogenii lososevykh ryb. *In:* Osnovy klassifikatsii i filogenii lososevidnykh ryb. Sbor. Nauch. Trudov Zool. Inst. Akad. Nauk SSSR, Leningrad, pp. 86–95.

DOROFEEVA, E. A., ZINOVEV, E. A., KLYUKANOV, V. A., RESHETNIKOV, YU. S., SAVVAITOVA, K. A. and SHAPOSHNIKOVA, G. Kh., 1980: Sovremennoe sostoyanie issledovanii filogenii i klassifikatsii lososevidnykh ryb. Vopr. Ikhtiol. **20:** 771–791.

DRENSKI, P., 1948: Synopsis and distribution of fishes in Bulgaria. Ann. Univ. Sofia **44:** 1–71.

DRYAGIN, P. A., 1933: Rybnye resursy Yakutii. *In:* Trudy Soveta po izucheniyu proizvoditel'nykh sil. Izd. Akad. Nauk SSSR, Leningrad, pp. 1–94.

DUDICH, E., 1958: Die Grundlagen der Fauna eines Karpaten-Flusses. Acta Zool. Acad. Sci. Hung. **3:** 179–200.

DULMAA, A., 1973: Zur Fischfauna der Mongolei. Mitt. Zool. Mus. Berlin **49:** 49–67.

DULMAA, A., 1979: Hydrobiological outline of the Mongolian lakes. Int. Rev. Ges. Hydrobiol. **64:** 709–736.

DYK, V., 1937a: Těžké hlavatky. Rybářský Věst. **17:** 179.

DYK, V., 1937b: Záchrana hlavatky na Podkarpatské Rusi. Krása našeho domova **29:** 53–55.

DYK, V., 1938: Malá náročnost plůdku hlavatek při rybničním chovu. Čs. Rybář **18:** 159–160.

DYK, V., 1952: Současný výskyt ryb v řece Moravici. Přírodověd. Sbor. Ostrav. Kraje (Suppl.) **13:** 1–24.

DYK, V., 1956: Naše ryby. 4th ed. SZN, Praha.

DYK, Z., 1926a: Hlavatka (*Hucho hucho* Linné) a její existenční poměry ve vodách naší republiky. Příroda **19:** 245–252.

DYK, Z., 1926b: Hospodaření na vodách Podkarpatské Rusi a jeho úprava u příležitosti vydání nového rybářského zákona. Čs. Rybář **6:** 93–96, 102–105.

EGOROV, A. G and MEIER, A. R., 1958: Lov ryby spinningovoi snast'yu na reke Angare. Irkutskoe knizhnoe izd., Irkutsk.

EHRLER, E. F., 1935: 50 Jahre Jagd- und Fischereischutzverein. Český Těšín.

ERGENS, R., GUSSEV, V. A., IZYUMOVA, N. A., and MOLNÁR, K., 1975: Parasite fauna of fishes of the Tisa river basin. Rozpr. ČSAV, Řada MPV **85:** 3–117.

FERIANC, O., 1948: Slovenské názvoslovie rýb Československej republiky a susediacich krajov. Spisy Prírodoved. Odb. MS **3:** 3–113.

FLASAR, I. and FLASAROVÁ, M., 1981: O rybách řeky Ohře. Monogr. Studie Kraj. Muz. v Teplicích 21. Krajské muzeum Teplice – Povodí Ohře Chomutov, Teplice.

FORTUNATOVA, A. F. and POPOVA, O. A., 1973: Pitanie i pishchevye vzaimootnosheniya khishchnykh ryb v delte Volgi. Izd. Nauka, Moskva.

FRANK, R., 1971: Date biometrice asupra lostriţei (*Hucho hucho* L.) din apele Maramureşului. Bul. Ştiinţ., Ser. B **3:** 91–95.

FRANK, R., 1972: Lostriţa (*Hucho hucho* L.) in apele Maramureşului. Ocrotirea Natur. **16:** 13–20.

GOVE, P. B. (Ed.), 1966: Webster's Third New International Dictionary of the English Language. C. and C. Marriam Co., Springfield.

GOZMÁNY, L., 1979: Septemlingual Dictionary of the Names of European Animals. Akadémiai Kiadó, Budapest.

GREENWOOD, P. H., ROSEN, D. E., WEITZMAN, S. H. and MYERS, G. S., 1966: Phyletic studies of teleostean fishes with a provisional classification of living forms. Bull. Amer. Mus. Natur. Hist. **131:** 339–456.

GREENWOOD, P. H., MYERS, G. S., ROSEN, D. E. and WEITZMAN, S. H., 1967: Names main divisions of teleostean fishes. Proc. Biol. Soc. Washington **80:** 227–228.

GRITSENKO, O. F., 1975: Sistematika i proiskhozdenie sakhalinskikh goľtsov roda *Salvelinus*. Trudy VNIRO **106:** 141–160.

GRITSENKO, O. F. and CHURIKOV, A. A., 1977: Issledovanie ekologii taimenya *Hucho perryi* (Brevoort) severnogo Sakhalina. Nauch. Otchet po Teme No. 2. VNIRO, Moskva (mimeographed).

GRITSENKO, O. F., MALKIN, E. M. and CHURIKOV, A. A., 1974: Sakhalinskii taimen' *Hucho perryi* (Brevoort) reki Bogatoi (vostochnoe poberezh'e Sakhalina). Izv. TINRO **93:** 91–100.

GRITSENKO, O. F., 1978: Diskussionnye voprosy sistematiki taimenei roda *Hucho* i vozmozhnye puti ikh resheniya. MS.

GRMAN, L., 1980: V chove hlavátky môžeme dokázať viac. Poľov. a Rybár. **32:** 24–25.

GÜNTHER, A., 1866: Catalogue of the Fishes in the British Museum. 6. Trustees. London.

H. M., 1962: Aklimatizacija głowacicy w Maroku. Gospod. Ryb. **14:** 15.

HAEMPEL, O., 1910: Über das Wachstum des Huchens (*Salmo hucho* L.). Ein Beitrag zur Alterbestimmung der Teleostier. Int. Rev. Ges. Hydrobiol. Hydrogr. **3:** 136–155.

HANKÓ, B., 1932: Ursprung und Verbreitung der Fischfauna Ungarns. Arch. Hydrobiol., **23:** 520–556.

HANNY, Š., 1965: Zriedkavý úlovok. Poľov. a Rybár. **17:** 20.

HARSÁNYI, A., 1982: Der Huchen. Paul Parey, Hamburg–Berlin.

HARTLIEB, R., 1948: Der Huchenfischer. Hubertusverlag, Wien.

HARTLIEB, R., 1961: Gute Huchenstandplätze (Huchenstände). Allg. Fisch. Ztg. **86:** 646.

HARTLIEB, R., 1969: Über Huchenköder. Allg. Fisch. Ztg. **94:** 74.

HARTLIEB, R., 1970: Über Huchenköder. Allg. Fisch. Ztg. **95:** 12–13.

HAVELKA, J. and IVAŠKA, S., 1954: Zvláštní případ onemocnění hlavatek furunkulosou. Sbor. ČSAZV, Řada B **27:** 307–314.

HAVELKA, J. and VOSTRADOVSKÝ, J., 1974: Výzkum podmínek rybářského obhospodařování revírů pod

Vltavskou kaskádou. Research report. Research Institute of Fisheries and Hydrobiology, Prague.
HECKEL, J. and KNER, R., 1858: Die Süßwasserfische der Östreichischen Monarchie. W. Engelmann, Leipzig.
HEINTZ, K., 1920: Der Angelsport im Süßwasser. München.
HENSEL, K., 1976: O hlavátkach (rod *Hucho* Günther, 1866). *In:* Hlavátka podunajská. Compiled papers. Príroda, Bratislava, pp. 32–39.
HENSEL, K., 1980: On the geographical distribution of the huchen – *Hucho hucho* (Linnaeus, 1758). MS.
HENSEL, K. and HOLČÍK, J., 1983: On the identity of *Hucho hucho* and *Hucho taimen* (Pisces: Salmonidae). Folia Zool., **32:** 67–83.
HENSEL, K. and PIVNIČKA, K., 1980: Notes to the morphology and taxonomy of *Hucho taimen* (Pallas, 1773) from Mongolia. MS.
HERMAN, O., 1887: A Magyar halászat könyve. 2. Budapest.
HIGASHI, H., 1961: Chapter 13. Vitamins in fish – with special reference to edible parts. *In:* Fish as Food. 1. Ed. G. Borgstrom. Academic Press, New York–London, pp. 411–486.
HIRTZ, M., 1956: Rječnik narodnich zooloških naziva. Kn. 3. Ribe (Pisces). Zagreb.
HOAR, W. S., 1958: The evolution of migratory behavior among juvenile salmon of the genus *Oncorhynchus.* J. Fish. Res. Board. Can. **15:** 391–428.
HOLČÍK, J., 1966a: Vývoj a formovanie ichtyofauny v Oravskej priehrade. Biol. Práce **12:** 5–75.
HOLČÍK, J., 1966b: Ichtyologický výskum Karpatského oblúka. 4. Ichtyofauna rieky Hornád so zreteľom na vybudovanie vodného diela Ružín. Biol. Práce **12:** 76–116.
HOLČÍK, J., 1970: The Klíčava reservoir. Biol. Práce **15:** 5–86.
HOLČÍK, J., 1977a: Changes in fish community of Klíčava reservoir with particular reference to Eurasian perch (*Perca fluviatilis*) 1957–72. J. Fish. Res. Board Can. **34:** 1734–1747.
HOLČÍK, J., 1977b: Z histórie o chove hlavátky. Poľov. a Rybár. **12:** 28–29.
HOLČÍK, J., 1978: Ryby ako limitujúci faktor biomasy zooplanktónu, so zreteľom na vodárenské údolné nádrže. Vertebr. Zpr. pp. 33–38.
HOLČÍK, J., 1982: Towards the characteristics of the genera *Hucho* Günther, 1866 and *Brachymystax* Günther, 1866 (Pisces: Salmonidae). Folia Zool. **31:** 369–380.
HOLČÍK, J., MIŠÍK, V., BASTL, I. and KIRKA, A., 1965: Ichtyologický výskum Karpatského oblúka. 3. Ichtyofauna Oravskej priehrady a jej prítokov. Acta Rer. Natur. Mus. Nat. Slov. **11:** 93–139.
HOLČÍK, J., HENSEL, K. and SKÁCEL, L., 1977: Hlavátka (*Hucho hucho* Linnaeus, 1758). Súbor poznatkov o jej biológii, využití a chove. Research report. Laboratory of Fishery Research and Hydrobiology, Bratislava.
HOLČÍK, J. and LEPIKSAAR, J., 1980: On the find of *Salmo salar* remains in a natural well of the Early Bronze Age at Gánovce (Czechoslovakia) with regard to the occurrence of this species in the upper Vistula basin. Věst. Čs. Spol. Zool. **44:** 94–100.
HOLLÝ, B., 1934a: Biológia a rozšírenie hlavátky na Slovensku a v Podkarpatskej Rusi. Rybář **1:** 64–67.
HOLLÝ, B., 1934b: Na záchranu hlavátky. Rybářský Věst. **14:** 149–151.
HOLLÝ, B., 1935a: Zpráva o výsledku vyšetrovania výskytu hlavátok – *Salmo hucho* L. – v rieke Hrone, jej prirodzených trlísk v Hrone a jej prítokoch a o vyšetrení vhodného miesta pre postavenie liahne pre umelé liahnutie ikier a vybudovanie rybníčkov pre chov mlade. Research report. State Research Station of Fisheries and Hydrobiology, Liptovský Sv. Mikuláš.
HOLLÝ, B., 1935b: Môže sa založiť umelý chov hlavátok v oblasti Hrona? Rybář **2:** 184–188.
HOMEI, V., 1956: Lostrița (*Hucho hucho* L.) in apele tarii noastra. Ocrotirea Natur. **2:** 101–109.
HSIEH, CHAI-YÜ, SHAN-WU HUANG and YUN-YÜ YÜAN, 1959: [Lenok and taimen, their natural hybrids in the Amur river basin.] Acta Hydrobiol. Sinica **2:** 215–220. (In Chinese with Russian summary.)
H-Ý., 1934: Hlavátka a spôsoby jej športového lovenia na Slovensku. Rybář **1:** 78–79, 118–120, 144–147.
IANCU, S. and DECEI, P., 1972: Valorificarea piscicole a principalelor lacuri de acumulare din zone carpațica a Romaniei (lacurile Bicaz si Vidraru – Arges). Hidrobiologia **13:** 199–207.
ISACHENKO, V. I., 1912: Ryby Turkhanskogo kraya, vstrechayushchiesya v r. Enisee i Eniseiskom zalive. Mater. Issl. r. Eniseya v Ryboprom. Otnosh. **6:** 1–111.
IVAŠKA, S., 1935: Pokus o umelé rozmnožovanie hlavátok. Rybář **2:** 83–85.

IVAŠKA, S., 1937: Umelý chov hlavátok a ďalšie moje skúsenosti. Rybářský Věst. **17:** 111–112.

IVAŠKA, S., 1946: Naše ťažkosti s hlavátkami. Čs. Rybář **1:** 163–164.

IVAŠKA, S., 1951: Hlavátka, jej lov a umelý chov. Tatran, Bratislava.

IVAŠKA, S., 1953: Hlavátka v slovenských vodách. Sbor. ČSAZV, Řada B **10:** 57–62.

IVAŠKA, S., 1959a: Pokusné osadzovanie hlavátky v menších vodách. Čs. Rybářství **14:** 83–84.

IVAŠKA, S., 1959b: Niekoľko poznatkov z výskytu hlavátok na Orave. Poľov. a Rybár. **11:** 22.

J. E., 1975: Rekordné úlovky. Poľov. a Rybár. **27:** 38.

JAGODITSCH, F., 1930: Der Huchen als Hochzeiter. Oesterr. Fisch. Ztg. **27:** 4–7, 15–16, 23–25, 45–47.

JANZA, L., 1958: O záchranu hlavátky na Slovensku. Ochr. Prír. **13:** 241–242.

JORDAN, D. S. and MCGREGOR. E. A., 1925: Record of fishes obtained by David Starr Jordan in Japan. Mem. Carnegie Mus. **10:** 3–145.

JORDAN, D. S. and SNYDER, I. O., 1902: A review of the salmonid fishes of Japan. Proc. U. S. Natur. Mus. **24:** 567–593.

JUNGWIRTH, M., 1977a: Huchen Hochzeit. Blinker **9:** 78–81.

JUNGWIRTH, M., 1977b: Der Huchen und seine Zucht. Österr. Fisch. **30:** 125–134.

JUNGWIRTH, M., 1978: The problem of farming and conservation of the Danube salmon (*Hucho hucho* L.). Envir. Biol. Fish. **3:** 231–234.

JUNGWIRTH, M., 1979: Ovulation inducement in prespawning adult Danube salmon (*Hucho hucho* L.) by injection of acetone-dried carp pituary (CP). Aquaculture **17:** 129–235.

JUNGWIRTH, M., 1980: Der Huchen – Derzeitiger Stand und Zukunftschanzen einer gefährdeten Fischart. Festschr. "Österreichische Fischereigesellschaft 1880–1980 – 100 Jahre Hege und Pflege", pp. 105–114.

JUNGWIRTH, M. and WINKLER, H., 1984: The temperature dependence of embryonic development of grayling (*Thymallus thymallus*), Danube salmon (*Hucho hucho*), Arctic char (*Salvelinus alpinus*) and brown trout (*Salmo trutta fario*). Aquaculture **38:** 315–327.

KAFKA, J., 1924: Hlavatka neboli losos dunajský. Rybářský Věst. **4:** 103–104, 121–123.

KALASHNIKOV, YU. E., 1978: Ryby basseina reki Vitim. Izd. Nauka, Sib. Otd., Novosibirsk.

KANEP, S. V., 1976: Printsipy, predlozheniya i analiz sistematiki i filogenii lososevidnykh ryb fauny SSSR (lososevye, sigovye, khariusovye). *In:* Ekologiya i sistematika lososevidnykh ryb. Materialy 1-ogo soveshchaniya po izucheniyu lososevidnykh ryb. Zool. Inst. Akad. Nauk SSSR, Leningrad, pp. 42–44.

KANG, Y. S. and PARK, E. H., 1973: Somatic chromosomes of the Manchurian trout, *Brachymystax lenok* (Salmonidae). Chromosome Inform. Serv. **15:** 10–11.

KARANTONIS, F. E., KIRILLOV, F. N. and MUKHOMEDYAROV, F. B., 1956: Ryby srednego techeniya reki Leny. Trudy Inst. Biol. Yakut. Fil. Sib. Otd. Akad. Nauk SSSR **2:** 3–144.

KARBE, L., 1970: Unterordnung Lachsenähnliche. *In:* Grzimeks Tierleben, Fische I. Kindler Verlag, Zürich, pp. 217–258.

KARPEVICH, A. F., 1975: Teoriya i praktika akklimatizatsii vodnykh organizmov. Izd. Pishchevaya promyshlennost', Moskva.

KASECKER, S., 1970: Es ist Huchenzeit. Allg. Fisch. Ztg. **95:** 5–8.

KAŠPAR, K., 1886: Ryby moravské a slezské. Čas. Vlasten. Spol. Muz. Olomouc **3:** 132–134.

KATTOŠ, L., 1966: Priehrada Nitrianske Rudno. Poľov. a Rybár. **18:** 20–21.

KENDALL, A. W., Jr. and BEHNKE, R. J., 1984: Salmonidae: Development and relationships. Amer. Soc. Ichthyol. Herpetol., Special Publ. **1:** 142–149.

KENNEDY, W. A., 1953: Growth, maturity, fecundity and mortality in the relatively unexploited whitefish, *Coregonus clupeaformis,* of Great Slave Lake. J. Fish. Res. Board Can. **10:** 413–441.

KHATKEVICH, V., 1973: Chevitsa – perspektivnyi ob'ekt lososevodstva. Rybov. i Rybolov. (1): 19.

KIFA, M. I., 1974: Materialy po ekologii neresta lenka (*Brachymystax lenok*) i taimenya (*Hucho taimen*) v basseine Amura. Issled. Biol. Ryb. Prom. Okeanogr. **5:** 105–108.

KIFA, M. I. and VDOVCHENKO, M. G., 1976: Opyt inkubatsii ikry lenka i taimenya na Bidzhanskom rybovodnom zavode Khabarovskogo kraya. Rybokhoz. Izuch. Vnutr. Vodoemov **17:** 38–42.

KIMURA, S., 1934: Description of the fishes collected from the Yangtze-kiang, China, by late Dr. K. Kishinouye and his party in 1927–1929. J. Shangai Sci. Inst., Section 3 **1:** 11–247.

KIMURA, S., 1966: On the life history of the salmonid fish, *Hucho perryi* (Brevoort) found in Nemuro, Hokkaido. Jap. J. Ichthyol. **14:** 17–25.

KING, L., 1967: Morfologiya Zemli. Izuchenie svedenii o rel'efe Zemli. (Russian translation of the Morphology of the Earth. A Study and Synthesis of World Scenery.) Izd. Progress, Moskva.

KIRILLOV, F. N., 1955: Ryby reki Indigirki. Izv. VNIORKH **35.**

KIRILLOV, F. N., 1962: Ikhtiofauna basseina reki Vilyuya. Trudy Inst. Biol. Yakut. Fil. Sib. Otd. Akad. Nauk SSSR **8:** 5–71.

KIRILLOV, F. N., 1964: Vidovoi sostav reki Aldana. *In:* Pozvonochnye zhivotnye Yakutii, Yakutsk.

KIRILLOV, F. N., 1972: Ryby Yakutii. Izd. Nauka, Moskva.

KIRILLOV, F. N., 1976: Morfo-ekologicheskaya kharakteristika taimenya *Hucho taimen* (Pallas) r. Anabar. Vopr. Ikhtiol. **16:** 165–167.

KIRILLOV, F. N., 1977: Rybokhozyaistvennoe osvoenie Vilyuiskogo vodokhranilishcha. Izv. GosNIORKH **116:** 24–36.

KIRKA, A., 1958: Deformity plutiev mladi hlavátok z umelých chovov. Čs. Rybářství **13:** 72.

KIRKA, A., 1963: Ein Beitrag zum Alter und Wachstum des Huchen (*Hucho hucho* Linnaeus) aus den Gewässern des Flusses Váh. Věst. Čs. Spol. Zool. **27:** 202–208.

KIRKA, A.,,1975: Výskum ichtyofauny rieky Poprad so zreteľom na rozšírenie hlavátky. Research report. Laboratory of Fishery Research and Hydrobiology, Bratislava.

KIRKA, A. and BASTL, I., 1980: Ichtyologický výskum chráneného náleziska hlavátky v rieke Turiec. Živoč. Výr. **25:** 835–845.

KIRKA, A., BASTL, I. and HOLČÍK, J., 1976: Ichtyocenózy povodia Váhu v oblasti nádrže Liptovská Mara, prognóza jej ichtyofauny a návrh na prvotné zarybnenie. Biol. Práce **24:** 9–79.

KIRPICHNIKOV, V. S., 1974: K voprosu o evolyutsii kariotipa ryboobraznykh i ryb. Usp. Sovr. Biol. **78:** 404–422.

KLEMM, M., 1973: Zoologisches Wörterbuch. Paläarktische Tiere. Paul Parey, Hamburg–Berlin.

KOHAUT, R., 1905: Halak. *In:* Az állatok világa (A. Brehm). Budapest.

KOLDER, W., 1964: Ichtiofauna dorzecza Czarnej Orawy. Wiad. wędkarskie **9:** 10–11.

KOLLER, R., 1907: Über die geografische Verbreitung des Huchens. Oesterr. Fisch. Ztg. **4:** 140–142.

KOMÁREK, J., 1955: Lovy v Karpatech. Orbis, Praha.

KORNHUBER, G. A., 1861: Beitrag zur Kenntniss der Fische in Waag-Gebiet. Verh. Ver. Naturk. zu Preßburg **5:** 40–44.

KOROVINA, V. M., 1978: O strukture semeistva lososevykh Salmonidae. Materialy po stroeniyu ikrinok i nekotorym osobennostyam morfogeneza. *In:* Morfologiya i sistemytika ryb. Sbor. Nauch. Trudov Zool. Inst. Akad. Nauk SSSR, Leningrad, pp. 40–52.

KOSORIĆ, D., KAPETANOVIĆ, N. and VELEDAR, I., 1975: Ribe Velikog Zlatarskog jezera. God. Biol. Inst. Univ. u Sarajevu **28:** 133–141.

KOSTOMAROV, B., 1937a: Několik úvah o hlavatce (*Hucho hucho* Linné) a jejím hospodářském významu u nás. Rybářský Věst. **17:** 94–95.

KOSTOMAROV, B., 1937b: Hlavatka (*Hucho hucho* L.) jako konkurent ve vzrůstu kapra. Čs. Rybář **17:** 125–128.

KOZÁK, J., 1972: Lovcem v tajze. Orbis, Praha.

KOWALSKA, K., REMBISZEWSKI, J. M. and ROLIK, H., 1973: Mały słownik zoologiczny. Ryby. Więdza powszechna, Warszawa.

KOZHOV, M., 1963: Lake Baikal and Its Life. Monographiae Biologicae XI. Dr. W. Junk, The Hague.

KÖTTL, R. and KAFKA, J., 1927: Doba tření hlavatky. Rybářský Věst. **7:** 66–68.

KRALL-KRALSBERG, E., 1930: 44 1b. Huchen caught in the Danube. Fish. Gaz. C (2776): 407.

KRATT, L. F. and SMITH, H. F., 1978: Breeding tubercles occur on male and female Arctic grayling (*Thymallus arcticus*). Copeia (1): 185–188.

KRAUSS, H., 1932: Mladica u Dravi. *In:* Vode i ribe Jugoslavije u slici i rječi. Zagreb.

KRAUSS, H., 1933: Der Drauhuchen, seine Lebensweise, Wanderung und die zu seinem Schutz notwendigen Maßnahmen. Oesterr. Fisch. Ztg. **30:** 2–4, 11–12, 17–19, 34–35.

KRIŠŠÁK, I., 1985: Výskyt hlavátky v Slovenskom raji. Poľov. a Rybár. **37:** 35.

KULMATYCKI, W., 1931a: O wzroscie głowacicy v Czeremosu. Rolnik **20:** 1–8.

KULMATYCKI, W., 1931b: Głowacica z punktu widzenia ochrony przyrody. Ochr. Przyr. **11:** 74–88.

KULMATYCKI, W. J., 1931c: Über das Vorkommen und die Biologie des Huchen im Czeremosz-Fluß. Verh. Int. Ver. Theor. Angew. Limnol. **5:** 354–396.

KULMATYCKI, W. J., 1935: Das Vergleichen des Wachstumstempo von *Hucho hucho* L. in den Gewässern Jugoslawiens und Polens. Verh. Int. Ver. Limnol. **7:** 308–312.

KUO, C. C., 1981: [Rectification of the original description of *Hucho bleekeri* Kimura.] Acta Zootaxon. Sin. **6:** 84. (In Chinese.)

KUX, Z. and WEISZ, T., 1964: Příspěvek k poznání ichtyofauny Slovenska. Čas. Morav. Muz., Přír. Vědy **49:** 191–246.

KVET, K., 1961: Tragédia hlavátky. Poľov. a Rybár. **13:** 22.

LACHNER E. A., ROBINS, C. R. and COURTENAY, W. R., 1970: Exotic fishes and other aquatic organisms introduced into North America. Smithsonian Contr. Zool. (59): 1–29.

LADIGES, W. and VOGT, D., 1965: Die Süßwasserfische Europas. Paul Parey, Hamburg–Berlin.

LEPIKSAAR, J., 1968: Fischreste der Burg-Grabung in Nürnberg. *In:* Grabungsbefunde des hohen und späten Mittelalters auf der Burg zu Nürnberg. Eds. P. Fehring and G. Stachel. Jb. Fränk. Landesforsch. **28:** 70–72.

LEPIKSAAR, J., 1973: Jungpleistozäne Fischreste aus der Brillenhöhle. *In:* Das Paläolithikum der Brillenhöhle von Blaubeuren. 2. Stuttgart, pp. 124–127.

LEVANIDOV, V. YA., 1951: Pitanie taimenya v predgornykh pritokakh Amura. Byull. MOIP, Otd. Biol. **56:** 31–38.

LEVANIDOV, V. YA., 1959: Pitanie i pishchevye otnosheniya ryb v predgornykh pritokakh nizhnego techeniya Amura. Vopr. Ikhtiol. (13): 139–155.

LI, S. C. et al., 1979: [Fishes of Xinjiang.] Xinjiang. (In Chinese.)

LINDBERG, G. U., 1972: Krupnye kolebaniya urovnya okeana v chetvertichnyi period. Izd. Nauka, Leningrad.

LINDBERG, G. U. and CHZHU, SH., 1964: Illyustrirovannyi slovar' nazvanii promyslovykh ryb zapadnoi chasti Tikhogo okeana na latinskom, russkom, kitaiskom, koreiskom, v'etnamskom, mongol'skom, yaponskom i angliiskom yazykakh. Pekin.

LINDBERG, G. U. and DUL'KEIT, G. D., 1929: Materialy po rybam Shantarskogo morya. Izv. Tikh. Nauch.-prom. Stan. **31:** 3–138.

LINDBERG, G. U. and GERD, A. S., 1972: Slovar' nazvanii presnovodnykh ryb SSSR na yazykakh narodov SSSR i evropeiskikh stran. Izd. Nauka, Leningrad.

LINDBERG, G. U. and LEGEZA, M. I., 1965: Ryby Yaponskogo morya i sopredel'nykh chastei Okhotskogo i Zheltogo morei. 2. Izd. Nauka, Moskva–Leningrad.

LINDBERG, G. U., HEARD, A. S. and RASS, T. S., 1980: Multilingual Dictionary of Names of Marine Food-Fishes of World Fauna. Izd. Nauka, Leningrad.

LINDSEY, C. C., 1964: Problems in zoogeography of the lake trout, *Salvelinus namaycush*. J. Fish. Res. Board Can. **21:** 977–994.

LINDSEY C. C., 1966: Body sizes of poikilotherm vertebrates at different latitudes. Evolution **20:** 456–465.

LINNAEUS, C., 1758: Systema naturae. Nantes Pisces. X. Holmiae.

LINKA, M., 1934: Hlavatka s bílou skvrnou na hřbetě. Rybář **1:** 165–170.

LOHNISKÝ, K., 1973: Aklimatizované a zavlečené druhy ryb v severovýchodních Čechách. Acta Mus. Reginaehrad., Ser. A, Sci. Natur. **14:** 121–126.

LOHNISKÝ, K., 1984: Změny rozšíření a druhové skladby ichtyofauny východních Čech v posledních desetiletích. Zprav. Kraj. Muz. Vých. Čech **11:** 29–106.

LÖBL, K., 1976: 64 Pfd. schon wieder österreichischer Rekord-Huchen. Blinker **3:** 1–9.

LUKIN, E. I., 1976: Piyavki presnykh i solonovatykh vodoemov. 1. Fauna SSSR. Izd. Nauka, Leningrad.

LUK'YANCHIKOV, F. V., 1967a: Promyslovo-biologicheskaya kharakteristika i sostoyanie zapasov promyslovykh ryb Bratskogo vodokhranilishcha v pervye gody ego sushchestvovaniya. Izv. Biol.-geogr. Nauch.-issled. Inst. Irkut. Gos. Univ. **20:** 262–332.

LUK'YANCHIKOV, F. V., 1967b: Ryby sistemy reki Khatangi. Trudy Krasnoyar. Otd. Sib. Nauch-issled. Inst. Ryb. Khoz. Krasnoyar. Kraev. Nauch.-tekhn. Obshch. Pishch. Prom. **9:** 11–93.

LUSK, S., 1976: Výskyt a vysazování hlavatky podunajské – *Hucho hucho* (Linnaeus, 1758) v moravských tocích. *In:* Hlavátka podunajská. Compiled papers. Príroda, Bratislava, pp. 65–74.
LUSK, S. and VOSTRADOVSKÝ, J., 1978: Ryby a rybářské hospodaření ve vodárenských nádržích. Vertebr. Zprávy, pp. 20–28.
M., 1924: Hlavatka, neboli losos dunajský. Rybářský Věst. **4:** 103–104.
MAHEN, J., 1926: Ještě něco o hlavatce. Příroda **19:** 303–305.
MAJEWSKI E., 1889: Słownik nazwisk zoologicznych i botanicznych polskich. Warszawa.
MARLBOROUGH, D., 1963: A supplement to the fishes of the London area. Lond. Natur. **42:** 62–70.
MARTIN, N. V. and OLVER, CH. H., 1980: The lake charr, *Salvelinus namaycush*. *In:* Charrs. Ed. E. K. Balon. Dr. W. Junk, The Hague, pp. 205–277.
MARTINKA, J., 1930: Slovenské rybárstvo. Súpis názvov rýb na Slovensku žijúcich. Sbor. MSS **25:** 65–102.
MATLAK, O., 1958: Historia badań nad głowacicą i próby jej hodowli w stawach. Gospod. Ryb. **10:** 8–10.
MATYAS, K., 1954: O možnostiach zarybnenia tekúcich vôd hlavátkou a poznatky z praxe. Čs. Rybářství **9:** 53–54.
MCALLISTER, D. E., 1968: Evolution of branchiostegals and classification of teleostome fishes. Nat. Mus. Can. Bull. **221,** Ser. Biol. **77:** 1–239.
MEDNYANSKY, A., 1844: Malerische Reise auf dem Waagflusse in Ungarn. 2nd ed. C. A. Hartleben. Pest.
MICHÁLEK, J., 1976: Setkání s královnou. Rybářství (9): 210.
MISHARIN, K. I. and SHUTILO, N. V., 1971: Tajmen', ego morfologiya, biologiya i promysl. Izv. Biol.-geogr. Nauch.-issled. Inst. Irkut. Gos. Univ. **24:** 58–105.
MONDEJAR-REYNA, C., 1981: Aclimatacion del salmon del Danubio *Hucho hucho* (L.) en el rio Tormes. Escalametria y crecimiento. MS.
MORAVEC, F., 1975: Reconstruction of the nematode genus *Rhabdochona* Raillet, 1916 with a review of the species parasitic in fishes of Europe and Asia. Studie ČSAV **8:** 5–104.
MORAVEC, F. and ERGENS, R., 1970: Nematodes from fishes and cyclostomes of Mongolia. Folia Parasit. (Praha) **17:** 217–232.
MORI, T., 1928: On the freshwater fishes from the Yalu river, Korea with description of new species. J. Chosen Natur. Hist. Soc. (6): 8–24.
MUNDA, A., 1925: Nekaj statističnih podatkov o sulčji lovi v Savi i Ljubljanici. Lovac (12): 1–8.
MUNDA, A., 1926a: Ribe v slovenskich vodah. Ljubljana.
MUNDA, A., 1926b: Kako uspevajo sulci v naših vodah. Glas. Muz. Družstva za Slovenijo. Reprint.
MUNDA, A., 1935: Die künstliche Zucht des Huchens in Jugoslawien und die hiebei gewonnenen biologischen Erfahrungen. Verh. Int. Ver. Limnol. **7:** 313–320.
NAGY, Š., 1976: Contribution to the knowledge of the food of the huchen (*Hucho hucho*) (Teleostei: Salmonidae). Zool. Listy **25:** 183–191.
NAVOZOV, N. P., 1912: Materialy k ikhtiofaune basseina r. Urala. Vest. Ryboprom., pp. 252–283.
NELSON, G. J., 1969: Gill arches and the phylogeny of fishes, with notes on the classification of vertebrates. Bull. Amer. Mus. Natur. Hist. **141:** 475–552.
NERESHEIMER, W., 1937: Die Lachsartigen (Salmonidae). 1. Teil. *In:* Handbuch der Binnenfischerei Mitteleuropas. 3. Systematik und Biologie. Eds. R. Demoll and H. N. Meier. Lfg. 5. A. Oelsschläger'sche Buchdruckerei, Wien, pp. 219–370.
NIKOL'SKII, G. V., 1956: Ryby basseina Amura. Izd. Akad. Nauk SSSR, Moskva.
NIKOL'SKII, G. V., 1971: Chastnaya ikhtiologiya. 3rd ed. Izd. Vysshaya shkola, Moskva.
NIKOL'SKII, G. V., 1974: Ekologiya ryb. 3rd ed. Izd. Vysshaya shkola, Moskva.
NILSSON, N. A. and SVÄRDSON, G., 1963: Donaulaxen. Svenskt Fiske **7:** 5–8.
NOAKES, L. G., 1980: Some histological observations on the skin of charrs. *In:* Charrs. Ed. E. K. Balon. Dr. W. Junk, The Hague, pp. 423–433.
NORDEN, O. R., 1961: Comparative osteology of representative salmonid fishes, with particular reference to the grayling (*Thymallus arcticus*) and its phylogeny. J. Fish. Res. Board Can. **18:** 679–791.
NOVÁK, I., 1969: 50 rokov rybárstva v Turci. MO SRZ, Martin.
NOVÁK, I., 1971: Fauna rýb rieky Turiec. Čs. Ochr. Prír. **12:** 5–25.
NOVÁK, J., 1932: O hlavatce řeky Oravy. Čs. Rybář **12:** 43–44.

NYMAN, L., 1967: Protein variation in Salmonidae. Rep. Inst. Freshwater Res. Drottningholm **47:** 6–38.

OHNO, S., 1970: Evolution by gene duplication. Springer-Verlag, Berlin–Heidelberg–New York.

OLIFER, S. A., 1977: Rybokhozyaistvennoe osvoenie Usť-Ilimskogo vodokhranilishcha. Izv. GosNIORKH **115:** 65–96.

OL'SHANSKAYA, O. L., 1965: Obzor ikhtiofauny r. Pyasiny. Vopr. Ikhtiol. **5:** 262–278.

OL'SHANSKAYA, O. L., VERSHININ, N. V., TOLMACHEV, V. A., MIKHALEVA, T. V., ROMANOVA, I. M., VOLKOVA, N. I. AND EREMEEVA, G. G., 1977: Rybokhozyaistvennoe ispol'zovanie Krasnoyarskogo vodokhranilishcha. Izv. GosNIORKH **115:** 97–138.

O-N, O., 1955: Vari sötvattensfiskers utländska namm. Svensk Fisk. Tidskr. (819): 119–120.

P. V., 1968: Le huchon dans les Usses. Bull. Franç. Piscicult. **40:** 112.

PALATKA, J., 1976: Hlavatka v revírech Dyje. Rybářství (9): 210.

PALLAS, P. S., 1773: Reise durch verschiedene Provinzen des Russischen Reichs. **2.** Acad. Sci., St. Petersburg.

PALLAS, P. S. [1814]: Zoographia Rosso-Asiatica. **3.** Acad. Sci., Petropoli.

PAPPENHEIM, P., 1909: Pisces (inkl. Cyclostomata), Fische. *In:* Süßwasserfauna Deutschlands. Eine Exkursionsfauna. 1. Ed. A. Brauer. Gustav Fischer Verlag, Jena, pp. 90–201.

PAVLÍK, I., 1955: Odchov hlavátky na jesienka a jeho vysadzovanie. Čs. Rybářství **10:** 117–118.

PAVLÍK, I., 1957: Umelý výter hlavátok. Čs. Rybářství **12:** 71.

PAVLOV, D. A., 1979: Sravniteľnyi analiz embryonaľno-lichinochnogo razvitiya evropeiskikh lososei roda *Salmo*. Zool. Zh. **53:** 674–684.

PAVLOV, D. A., 1980: Osobennosti embryonaľno-lichinochnogo razvitiya atlanticheskikh i tikhookeanskikh lososei roda *Salmo* v svyazi s ikh evolyutsiei. Zool. Zh. **54:** 569–575.

PAŽUR, K., HABEKOVIČ, D. and POPOVIČ, J., 1982: Dinamika rasta mladice (*Hucho hucho* L., 1758) u vodama SR Hrvatske. Ichthyologia **14:** 161–169.

PEŇÁZ, M. and PŘÍHODA, J., 1981: Reproduction and early development of the Danube salmon, *Hucho hucho* (Linnaeus, 1758). Acta Sci. Natur. Acad. Sci. Bohemoslov. Brno **15** (N. S.): 3–33.

PICKERING, A. D., 1977: Seasonal changes in the epidermis of the brown trout *Salmo trutta* (L.). J. Fish Biol. **10:** 561–566.

PIROZHNIKOV, P. L., 1955: Materialy po biologii promyslovykh ryb reki Leny. Izv. VNIORKH **35:** 61–128.

PODLESNYI, A. V., 1958: Ryby Eniseya, usloviya ikh obitaniva i ispoľzovaniya. Izv. VNIORKH **44:** 97–178.

PODYAPOL'SKAYA, V. P. and KAPUSTIN, V. F., 1950: Glistnye zabolevaniya cheloveka. Izd. AN SSR, Moskva.

POLÁŠEK, J. N., 1936: Hlavatky na Valašsku. Rybářské Listy **1:** 77–78.

POLSKI ZWIĄZEK WĘDKARSKI, 1977: Regulamin sportowego polowu ryb. Warszawa.

POPOVA, O. A., 1967: The "predator-prey" relationship among fish. *In:* The Biological Basis of Freshwater Fish Production. Ed. S. D. Gerking. Blackwell Scientific Publ., Oxford–Edinburgh, pp. 359–376.

PÖLZL, F., 1905: Zum Kapitel „Huchenzucht". Oesterr. Fisch. Ztg. **2:** 193–195.

PÖLZL, F., 1910: Der Donauhuchen in England. Oesterr. Fisch. Ztg. **7:** 291.

PÖLZL, F., 1929: Ein Beitrag zur Frage: Soll man den Huchen künstlich züchten? Oesterr. Fisch. Ztg. **26:** 134.

PRAVDIN, I. F., 1949: Taimen' – *Hucho hucho* (Pallas). *In:* Promyslovye ryby SSSR. Opisaniya ryb. Eds. L. S. Beg, A. S. Bogdanov, N. I. Kozhin and T. S. Rass. Pishchepromizdat, Moskva, pp. 205–207.

PRAWOCHEŃSKI, R. and KOLDER, W., 1968: Synopsis of biological data on *Hucho hucho* (Linnaeus, 1758). FAO Fish. Synopses **22** (Suppl. 1): 1–14.

J. G. P. (Preudhomme, J. G.), 1953: Importation de nouvelles espèces sportives. Fishing-Club de Fes et Truite de l'Atlas **13:** 13.

PREUDHOMME, M. J. G., 1955: Acclimatation d'espèces nouvelles au Maroc. Fishing-Club du Moyen-Atlas **2** (N. S.): 23–24.

PREUDHOMME, J. G., 1959a: Importation d'œufs de huchons en 1959. Fishing-Club du Moyen-Atlas **5** (N. S.): 37.

PREUDHOMME, J. G., 1959b: Les premiers œufs de huchon marocain. Fishing-Club du Moyen-Atlas **5** (N. S.): 37.

PREUDHOMME, M., 1961: Rapport annuel 1960 sur la pêche dans eaux continentales et la pisciculture. Bilan et enseignements de 25 années d'acclimatations piscicoles. Fishing-Club du Moyen-Atlas **6** (N. S.): 1–15.

PRIVOL'NEV, T. I., 1967: K voprosu o proiskhozhdenii lososevykh i sigovykh ryb. Izv. VNIORKH **62:** 31–38.

PUSHKINA, R. G., 1977: Osnovnye napravleniya povysheniya ryboproduktivnosti Bratskogo vodokhranilishcha. Izv. GosNIORKH **115:** 55–64.

R., 1970: Zur Ernährung des Huchens. Allg. Fisch. Ztg. **95:** 801.

R., 1970: Was frißt der Huchen? Allg. Fisch. Ztg. **95:** 313.

RÁB, P., 1981: Karyotypy hlavatky podunajské (*Hucho hucho* L.) ze Slovenska a Jugoslávie. Reprodukce, genetika a hybridizace ryb. Slov. Zool. Spol., Vodňany: 190–193.

RÁB, P. and LIEHMAN, P., 1982: Chromosome study of Danube salmon *Hucho hucho* (Linnaeus, 1758) (Pisces: Salmonidae). Folia Zool. **31:** 181–190.

RANDÍK, A., 1976: Problémy s ochranou hlavátky podunajskej (*Hucho hucho* L.) v strednej Európe. *In:* Hlavátka podunajská. Compiled papers. Príroda, Bratislava, pp. 40–50.

RANDÍK, A., 1978: Výskum niektorých vzácnych a chránených druhov stavovcov na Slovensku. Čs. Ochr. Prír. **1:** 89–175.

RASS, T. S. (Ed.), 1971: Zhizn' zhivotnykh. Izd. Prosveshchenie, Moskva.

REES, H., 1964: The question of polyploids in the Salmonidae. Chromosoma **15:** 275–279.

REMBISZEWSKI, J. M. and ROLIK, H., 1975: Krągłouste i ryby. Cyclostomata et Pisces. Katalog Fauny Polski. 38. Państwowe Wydawnictwo Naukove, Warszawa.

RENZ, H., 1969: Entwicklung des Huchens in der Saane. Fischerei (Schw.) **9:** 18–21.

RESHETNIKOV, YU. S., 1980: Ekologiya i sistematika sigovykh ryb. Izd. Nauka, Moskva.

REVNIVYKH, A. R., 1939: O pitanii taimenya v reke Loz've. Trudy Ural'. Otd. VNIORKH **1:** 246.

REŽNÝ, J., 1951: Ako sme vytierali hlavátky. Čs. Rybář **6:** 18.

REŽNÝ, J., 1968: Vydarený pokus. Poľov. a Rybár. **20:** 20.

RIBIŠKA ZVEZA SLOVENIJE, 1966: Informativni bilten o možnostih ribolova. Ribič **25:** 213–218.

RICKER, W. E., 1962: Russian-English glossary of names of aquatic organisms and other biological and related terms. Fish. Res. Board Can., Circular **65:** 1–173.

RICKER, W. E., 1975: Computation and interpretation of biological statistics of fish populations. Fish. Res. Board Can. Bull. **191:** 3–382.

ROBERTS, R. J. and SHEPHERD, C. J., 1974: Handbook of Trout and Salmon Diseases. Fishing News (Books) Ltd., West Byfleet, Surrey.

ROBIDA, J., 1902: Der Huchen und sein Fang mit der Angel. Kleinmayer und Bamberg, Laibach.

ROITMAN, V. A., 1967: Ekologo-geograficheskaya kharakteristika gelmintofauny taimenei (rod *Hucho*) i lenka (rod *Brachymystax*), obitayushchikh v vodoemakh SSSR. *In:* Problemy parazitologii. Izd. Naukova Dumka, Kiev, pp. 195–197.

ROSEN, D. E., 1974: Phylogeny and zoogeography of salmoniform fishes and relationships of *Lepidogalaxias salamandroides*. Bull. Amer. Mus. Natur. Hist. **153:** 265–326.

ROTHSCHEIN, J., 1976: Prognóza zmien kvalitatívneho režimu Dunaja po výstavbe vodných diel. Práce a Štúdie (82): 7–89.

ROTHSCHEIN, J., 1980: S hlavátkou účelnejšie. Poľov. a Rybár. **32:** 32–33.

ROUNSEFELL, G., 1962: Relationships among North American Salmonidae. Fish. Bull. and Wildl. Serv. **62:** 235–270.

ROUSEK, J., 1934: Losos všem. Rybář **1:** 253–254.

ROWLAND, W., 1881: Beitrag zur Kenntniss unserer ichthyologischen Verhältnisse. Jb. Ung. Karpathen-Ver. **8:** 38–62.

RUDOLF, J., 1964: Návštevou u ryb v slovenských vodách. Čs. Rybářství **19:** 140–141.

SABANEEV, L. P., 1911: Ryby Rossii. 3rd ed. Izd. Kartsev, Moskva.

SABIONCELLO, I., MARKO, S. and PAŽUR, K., 1970: Bioekološka ispitivanija salmonida u SR Hrvatskoj. Ribar. Jugosl. **25:** 29–37.

SCHÄFERNA, K., 1928: Obrovitý pstruh a úvaha o růstu ryb. Rybářský Věst. **8:** 110–111.

SCHINDLER, O., 1935: Die Brut der mitteleuropäischen Süßwasserfische. 3. Der Huchen *Salmo hucho* L. Allg. Fisch. Ztg. **60** (N. F. 50): 306–309.

SCHM., 1970: Fischmord im Inn. Allg. Fisch. Ztg. **95:** 823.

SCHNAKENBECK, W., 1940: XV. Untersuchungen über die Entwicklung von Süßwasserfischen. Z. Fisch. **38:** 269–321.

SCHULZ, N., 1985: Das Wachstum des Huchens (*Hucho hucho* L.) in der Drau in Kärnten. Österr. Fisch. **38:** 131–142.

SCHULZ, N. and PIERY, G., 1982: Zur Fortpflanzung des Huchens (*Hucho hucho* L.) – Untersuchung einer Laichgrube. Österr. Fisch. **35:** 241–249.

SCHULZE, E., 1890: Fauna Piscium Germaniae. Verzeichnis der Fische der Stromgebiete der Donau, des Rheines, der Ems, Weser, Elbe, Oder, Weichsel, des Pregels und der Memel. Jber. u. Abh. Naturwiss. Ver. Magdeburg 1889, pp. 137–213.

SCOTT, W. B. and CROSSMAN, E. J., 1973: Freshwater fishes of Canada. Fish. Res. Board Can. Bull. **184:** 1–966.

SEDLÁR, J., 1969: Súčasný stav zarybnenia povodia rieky Nitry. Biol. Práce **15:** 5–78.

SEDLÁR, J. (Ed.), 1970: Príručka športového rybára. Príroda, Bratislava.

SEDLÁR, J. and STRÁŇAI, I., 1975: Výskum ichtyofauny povodia rieky Hron. Research report. Faculty of Agronomy, Agricultural University, Nitra.

SEEZ, R., 1939: Über den Huchen. Allg. Fisch. Ztg. **64:** 272–276, 279–283, 287–288.

SERVICE CANTONAL DE LA CHASSE ET DE LA PÊCHE FRIBOURG, 1977: Resumé des expériences faites avec les huchons (*Hucho hucho* L.). Report. Fribourg.

SHAPOSHNIKOVA, G. KH., 1964: Biologiya i raspredelenie ryb v rekakh ural'skogo tipa. Izd. Nauka, Moskva.

SHAPOSHNIKOVA, G. KH., 1967: O sistematicheskom polozhenii rodov *Hucho* Günther i *Brachymystax* Günther. Zool. Zh. **46:** 254–257.

SHAPOSHNIKOVA, G. KH., 1968: Sravnitel'no-morfologicheskoe izuchenie taimenei i lenka. Vopr. Ikhtiol. **8:** 440–464.

SHAPOSHNIKOVA, G. KH., 1975: Sistematicheskie otnosheniya nekotorykh predstavitelei semeistva Salmonidae. Zool. Zh. **54:** 575–582.

SHMIDT, P. YU., 1947: Migratsii ryb. Izd. Akad. Nauk SSSR, Moskva–Leningrad.

SHMIDT, P. YU., 1950: Ryby Okhotskogo morya. Izd. Akad. Nauk SSSR, Moskva–Leningrad.

SHNAREVICH, I. D. and MOSHUK, K. D., 1957: K izucheniyu ekologii dunaiskogo lososya i ego rybokhozyastvennogo ispol'zovaniya na Bukovine. Nauch. Ezhegod. Chernovitskogo Gos. Univ. **1:** 107–112.

SIEBOLD, C. TH. E., von, 1863: Die Süßwasserfische von Mitteleuropa. W. Engelmann, Leipzig.

SIGUNOV, P., 1972; Taimen'i istorii. *In:* Ozherelie Dzhekhangira. Izd. Nauka, Moskva, pp. 71–92.

SKÁCEL, L., 1963: K diskusii o hlavátke. Poľov. a Rybár. **15:** 13.

SKÁCEL, L., 1976: Súčasný stav a rozšírenie hlavátky podunajskej (*Hucho hucho* L.) a perspektíva aklimatizačných pokusov na Slovensku a v zahraničí. *In:* Hlavátka podunajská. Compiled papers. Príroda, Bratislava, pp. 11–21.

SKALIN, B., 1972: Pokus vzreje lipana i sulca v ribniku i potoku. Ribič **31:** 248–249.

SKALIN, B., 1976: Nekaj misli o sulcu. Ribič **35:** 102–103.

SKALIN, B., 1983: Technologija održavanja populacije mladice (*Hucho hucho* L.) u vodama Slovenije. Agric. Consp. Sci. **63:** 619–634.

SMÍŠEK, J., 1953: Hlavatka do českých řek. Čs. Rybářství **8:** 24.

SMITH, H. M., 1945: The freshwater fishes of Siam or Thailand. Bull. U. S. Natur. Mus. **188:** 3–622.

SMOL'YANOV, I. I., 1961: Razvitie lenka *Brachymystax lenok* (Pallas). Vopr. Ikhtiol. **1:** 136–148.

SOFRADŽIJA, A., 1979: The chromosomes of *Hucho hucho* (L.). Third European Ichthyological Congress, Warszawa. Abstracts, p. 166.

SOIN, S. G., 1980: O tipakh razvitiya lososevidnykh i ikh taksonomicheskom znachenii. Vopr. Ikhtiol. **20:** 65–72.

SOKOLOV, V. E. et al., 1983: Ryby Mongol'skoi narodnoi respubliki. Izd. Nauka, Moskva.

SPILLMANN, C. J., 1961: Poissons d'eau douce. Faune de France 65. Librairie Fac. Sci., Paris.

STEINMANN, P., 1948: Sollen wir in der Schweiz Huchen einsetzen? Fisch. Ztg. **56:** 31–33.

STOKŁOSOWA, S., 1966: Sexual dimorphism in the skin of sea-trout *Salmo trutta*. Copeia (3): 613–614.

STOKŁOSOWA, S., 1970: Further observations on the sexual dimorphism in the skin of *Salmo trutta trutta* in relation to sexual maturity. Cooeia (2): 332–339.

STORK, H., JR., 1908: Huchen-Rekord. Allg. Fisch. Ztg. **33:** 52–53.

STÖLZLE, A., 1910: Zur Frage der Huchenwanderung. Oesterr. Fisch. Ztg. **7:** 276–278.

STRAŠKRABA, M., ČIHAR, J., FRANK, S. and HRUŠKA, V., 1966: Contribution to the problem of food competition among the sculpin, minnow and brown trout. J. Anim. Ecol. **35:** 308–311.

STRAŠKRABA, M. and STRAŠKRABOVÁ, V., 1980: Eutrofizace vodních nádrží. Život. Prostredie **14:** 188–190.

STRELKOV, YU.A. and SHUL'MAN, S. S., 1964: Itogi rabot Amurskoi ikhtioparazitologicheskoi ekspeditsii 1957–1959. Vopr. Ikhtiol. **4:** 162–177.

SVÄRDSON, G., 1945: Chromosome studies on Salmonidae. Meddelenden fran Steten undersöknings-och försökanstalt för sötvattenfisket **24:** 3–151.

SVETINA, M., 1962: Étude synoptique sur le biologie du huchon *Hucho hucho* (Linné, 1758). FAO Fish. Biol. Synopses **22:** 1–30.

SVETINA, M., 1964: Repport sur les possibilités de la transplantation du huchon – *Hucho hucho* (L.) dans le Sarine (Canton Fribourg). Research report. Fribourg.

SVETINA, M., 1966: Kako se je naselil sulec v srednejevropejske vode. Ribič **25:** 138–143.

SVETINA, M., 1967: Der Huchen hat sich in Frankreich gut eingebürgert. Allg. Fisch. Ztg. **92:** 337–338.

SVETINA, M., 1968: Dinamika rasti sulca – *Hucho hucho* (Linnaeus, 1758). Biol. Vest. **16:** 103–114.

SVETINA, M., 1970: Wachstums-Dynamik des Huchens. Allg. Fisch. Ztg. **95:** 133–137.

SVETOVIDOV, A. N., 1978: Tipy vidov ryb, opisannykh P. S. Pallasom v "Zoographia Rosso-Asiatica". Izd. Nauka, Leningrad.

SVETOVIDOV, A. N., DOROFEEVA, E. A., KLYUKANOV, V. A. and SHAPOSHNIKOVA G. KH., 1975: Morfologicheskie osnovy klassifikatsii lososevykh ryb. Zool. Zh. **54:** 559–574.

SVETOVIDOV, A. N., DOROFEEVA, E. A., KLYUKANOV, V. A. and SHAPOSHNIKOVA, G. KH., 1976: Morphological bases of the salmonoid fishes. Ichthyologia **8:** 135–154.

SVETOVIDOVA, A. A., 1960: Materialy po ikhtiofaune, vozrastnomu sostavu i tempu rosta ryb ozera Dalainor (Kitaiskaya Narodnaya Respublika). Zool. Zh. **39:** 250–262.

SYCH-AVERINTSEVA, N. V., 1933: O meristicheskikh priznakakh nekotorykh predstavitelei lososevykh r. Leny. Trudy Yakut. Rybokhoz. Stan. (quoted after Prirozhnikov, 1955).

SÝKORA, S., 1957: Výsledky pokusného vysazování hlavatky do Moravice. Čs. Rybářství **12:** 88.

SYROECHKOVSKII, E. E. and ROGACHEVA, E. V., 1975: Zhivotnyi mir SSSR (geografiya resursov). Izd. Mysl', Moskva.

SZABÓ, J., 1976: Kráľovná z Oravy. Poľov. a Rybár. **28:** 32–33.

SZCERBOWSKI, J., 1977: Próba określenia kryteriów oceny tempa vzrostu ryb. Gospod. Ryb. **29:** 8–11.

ŠIMEK, Z., 1946: Chytání pstruhů, lipanů a hlavatek. A. Neubert, Praha.

ŠIMEK, Z., 1954: Rybářství na tekoucích vodách. SZN, Praha.

ŠIMEK, Z. and KOPÁČIK, L., 1959: Športové rybárstvo. SVPL, Bratislava.

ŠIVIC, I. and BAJKOV, A., 1925: Umelý chov hlavatice (*Hucho hucho*). Rybářský Věst. **5:** 100–102.

ŠUSTA, V., 1898: Fünf Jahrhunderte der Teichwirtschaft zu Wittingen. Herrcke und Lebelling, Stettin.

TT., 1970: Kapitaler Huchen. Allg. Fisch. Ztg. **95:** 58.

TALER, Z., 1953: Rasprostranjenje i popis slatkovodnich riba Jugoslavije. Glas. Prir. Muz. Srpske Zemlje, Ser. B **56:** 425–455.

TARANETS, A. YA., 1936: Presnovodnye ryby basseina severo-zapadnoi chasti Yaponskogo morya. Trudy Zool. Inst. Akad. Nauk SSSR **4:** 483–537.

TARLING, D. H. and TARLING, M. P., 1972: Continental Drift. A study of the Earth's Moving Surface. Penguin Books, Harmondsworth.

TCHANG, T. L. and LIU, C. H., 1957: A preliminary survey of the fishes of the Min river with some notes on their distribution. Szechwan Univ. Sci. J. **2:** 221–246.

TCHERNAVIN, V. V., 1939: The origin of salmon. Is its ancestry marine or freshwater? Salmon and Trout Mag. **95:** 1–21.

TEJČKA, J., 1934a: Hlavatka či losos dunajský. Rybář **1** (3): 40–43.

TEJČKA, J., 1934b: Rybářský sport. A. Neubert, Praha.

TEPLOV, V. P., 1951: O pitanii tajmenya (*Hucho taimen* Pallas) i ego vzaimootnosheniya s molodyu semgi (*Salmo salar*). Zool. Zh. **30:** 641–643.

THALLER, Z., 1932: Vode i ribe Jugoslavije. Zagreb.

TRET'YAKOV, D. K., 1947: Vyznachnyk kruglorotykh i ryb URSR. Izd. Akad. Nauk URSR, Kyiv.

TSEPKIN, E. A., 1976: K istorii promyslovoi ikhtiofauny i rybolovstva v basseine oz. Baikal. Byull. MOIP, Otd. Biol. **81:** 65–73.

TSEPKIN, E. A., 1979: Pozdnechetvertichnaya ikhtiofauna reki Aldan (bassein Leny). Vopr. Ikhtiol. **19:** 28–32.

TSEPKIN, E. A., 1980a: K istorii promyslovoi ikhtiofauny i rybolovstva v srednom techenii Angary. Vopr. Ikhtiol. **20:** 543–545.

TSEPKIN, E. A., 1980b: K istorii vozniknoveniya i razvitiya rybolovstva v kontinentaľnykh vodoemakh aziatskoi chasti SSSR. Vopr. Ikhtiol. **20:** 806–813.

TUGARINA, P. YA., 1977: Irkutskoe vodokhranilishche i produktivnost' ego ikhtiotsenozov. Izv. GosNIORKH **115:** 44–54.

TUGARINA, P. YA. and KUPCHINSKAYA, E. S., 1977: Pitanie i pishchevie vzaimootnosheniya ryb Baikalo-Angarskogo basseina. Izd. Nauka, Sib. Otd., Novosibirsk.

TUŠA, I., 1968: On the feeding biology of the brown trout (*Salmo trutta* m. *fario* L.) in the Loučka Creek. Zool. Listy **17:** 379–395.

TUTUROV, YU. A. and OMEL'CHENKO, V. T., 1976: Opredelenie gomologii DNK nekotorykh predstavitelei Salmoniformes kineticheskim metodom. *In:* Ekologyiya i sistematika lososevidnykh ryb. Materialy 1-ogo soveshchaniya po izucheniyu lososevidnykh ryb. Zool. Inst. Akad. Nauk SSSR, Leningrad, pp. 94–98.

VARPAKHOVSKII, N., 1899: Dannye po ikhtiologicheskoi faune basseina reki Obi. 2. Ezhegod. Zool. Muz. Akad. Nauk **4:** 325–374.

VASIĽEV, V. M., 1960: Nastoľnaya kniga rybolova-sportsmena. Gos. izd. Fizkultura i sport, Moskva.

VASIL'EV, V. P., 1980: Khromosomnye chisla ryboobraznykh i ryb. Vopr. Ikhtiol. **20:** 387–422.

VASIL'EV, V. P., 1983: Nekotorye aspekty khromosomnoi differentsiatsii ryb. *In:* Biologicheskie osnovy rybovodstva: Problemy genetiki i selektsii. Izd. Nauka, Leningrad, pp. 166–180.

VASILIU, G. D., 1959: Peşti apelor noastre. Editura Ştiinţifică, Bucureşti.

VERCE, F., 1973: Pogled na naše ribištvo skozi prizmo statistike. Ribič **32:** 98–100.

VIKTOROVSKII, R. M., 1976: Sravniteľnaya kariologiya, evolyutsiya i vidoobrazovanie u goľtsov roda *Salvelinus*. *In:* Ekologiya i sistematika lososevidnykh ryb. Materialy 1-ogo soveshchaniya po izucheniyu lososevidnykh ryb. Zool. Inst. Akad. Nauk SSSR, Leningrad, pp. 16–19.

VIKTOROVSKII, R. M., MAKOEDOV, A. N. and SHEVCHISHIN, A. A., 1985: Khromosomnye nabory lenka i sibirskogo taimenya i divergentsiya rodov lososevykh. Tsitologiya **27:** 703–708.

VIVIER, P., BLANC, L. and SVETINA, M., 1964: Le huchon et son acclimatation en Haute-Savoie. Bull. Franç. Piscicult. **36:** 77–85.

VLADYKOV, V., 1925: Hlavatka neboli losos dunajský. Rybářský Věst. **5** (3): 35, (4): 50–52, (5): 66–67.

VLADYKOV, V. D., 1926: Ryby Podkarpatskoi Rusi i ikh glavneishie sposoby lovli. Uzhgorod.

VLADYKOV, V., 1929: Umělý chov hlavatek (*Hucho hucho* L.) na Podkarpatské Rusi. Rybářský Věst. **9** (8): 118–121, (9): 136–138.

VLADYKOV, V., 1931: Poissons de la Roussie-Sous-Carpathique (Tchécoslovaquie). Mém. Soc. Zool. France **4:** 217–374.

VLADYKOV, V. D., 1954: Taxonomic characters of the eastern North American chars (*Salvelinus* and *Cristivomer*). J. Fish. Res. Board Can. **11:** 904–932.
VLADYKOV, V. D., 1963: A review of salmonid genera and their broad geographical distribution. Trans. Roy. Soc. Can. **1:** 459–505.
VLADYKOV, V. D., 1970: Pearl tubercles and certain cranial peculiarities useful in the taxonomy of coregonid genera. *In:* Biology of Coregonid Fishes. Eds. C. C. Lindsey and C. S. Woods. Univ. Manitoba Press, Winnipeg, pp. 167–193.
VLADYKOV, V. D. and GRUCHY, C. G., 1972: Comments on the nomenclature of some subgenera of Salmonidae. J. Fish. Res. Board Can. **29:** 1631–1632.
VLASOVA, E. K., 1956: Materialy po ikhtiofaune Zakarpat'ya. Nauch. Zap. Uzhgor. Gos. Univ. **16:** 3–38.
VLASOVA, E. K., 1959: Lososevye ryby (Salmoidei) rek Zakarpat'ya. Nauch. Zap. Uzhgor. Gos. Univ. **40:** 89–100.
VOKAČ, O., 1959: Sulci v Dravi. Ribič **18:** 10–13, 34–35.
VOKAČ, O., 1973: Sulec v Dravskom bazenu. Ribič **32:** 285.
VOLF, F. and HAVELKA, J., 1958: Rybářská zdravověda. SZN, Praha.
VOOREN, C. M., 1972: Ecological aspects of the introduction of fish into natural habitats in Europe, with special reference to the Netherlands. J. Fish Biol. **4:** 565–583.
VOSTRADOVSKÝ, J., LEONTOVYČ, I. and VOSTRADOVSKÁ, M., 1973: Ichtyofauna pražské Vltavy v letech 1970–1972. Bul. VÚRH Vodňany (2): 19–26.
VOSTRADOVSKÝ, J. and NOVÁK, M., 1959: Několik poznatků z Lipenské údolní nádrže v roce 1958. Sbor. ČSAVZ – Živoč. Výr. **4:** 877–888.
VOTRUBEC, J., 1932: Vzpomínky na slovenské a podkarpatoruské rybářství. Čs. Rybář **12:** 84–85.
VUTSKITS, G., 1918: Classis Pisces. *In:* Fauna Regni Hungariae. 1. Vertebrata. Regia Soc. Sci. Natur. Hungarica, Budapest.
W., 1970: Kapitaler Huchen. Allg. Fisch. Ztg. **95:** 26.
WAJDOWICZ, Z., 1964: Rozwój ichtyofauny zbiorników zaporowych o małych wahaniach stanu wody. Acta Hydrobiol. **6:** 67–69.
WEBER, E., 1979: Wechselwirkungen zwischen der Donau und ihren Nebengewässern in Österreich. XXI. Arbeitstagung der SIL, Übersichtsreferate, Novi Sad, pp. 44–59.
WEITZMAN, S. H., 1967: The origin of the stomiatoid fishes with comments on the classification of salmoniform fishes. Copeia (3): 507–540.
WENDLAND, H., 1965: So fängt man Huchen. Paul Parey, Hamburg–Berlin.
WHEELER, A., 1969: The fishes of the British Isles and North West Europe. Michigan State University Press, East Lansing.
WHEELER, A., 1979: The Tidal Thames. Routledge and Kegan Paul, London–Boston–Henley.
WHEELER, A. and MAITLAND, P. S., 1973: The scarcer freshwater fishes of the British Isles. 1. Introduced species. J. Fish Biol. **5:** 49–68.
WILEY, M. L. and COLLETTE, B. C., 1970: Breeding tubercles and contact organs in fishes: Their occurrence, structure and significance. Bull. Amer. Mus. Natur. Hist. **143:** 147–216.
WILSON, J. G. M., 1976: Immunological aspects of fungal diseases in fish. *In:* Recent Advances in Aquatic Mycology. Ed. E. B. Gareth. Paul Elek Ltd., London, pp. 573–601.
WILSON, M., 1977: Middle Eocene freshwater fishes from British Columbia. Roy. Ontario Mus. Life Sci. Contr. **113:** 1–61.
WITKOWSKI, A. and BŁACHUTA, J., 1980: Występowanie *Basanistes huchonis* (Schrank, 1786) (Copepoda parasitica, Lernaeopodidae) na głowacicy *Hucho hucho* (L.) v Dunajcu. Przegl. Zool. **24:** 463–467.
WITKOWSKI, A. and KOKUREWICZ, B., 1981: The embryonal and postembryonal development of the Danube salmon *Hucho hucho* (L.) (Pisces: Salmonidae). Acta Hydrobiol. **23:** 85–94.
WITKOWSKI, A., KOKUREWICZ, B. and KOWALEWSKI, M., 1984: Early scale development in the Danube salmon *Hucho hucho* (L.) (Pisces: Salmonidae). Acta Hydrobiol. **25/26:** 215–223.
WITKOWSKI, A. and KOWALEWSKI, M., 1980: Aklimatizacija i rozsiedlenie głowacicy w Polsce. Gospod. Ryb. **32:** 6–9.

Witkowski, A. and Kowalewski, M., 1982: Uwagi o odżywieniu sie głowacicy introdukowanej do Dunajca. Gospod. Ryb. **34:** 10–11.

Witkowski, A. and Kowalewski, M., 1984: Food of the Danube salmon *Hucho hucho* (L.) introduced into the river Dunajec. Acta Hydrobiol. **25/26:** 205–214.

Witkowski, A., Kowalewski, M. and Błachuta, J., 1985: The growth of the Danube salmon (*Hucho hucho* L.) (Salmonidae) introduced into the river Dunajec. Acta Hydrobiol. **27:** 113–125.

Wu, H. W. (Ed.), 1979: [Economic Fauna Sinica: Freshwater Fishes.] 2nd ed. Beijang. (In Chinese.)

Wu, Y. F. and Chen, Y., 1979: [Notes on fishes from Golog and Yushu region of Quinghai Province, China.] Acta Zootax. Sin. **4:** 287–295. (In Chinese.)

Wunder, W., 1936: Physiologie der süßwasserfische Mitteleuropas. Sonderabdr. aus dem Handbuch der Binnenfischerei Mitteleuropas. Bd. 2. E. Schweizerbart'sche Verlagsbuchhandlung, Stuttgart.

Yakovlev, V. N., 1961: Rasprostranenie presnovodnykh ryb neogena Golarktiki i zoogeograficheskoe raionirovanie. Vopr. Ikhtiol. **1:** 209–220.

Zakhvatkin, V. O., 1951: Parazity ryb vodoemov Zakarpatskoi oblasti. Nauch. Zap. L'vovskogo Nauch. Prirod. Mus. Akad. Nauk USSR **1:** 119–149.

Zaplata, R. and Taler, Z., 1932: Ribe Sarajeva i okoline. Glas. Zemal. Mus. Bosni Herceg., Prir. Nauke **44:** 1–32.

Zbořil, J. and Absolon, K., 1916: Zoologická pozorování z okolí Hodonínského. Čas. Morav. Mus. Zem. **15:** 3–12.

Zenkevich, L. A., 1933: Nekotorye momenty zoogeografii severnogo poľarnogo basseina v svyazi s voprosom o ego paleogeograficheskom proshlom. Zool. Zh. **12:** 17–34.

Žitňan, R., 1974: Pisces. Zbor. TANAP **16:** 209–213.

Žitňan, R., 1976: Príspevok k poznaniu parazitov hlavátky v tečúcich vodách Slovenska. *In:* Hlavátka podunajská. Compiled papers. Príroda, Bratislava, pp. 56–64.

Index of Scientific Names

Index of Geographical Names

r. – river, res. – reservoir, s. – stream, Mts. – mountains.

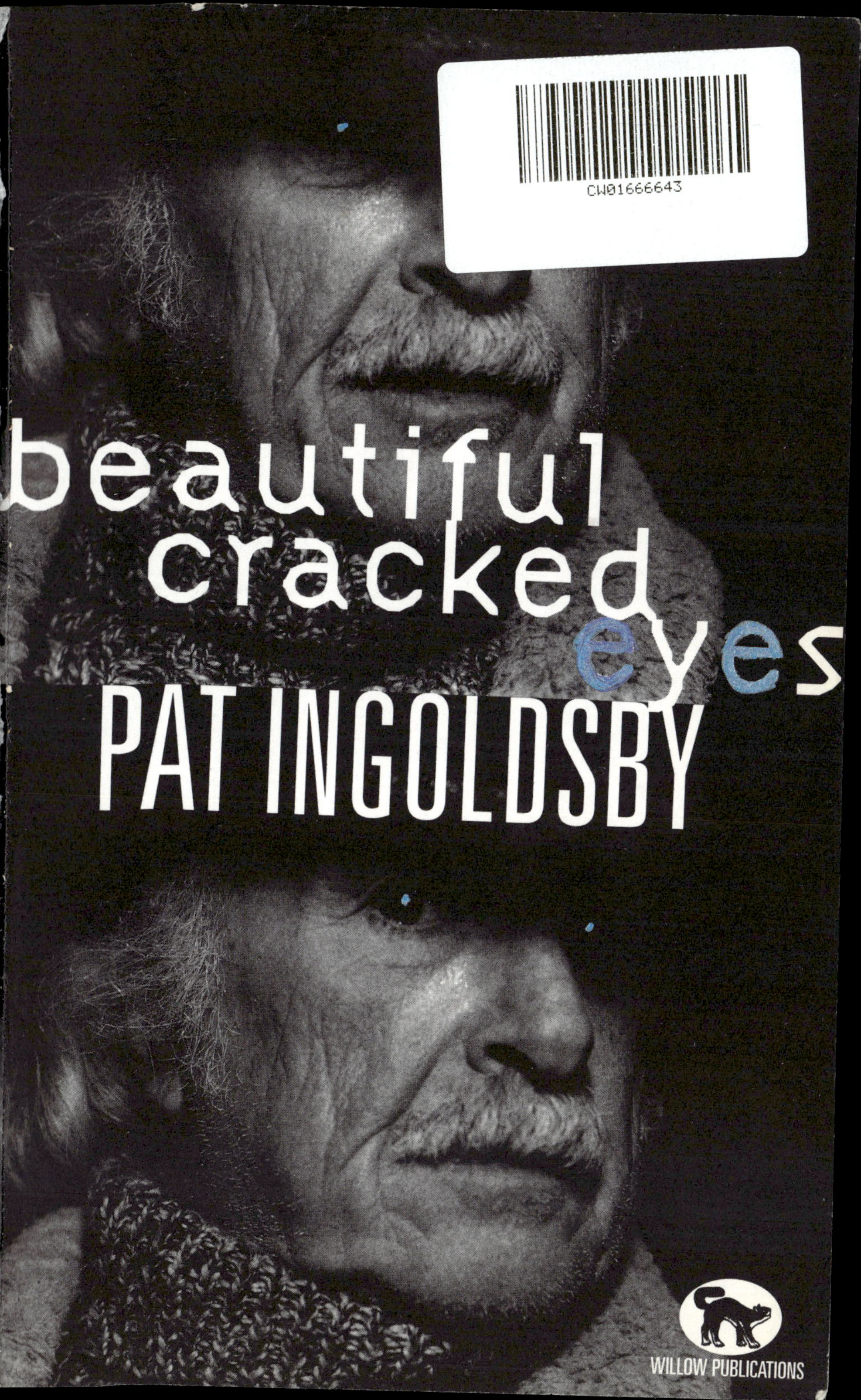

beautiful cracked eyes

PAT INGOLDSBY

WILLOW PUBLICATIONS

Other books by Pat Ingoldsby.

POETRY
You've Just Finished Reading This Title
Rhyme Doesn't With Reason
Up The Leg Of Your Jacket
Welcome To My Head (Please Remove Your Boots)
Salty Water
Scandal Sisters
How Was It For You Doctor?
Poems So Fresh And So New – Yahoo!
If You Don't Tell Anybody I Won't
See Liz She Spins
Half A Hug

PROSE
The Peculiar Sensation Of Being Irish
Laugh Without Prejudice

FOR CHILDREN
Zany Tales
Tell Me A Story Pat (Tape)

For Reg

with high regard
and delight
you're still around

A delight to meet you Noreen!
Love
4/3/2004

First Published in 1999
by Willow Publications (Dublin),
that's me, Willow and Hoot,
3 Vernon Court,
Seafield Road West,
Clontarf, Dublin 3,
EIRE.

love my poems, hate them, have fun with them,
enjoy them, laugh with them, cry with them,
put them on toast and eat them, do lots and lots
of sparkly things with them and they will dance

study them and the life will go out of them

thanks to my good and true friend Steve Averill
who more than anyone has made the difference
the superb cover design and layout is his

thanks to Diarmuid Doran for the lovely new look
the photos on the front and the back are his

thanks to Mark Curran who has been a good friend
to Willow Publications down the years
the photos on the title page are his
I'm delighted that you're in this one Mark.

Thanks to Willow and Hoot for keeping me warm at night,
dipping their tails into my coffee, sticking their whiskers
up my nose, letting me open and close doors,
for being the best of friends.

*honorary human directors: Steve Averill & Mike Moloney

Assets: 20p

ISBN: 0-9523052 5 9
Printed in Ireland by Colour Books Ltd.

Serene

The sea receives purple light
when day is ending
and lets it in deep.

It never says look at me.

The sea smashes high
against black rock
and hisses back
into itself.

It never says look at me.

The sea leads shells
into perfect spirals
and puts
mother of pearl
into them.

It never says look at me.

It is Finished

Every afternoon when school was over
the tall young man and two little boys
waited on the railway station at Raheny.
The man had jet black hair
brylcreemed and brushed straight back.
He had a black moustache.
The two little boys clambered
up and down the grassy bank.
They chased up and down the platform.
The man watched them and smiled.
The man was their teacher.
He was also their da.
When the train came into sight
around the corner from Killester
under the bridge
and into the straight
Paddy the stationmaster
rushed out of his office
jumped down onto the tracks
and scampered across the rails
little jumping steps to make him faster
one hand on his coat
one eye on the train.
The little boys watched him cut it fine.

When the train approaches Howth Junction
it can go two ways – right towards Howth
or straight on to Malahide
the way that me, Michael and Da
went every day
home to the little house
in the heart of old Malahide village
so close to the sea that high tides
flooded underneath the door
so close to the village green
that circus horses pulled grass with their teeth
and we heard them.

When we lay in bed at night
we heard Da tip-tapping the typewriter downstairs
we heard the sea worrying the edges of the village green
we heard the sign over Long's Garage creaking
we heard the wind coming down the chimney
we heard trains rushing towards the Twelve Arches
we heard a far-away foghorn saddening the night.

The time old Mr. Daniels lay dead next door
we heard men and women's voices together
rising and falling with the rosary.

Long after my father died in 1978
every time I went out to Howth
and my carriage swung off to the right
I would look away down the other line
the way that me and Michael and Da
went together.

All the old sounds were still down there
at the end of going home
waiting safe
for me

The day after my mother died in 1997
I was travelling out to Howth
and I looked down the line.
There was nothing there at all.

A cold door had slammed shut
and frightened all my sounds to death.

About Hoot (One)

I gave Hoot a dish filled with lovely crunchies.
On the box it said – WHISKAS COMPLETE
WITH CHICKEN, GOOSE, TURKEY AND DUCK.

After munching for a couple of seconds
he said – "There's no bloody goose in this!"

I do not feel sufficiently qualified to argue.

About Hoot (Two)

Hoot's idea of bliss is to waken
deep in the heart of the beanbag
deep in the heart of Willow,
haul himself ashore,
rob Willow's dinner
and lower himself
back into sleep again

About Hoot (Three)

When I called Hoot for his breakfast this morning
he handed me the following prepared statement –

"Until I instruct you otherwise
I am to be addressed as
– *'The Pussy Cat formerly known as Hoot.'*

American papers please copy.

It's all the one

A man who spends much of his time standing in O'Connell Street
told me – "I saw this fellah with black gloves walking past
and the next time I looked he was a gorilla."

I told him that my sitting room walls, ceiling and floor
changed shape so much last week that my stomach fell down.

"My psychiatrist wants to take me in again" he said.

I haven't got one.

Aigne Cham

Somedays more than others
when you look out to sea
Lambay Island
appears to be much nearer
to the shore.
This is because it floats in
during the night.

It knows that none of us
will ever realise the truth.

It knows that people
will talk to one another
about
    tricks of the light
    atmospheric conditions
    and other arid stuff like that.

Celestín Mac na hAigne Caime
told me about it.
But nobody
will ever believe him
because they say he is mad.

Two Disprin in a hurry please

In the window
of the Irish Aviation Authority on Burgh Quay
is a brightly coloured poster
captioned *AIR TRAFFIC CONTROL.*
It shows a young man and young woman
wearing headphones
concentrating deeply
and looking for all the world
like they are in the throes
of a blinding migraine headache
– zig-zagged lines are whizzing around
all over the place.
Even thinking about it gives me nausea.

I'll get the ferry thanks.

Going Going Going

How happy I was to go driving with my sister Brig
when David was little
and loved to watch church spires and gasometers
sinking down behind buildings.
Nobody realised what he was doing until he asked
– "Mammy – where is that church going"?

So we would drive down towards Holles St. Hospital
chanting – "Going going going......GONE!!!"
as the Pigeon House gasometer completely vanished.
You'd have to go into reverse to bring it back up again.

MÁTHAIR

It amazes me how
Eileen
a very old lady
skeletal
feeble
confused
and
utterly dependent
represented to me
a safety net
under
my worst fears
even though
I seldom went
to see her.

Diminishing Returns

During my very darkest times
when people looked in at me
through a peephole in a door
and I didn't know
who was looking
and cared less
my grandmother
came to see me
and said –

"Think of people
who
have only
got
one foot."

She went on to visit my grandfather
who had just had his right leg removed
and probably told him
to think about people
who have no legs at all.

Rock on granny
– two arms and a leg to go.

A Treble Dare

The next time you are passing a gang of workmen
hacking the road up
with picks, shovels, drills and diggers,
call them together
point lovingly at the sign which reads

WE APOLOGISE FOR ANY INCONVENIENCE
CAUSED BY THESE ROADWORKS

and say –

"It's alright boys – carry on.
Please don't worry about it.
I forgive you."

See what happens.

YOUR BEST FRIEND

I think it's awful
the way that we take dogs
and train them to sit
whenever we say so.

It's no wonder
they go around the place
biting postmen.

Parents Please Note

When it came to dying
both of my parents
had a terrible sense
of timing.

My father died on Christmas Eve.
My mother died on New Year's Eve.

I think there should be
a very unspecial insignificant date
– like say, the 15½ of May
or the 9⅔ of September
which doesn't celebrate
or signify anything at all.

If mothers and fathers
have to go and die
they can do it
on one of these.

Separately.

Little Life

When winter hit its deepest dark

rabbits tied their legs together
and rolled off river banks

steep birds climbed as high
as the sky would take them,
shipped their wings
and dived headfirst down

cows held their breath
till it stopped

rich trees withered

old folk shuffled into cupboards
and forgot

On the first day of Spring

a giddy flea danced.

The day I went out to Jurys to meet Robert
29-1-98

16.25 The Gents

I find it a deeply spiritual experience
to be doing my wee wee
and listening to "The Lord is my Shepherd"
at one and the same time.

16.30 The Coffee Dock

Eating a omelette and chips
and looking down at a footrail
which is being supported by four brass horses.
Even though their faces all look the same
I can tell at a glance which one is not taking
its fair share of the weight.
After working as a stage-hand
lifting heavy scenery in The Abbey
I can spot chancers like that a mile off.

17.10 The Front Lobby

She is wearing a short tartan skirt
black tights
her legs are crossed

I know excactly what you are doing
man in the armchair

now you know that I know
because you are standing up
and walking away

you had the better view

YOUR PLACE

In my mother's time
you sat beside your husband
in the passenger seat of the car
and you went wherever
The Lord and Master drove you.
When he went into the shop
to get the paper
you sat there in silence
and you waited.
When The Master returned
he dropped the paper into your lap.
You could read it till you got home.

Ah Here

"Would you send down a coffee to my room please?"

"No problem."

"With Jersey milk please."

"No problem."

"From a hand-milked cow please."

"No problem."

"Called Ermintrude."

"No problem."

"Ermintrude McGinty."

"No problem."

"Direct descendant of the Táin Bó Cuailgne."

"Would you ever go and fuck yourself."

Letter to a Greyhound on the Night Me and Brendan Each Placed Bets on Ten Consecutive Races and Didn't Win a Penny

Shelbourne Park Wednesday night 4th February 1998

Dear Glenbrien Buzz,

Do you know why I put a pound on you?
I don't know the first thing about greyhounds, bets, form,
weights, or any of that other stuff.
It's just when I saw what they had written about you
on the racecard I said to myself
"There's the one for me."
It was like reading the story of my life.

– *Old stager who is not well drawn and could struggle* –

You didn't win the race and I couldn't care less.
The way you wagged your tail before and after
was simply brilliant.
You got out there underneath the lights
and you had yourself a ball.
The night air, the scent of the grass, the good buzz,
you missed nothing.
Fair play to you.
I have already forgotten all the other dogs I backed.
I'll remember you.

Thundering Bunnies

The man looked into the traps
liked what he saw
waved his white paper
and
ring ring
"Here comes the hare!!"
Jerking speeding
riding the rail
"Here comes the hare!!"
The dogs
ready steady primed
programmed to burst
when the traps go slam
and the hare in his coming
starting to grow
like nobody in their right mind
had ever seen before
bit by bit
bigger and bigger
swelling up huge and huge
rounding and plumping
into the most gigantic
thundering bunny
that ever shot the fear of God
through your granny
lumbering on and on
rocking and rolling the rail
hurtling by the traps so fast
the dogs shot out
before they looked
and saw and said – "Holy shit –
that thing'd eat ye!"
Back they went the way they came
screaming through reverse
pulling the doors panic-shut
behind them
and every hare that ever fled
went: "Bah jeerie bah –
get your mammies to mind you!!"

SHOUT AND RUN

Under the street light
four young teenage lads
are watching me
crossing over to my house
from the shops.

One of them shouts after me

"Pat – you're only a bollix!"

They are a safe distance away.

I thought to myself
– I've got a pain in my arse
with conveyor-belt trainee-sheep
who feel the need to shout abuse
at anyone who doesn't trudge
the same predictable boring path
as themselves.
So I started to walk towards them.
I wanted to see what they looked like.
I genuinely did.
Usually when people feel the
need to shout insults at me
they are much too far away
for me to see them properly.

The four lads started walking.
I walked after them.
They walked faster.
So did I.

They turned right at the end of my road
and disappeared up Castle Avenue
in a hell of a hurry.

Running.

Caught short at Clontarf Road

If by chance I wander into the deep gateway
at the side of the railway station,
put up two chairs and a table,
and lay out my draughts set
I can sit there for the rest of the week
and wait for a partner.

If I put up a sign which reads

– *"For the next hour I am giving £250*
to anyone who stops beside me"

I will not see a living soul.

If an asthma attack strikes
I may puff and gasp till I die.

IF HOWEVER

I stand up against the wall
open my zip
and get into position
for a rattling good wee wee

a train will stop directly above me
on the embankment
and line itself with faces

a helicopter will hover inches above my head
and ask me for change of a pound.

A marching Fife, Drum and Spit band
who have lost their way
will mark time
and ask for the loan of a compass.

Knot your legs till you get home.

Sort of awkward but Bartley will think of something

Bartley Gavigan and his sister Myra
are pleased to announce the engagement
of their goldfish Lucy
to a table-lamp
which lives next-door.
It hasn't got a name
because it isn't a goldfish.

Seen and heard outside Penneys in O'Connell Street

Tuesday 27th January 1998   13.20

A young teenage guy
                blonde
                handsome
                vulnerable
speaking to
a hungry woman
his mother's age
expensively dressed
leaning forward
all her magnets
working
she'd love
to eat him whole

he says – excited

"I'm reading Ulysses
at the moment"

she says – husky

"Oh! WOW!!"

Grace before meals.

Down the Steps

I once asked the River Liffey
in a radio play
dark years ago

"Can I come down
                    down
the steps

While you come up
                around me

to keep me safe

but only
till the morning.

Can you do that?"

But Liffey said

                not.

Jesus Liffey
it's 30 years on
and the pain
is cutting
my stomach out.

I am asking you
again.

"Can I come down
                    down
the steps

while you come up
                around me?

                forget about
                the morning.

Can you do that?

Can you do it?"

Dormitory first time ever on a ship

When I went deep down inside
the M.V. Munster
steep steps down
and down again
stepping up
and over doorways
until I reached the steerage
the man took my ticket
and pulled out a heavy grey blanket.
"This way son" he said.
Men were sitting on bunks
swinging down their legs
lying on bunks
smoking towards the ceiling
looking down
looking up
dealing cards
laughing coughing
clinking bottles

high up on my bunk
with a curtain
and voices all around me
through the long night
I have never felt safer
waking
sleeping
waking
hearing the men
the silver coins
and every wall around me
throbbing with the ship.R44EYHUUU7Y666

(I wrote this down as far as – "throbbing with the ship."
Then Hoot walked across my keyboard and wrote the last bit.
He probably knows exactly what it means.
I haven't got the faintest idea.)

Me Bollox and Bartley Gavigan

The only reason that people get away
with making boring speeches
is that other people stand and listen.

An Cigire Barnaby Plop

"When I become Minister for the Marine"
pledged Barnaby Plop.
"I shall appoint a cigire
with responsibility for séimhiús and síneadhs
on the backs of trawlers.
Whenever he finds a síneadh out of place
he shall be authorised
to impound the boat
eat the catch raw
and sell the crew into slavery."

That should keep up standards.

Augustín Plop (his brother)

"When I am appointed Harbour Master in Howth"
pledged Augustín Plop.
"I shall make it an offence
to walk down the East Pier
carrying a sand-wedge and a small child.
Too many people are chipping youngsters
over onto Ireland's Eye
to save the fare."

The Plop Family.
A credit to The Brothers and the bata mór.

Toys I Remember

I remember my tin trolley-bus
with the cut-out conductor
standing forever on the platform
his flat front criss-crossed
by the straps
of his money-bag
and his ticket-clicker.

I remember my lead soldiers
with red tunics and black busbies
some of them down on one knee
pointing a rifle
or what remained of it.
Some of them marching
rifles over shoulders.
Some of their heads broken
and wobbled back on again
with matchsticks.

I remember my plastic cowboys
with bandy legs curved so wide
they looked ridiculous
until you pushed them down
onto a horse
and they clicked into place.

I remember my horse
with four legs deliberately loose.
When you stood him on a slant
he walked all the way to the bottom.

I remember Tipsy Joe
top-hatted and dressed in tails
holding a bottle.
Everytime you stood him up
he toppled over.
If you held him upside-down

for a while
then stood him up again
he didn't fall for ages.

I remember my other horse
with spindly thin legs
standing on a circular base.
When you pushed
the bottom of it in and out
the horse bent its legs
this way and that
and courtesied to you.

It seems like
I put each of them down
a couple of minutes ago
and now they are gone.

Don't Look at Me

When the wind blew the nest
out of the tree
and shattered it

the tree said –

"It's not my fault
I blame the birds
they didn't build it properly"

the wind said

"Don't look at me
it's not my fault
I blame the tree
it didn't hold it properly."

And while they argued high above
the birds gathered up
their bits and pieces
flew away
and built a new one.

It's all gone Mrs. Condron all gone

I met Mrs. Condron trudging up New Street
leaning forward heavy on her stick
stopping and resting
and trudging again
brown bandages wrapped elastic-tight
around her legs
"Where's Howard's Yard?" she asked.
"Gone" I said. "All gone."
"And Bertie Boyle's?"
"Gone too. All gone."
"Long's Garage?"
"Gone."
"The Boat Yard?"
"It's a marina."
"Ma-sie Walshe's?"
"A supermarket."
"Howard's cows?"
"God knows."
I left her before she got to where
her little sweet shop used to be.
When she finds a window
filled with antiques
she will shuffle back down New Street
around the corner
and on towards Church Street
where home was
and she could make a cup of tea.

She will go back to the graveyard.

The Hidden Depths of Gavigan Charles Whipsnade McGinty

"The most adventurous thing he ever did
was a silent fart during a history exam."

"Ah yeah – but that was until yesterday."

The first time ever I felt lonely for my Books

God – I felt so sad leaving my books behind
in the shop
even though Tara who signed for them
is lovely.
"That's the section I look after" she said
"I'll keep them nice and prominent."
Then I turned to walk away
and a deep wave of lonely
came up through me.
It took me completely by surprise
I didn't want to walk out of the shop
I kept turning around and looking back.
For a second I even thought about
sleeping outside on the pavement.

We came a long journey today.
I carried them on my back this morning early
across the road to the 130 bus
and on down to the DART at Clontarf Road.
Then rail to Tara Street.
A 90 bus down to Heuston.
Train to Clara.
Bus to Ballinasloe.
Train to Galway.
And I signed each one on the way.

Across Eyre Square in my blue sack
and on down to Easons
where Tara signed for them.

Now I feel lonely pushing up through me
and I want to cry.

I told Brian all about it.
He was selling The Big Issues outside the shop.
He smiled a huge smile and said –

"They'll meet loads of new books in there tonight."

On my way back to the station
I saw a plaque in a butcher's window

– 'WHITE PUDDING OF THE YEAR' –

and I felt better again.

Other white puddings probably hitch in after dark
from all over Galway just to have a look at it.

It's not so bad really.

THE WORLD MAY WELL BE GETTING SMALLER BUT I SEEM TO BE SPENDING MORE AND MORE OF MY TIME TRUDGING AROUND THE PLACE TRYING TO GET TO PACKAGES BEFORE THEY'RE SENT WHIZZING BACK THE WAY THEY CAME

Our courier called at...11.35...today
but was unable to effect delivery
of a parcel addressed to you.
You have got exactly one hour
from the time you read this
to locate our depot
which is tucked away
in the arsehold of nowhere
and claim your parcel
which looks to us like
very expensive stuff indeed.
Failing this
we will either raffle it
or give it Gerry's mammy
who has been having
a very hard time
with the other thing.

Assuring you of our best attention
at all times

IZZ INTERNATIONAL

TRAVEL BACKWARDS AND THEY'LL FLOW AROUND YOU

After travelling at high-speed by train
from Dublin to
                Belfast
                Cork
                Galway
                Waterford
                Dundalk
in the space of six days
I now know for certain
that the fields are moving.

Don't tell me they're not.
I saw a bread-van parked
directly in their way
outside Naas.
They whizzed around it.

The wires go up and down.

Lavender

I'd be much more inclined towards Ruby's snuff
myself.
The lavender-scented scorching powder
of Rossnowlagh

– making eyes water with a suddenness
which drowns kittens

– firing thrills up legs
and down the backs of them
with hair-crackle

– rising the top of your head infinitely higher
than it is supposed to go
and hovering it there for two or three days
while the wind whips out your dead straw
and cools your dereliction.

Then
when you have forgotten
that you had a top to your head at all
– bringing it down again
just like Ruby said it would.

Or Without You

At nighttime
in the forest
secret
rustling things
scurries
cracks
twigs
and creaks

deep ditches

a hundred thousand
hidden things
happening
long before
you ever were
and
happening
long after.

Ciúin

Quiet and quiet
as quietness be

swishing

ease

incoming sea

loosening gravel

little shells

free

whispering
love things

for you
and for me.

Codladh Fada

Just once in every man, woman and child's lifetime
– while they are sleeping alone
with nobody around to observe,
film, or take photographs
just once, for fifteen seconds
they will become Bartley Pipes McElligott
with all of his facial hair, bodily clicks,
porcelain ears, telescopic toes,
musical ribs and sulphurous digestion.
Then they will change back into themselves again
and never know a single thing about it.
In this mysterious, wonderful way
Bartley Pipes McElligott
has ever been
and
always will.

He has never been awake.

Professionally Speaking

"My daughter *the doctor* was only saying
to my son *the engineer*

– 'Isn't Mammy an awful snobby waggon
all the same' –

She certainly didn't get that from me!"

Conversation between two workmen out in the middle of the countryside a long long time ago building the railway to Cork

"Do you know something – at 12.30 p.m. on the 20th of
February 1998 a guy called Pat Ingoldsby will go
whizzing past this very spot on a train and he'll say to
the person sitting opposite –

'Did you ever wonder about the men who built this
railway?
Did you ever wonder what sort of things they talked about?'

And the person sitting opposite will say –

'Now you know.' "

And three Hail Marys...oh my God

There are legions of half-cracked
gap-toothed oul ones
pacing around all over Ireland
short cut
scraggeldy hair
stained-glass eyes
long coats
long skirts
flat shoes
ankle-socks
utterly grey
the fear of God and the devil
hunting them up and down
railway stations

Hail Glorious Saint Patrick.

A HEAVY LOT OF US

For a total of three years
and nine months of her life
my mother was walking around
with either Michael, me, Ann,
David or Brigid inside her.

Thanks Ma.

Nina and Bloody Frederik

The supermarket where I shop
never plays Jimi Hendrix.
They play jumpery men singing
– "Michael Row The Boat Ashore"
and "Puff the Magic Dragon."

I don't think we'll ever have an orgy.

God if I could look out now

My brother Michael said to me

"Do you remember
how we used to try and stay awake
to see the circus going in the morning?
But we always fell asleep.

When we woke up it was gone."

I shook my head.
I couldn't remember at all.

I can picture perfectly in my mind
how it must have been.
Two little boys
trying to keep their eyes open.
Eager to kneel on the morning-sill
and watch the horse-drawn wagons
creaking off the green.

I could draw you a picture
of how it must have been.
I'd love a second chance
to remember.

TELL ME ANYTHING YOU LIKE EXCEPT GOOD

The man had just had his hair cut.
A total change.
Very short.
I thought he looked brilliant.
"That's really great" I said to him.
"Makes you look years younger."

He looked surprised.
Very very unsure.

"All the lads are slaggin'" he said.
"Givin' me a real hard time."

"No – it's great...really suits you."

"You're not just sayin' that."

"No – I really mean it."

Suddenly he realised I was sincere.
He clicked into defensive mood.
Fluttered his right wrist.
Imitation camp.

"Ooooh – you say all the right things."

HIGH AND DRY

The only time I go into churches now
is for weddings and funerals.
When they reach the communion part
everybody in the place gets up
row by row
and approaches the altar.
I always feel very lonely sitting
on my own.

Just once before I die
I would like to be
in a packed Pro-Cathedral.
When they get to the communion
I will go up to the altar
with my head bowed.
Everybody else will sit where they are
saying – "My Jesus – isn't he very holy."

Paul Street Shopping Centre

Friday afternoon 21-2-98 Cork

10p for a wee-wee.
If that's the best
the Celtic Tiger
has to offer
I will gladly go
piss against the wall

What "the daffodils" looked like before William Wordsworth wrote it

SPECIAL IS RIGHT

The sign in the window read

– "PLASTIC RULERS...*SPECIAL*" –

Sebastian Shuffle liked the way
the little bell over the door tinkled
as he went in.

"Eh – a plastic ruler please."

"Special?" asked Gansey
the enterprising stationer.

"What's the difference?"

"Slightly dearer
but much much better."

"Right you be so."

"Who's been a naughty boy then?"
demanded Gansey suddenly
with wild swishes.
"To the back room with you
for a jolly good whacking."

"Oh my God!!!!
Give me an ordinary one so."

"That'll be sixty pence...
eh – can I give you some nice
springy elastic bands?
They're on special as well."

"I'm ringing the guards."

Keep It

I would like to go into the sea
and fall asleep peacefully
without any struggle
but drowning makes you fight.
I am very tired of that.

My Friend the Leaf

Something lovely made me very happy last Thursday morning. It was a sharp clear cold day and I was going across the road to the shops. It was one of those mornings when I was taking my time and looking at silver marks on concrete and red berries on bushes and frost white on black twigs. And then I noticed it. High up in the tree outside my house. One leaf. All by itself. I have never in my life seen anything like it before. A leaf completely alone. All the branches around it were stiff and frozen and empty. Nothing left. And there it was high up. One leaf. The only one on the tree. I stood for ages just staring up and saying to myself – "God – would you look at that" – and I was smiling to myself with a mixture of delight and wonder and admiration. I was saying to the leaf – "You are absolutely brilliant – I don't know how in the name of God you managed it...all the other leaves are gone and you are still there...amazing...that's what you are...amazing. The very best of luck to you...hang on...as long as you can...hang on."

Coming back across the road from the shops I looked up and grinned. It was still there.

Two or three times during the afternoon I stopped whatever I was doing and suddenly thought..."I hope it's OK." – and I went out to check. Half of me didn't want to look in case the tree was completely bare. I felt a lovely warm joy each time I discovered that my leaf was still there.

That night there was an angry wind growling around outside my house and I lay awake in bed listening. "That's it" I thought. "It's gone now anyway...no leaf in the world could survive a wind like that." And I fell asleep feeling very sad and lonely.

Next morning it was still there. Proud and defiant at the top of the tree. God – I felt so happy. And the next morning. And the one after that. I could not believe it. It was the first thing I checked every time my bus swung around the corner into Vernon Avenue. It was the first thing I checked when I woke. Time after time that one brown surviving leaf filled me up with delight and wonder and courage.

For five days it hung on. Five days of wind and rain and bitter cold. I travelled down to Cork on the third day and when I got back late the following night I was thrilled to see it was still there. Part of me was hoping that maybe it would last the winter but I knew that it couldn't.

On the sixth day it was gone.

D 285 MFV Lemarin Slán

East Pier 21-3-98

They towed her in
green and silent and slow
her wheelhouse
her masts
already gone.
They towed her in flat
her body cut away
and tied her to the pier
the living parts of her
obscenely open
to wind or rain
or anybody walking
who wished
to look down
and see her secrets

men walked up and down on her
saying with loud voices
the best way to cut her into pieces.
Then they drove away
and left her tied.

Late into the evening
the gulls
were making loud crying
around her.

Heavy equipment
is coming in the morning.

Putting the quiet into me

The grass is loving my feet
the grass is loving my feet
the grass is loving my feet
and I can feel it.

The wind is loving my skin
the wind is loving my skin
the wind is loving my skin
and I can feel it.

The rain is loving my hair
the rain is loving my hair
the rain is loving my hair
and I can feel it.

The cold wall is loving my back
the cold wall is loving my back
the cold wall is living my back
and I can feel it.

The silence is loving my mind.

Cold Harbour Empty Mirror

East Pier Howth   Friday morning   20-3-98

Silence

empty

a dog barking
onto it

silence again

silence

empty

shrieks drowning
onto it

silence again

silence

empty

frightening itself

silence again.

Looking off the back of the East Pier

late afternoon 20-3-98

Four black heads
motionless on a flat calm

a distance from the shore

making you wonder
for a long time
whether or not
they are seals

then
each one
in its own good time
proving to you
suddenly
that they are!!

Punt ar an mBradán

Using a highly trained herring in a furry suit
and six greyhounds who love to swim
an t-Uasal Macdara Haverty
has put great racing into the harbour.

Two chances.

She didn't have to tell me

Once when I was little
I floated a stick
in the river of rain
which was racing
down the edge
of our road
and I watched it
spinning
away
towards China.
Ber Howard's sister
brought it back
an hour later
and said –

"It got stuck
outside Brady's."

WHERE ARE YOU ALL GOING?

All of a sudden
so many good people in my life
are going out of churches
backwards on wheels
that I am confused with the shock
and with the sadness.
They are almost indecent in their haste
to get out of here.
Five good people gone since Christmas
and another one I know of
on the way.

I followed them all from the church
to see what the hurry was about.
Knowing them all as I did
I felt sure they were hastening to
a groooooovvvvvy place indeed.

Four went straight into the ground.
One into an oven.
They don't fool me for a minute.
They're up to something.
Especially Maureen.

PLATFORM 2 HEUSTON
11.00 a.m. 25-3-98

The station announcer echoes her mysteries
around and around the cavern
mystifying everyone except the locomotives
who translate them into loud pishes
and hiss them to one another.
The checker at the gate
continues to spread ticket-confetti
for the wedding which will never happen
because nobody can understand the pishes.

Off with their heads
as well

The old lady barely tethered this side of connection
indicated her dog and confided –
"Jessie is the winkiest cat in the whole world.
Whenever a hornpipe comes upon her
she bounces it out through the marigolds."

An hour later I met her again.
She was waving a letter at me.
"I have just been summoned for jury duty."

I went straight down and bought a television licence.

AND THE LANDLORD MAYBE
as long as you've got the rent

I have never seen a sign which says

– POET WANTED –

turnip trimmers
forklift drivers
dog handlers
roustabouts
wranglers
finishers
riveters
tanners
extras

yes

all these and more

wanted

poets

never

I suppose it is because
apart from their mammies
nobody really does.

Extra-large Eggs
memo to farmers

Sudocrem

sound-proof the farm

THEY WERE HERE BEFORE YOU

I would feel infinitely kinder
towards huge big shopping centres
if they let the pigeons in when it rains.

Don't tell anyone

There was a dog going down the pier
which looked so much like
a curranty bun
that a crow spread butter over it
flew away to get a bit of jam
and only made it back in time
for the crumbs.

Sign on the back of a bus

"JERVIS CENTRE

750 spaces in the heart of the city."

Please add –

"They ripped it out to make room for them."

Titanic

The next expedition
is simply going down
to switch off the engine.

Let that be an end to it.

THANKS LITTLE BIRD

Malahide – early morning   2nd April 1998

In the quiet of the Protestant graveyard this morning
a single bird
hidden
sang to my tired body

– "If you were to sit down here and make a list
of all the things which worried these people
there would not be enough paper in the world
for you to write them on –
yet for each person, only one of the things
definitely happened" –

I walked down the hill towards the village
feeling infinitely lighter.

REGARDING THE HEADS AND LEGS OF ELEPHANTS MOSTLY

While I was attending O'Connell Schools
during the middle fifties
I learned how to say –

"Elephants have got big heads and little legs"

in Latin.

I have never had an opportunity to include this fact
naturally in a conversation since.
The one single chance I used to get
was when Brother Allen would ask –

"Ingoldsby – how would you say in Latin –
'Elephants have got big heads and little legs'?"

and even thought he knew already
I would tell him –

"Elephanti habent magna capita et parva crura."

Sometimes he would go on to ask the entire class
individually
– 37 boys –
how they would say it.
He would keep at it until we all got it right.
On a bad day it might take hours.

Then the subject was closed.
We might not discuss the heads and legs
of elephants again for weeks.

I have lived my life in waiting ever since.
Eternally on stand-by.
Ready for the right person to rush up and ask –

"What can you tell me in Latin
about the relative dimensions of elephants' heads and legs?"

I have waited so long at this stage
it can even be a burglar
who comes crashing in through my window
at three o'clock in the morning.
As long as he went to the Brothers
and has got the right question
I couldn't give a shit.

UAIGNEACH

There is a huge big Dodo lives behind the hill at Howth.
A very long time ago when all the other Dodos became extinct
this one didn't.
It became big and huge instead and came to live in Howth.
If you saw it from a great distance
you would think it was six gigantic elephants
with their legs wrapped around each other.
But it's not.
It's a huge big Dodo which should be extinct.

Sometimes it pretends to be a very large bus stop
and stands at the side of the road for ages without moving at all.
People often wait beside it until their bus comes.
Sometimes when it is feeling sad
it lies down and pretends to be dead for ages.
Sometimes it stares out to sea and hopes for another Dodo.

It never comes.

A SAFE PLACE INDEED

I would feel safe
going deep into the woods
with my father in winter
I would feel very safe indeed
looking at the sawdust
and white chippings
on the floor
where the foresters
had been cutting with axes
slicing round with saws
and putting woodsmoke
on the air

I would feel safe
when we reached the clearing
where Reburns lived
I would feel very safe indeed
loving the turnstile in the wall
the cold feel of it
the way it squeaked
when we went through

I would feel very safe
in Reburn's kitchen
with my father
big delph plates
rolling on the dresser
bordered round in blue
plump toby jugs
with firelight on their smiles
Mrs. Reyburn
splashing milk with bubbles
for me to drink
her hair tight-bunned
pockets deep enough
in her apron
to drop a clothes peg
and never hear it fall

and the shy girl in her silence
white in the shadows
making me shy just to look at her
making me wonder silent things
all the way home.

Malahide Castle Bell

The castle bell would come ringing at eight o'clock
when there was nothing on the morning
nothing on the air
a single ring
over and over
                calling through the trees
                the fields
                across The Lady Acre
a slow space between each call
up The Hill
down the village
over Yellow Walls
across La Mancha
everybody knew it was time
and the men were starting work

At twelve
and one
and six

stopping the men
and starting them

the bell would come again.

Two years since I seen you

A young man now – still whistling on the street
still running when they chase you

this you told me

April 1998 near Nassau Street

"Two of them took off their caps
an' I lept up in the air
higher than the both of them
an' they took off after me
round an' round the statue
an' I ran into that brown place
where you got your coffee or somethin'
an' they came in flyin' after me
an' I ran into the pool hall
an' hid under the table
an' they said – 'He's gone'

an' I was."

Good health to your lungs.
Good health to your legs.
May they never catch you.

The country diary of an Irish woman who spends most of her week watching birds

Wednesday 3rd Jan:	*Cock blacknap on kitchen window-sill several times* *Snow remains on trees and bushes in garden*
Thursday 4th Jan:	*Cock blacknap on window-sill* *Ice still on drive*
Friday 5th Jan:	*Dark thrush seen* *One coal tit eating fat scraps off kitchen window-sill*
Saturday 6th Jan:	*Severe frost all day*
Sunday 7th Jan:	*All shrubs and trees covered with frost again*
Monday 8th Jan:	*K. came today* *Together we bit, chewed, licked, screwed, humped, bumped, rucked, fucked, moaned, groaned and generally drilled the living shite out of one another* *2" of rain in Blackrock*
Tuesday 9th Jan:	*Solitary redwing in trees at edge of wood* *collared dove and 2 robins in garden*
Wednesday 10th Jan:	*Severe frost all day* *cock blacknap feeding again*

Cut To Bits On Easter Sunday

East Pier   April 1998

Today at 12.35 I reached breaking point.
A deep black cloud paused over Malahide
and took a precise bearing on Howth.
Then it began to move across the water
making everything dark underneath.
When it reached the entrance to the harbour
it paused again and lined itself up
for a clear run at the East Pier.
When it was in place directly over me
it cut loose with a wicked spitting fusillade
of hailstones which cut, stung, prickled
and pin-pointed me back against the wall.
27 particularly bitter little bits of hail
riddled into my basket
and ate one of my books whole
– a copy of 'See Liz She Spins" –
the only book I shifted all day.

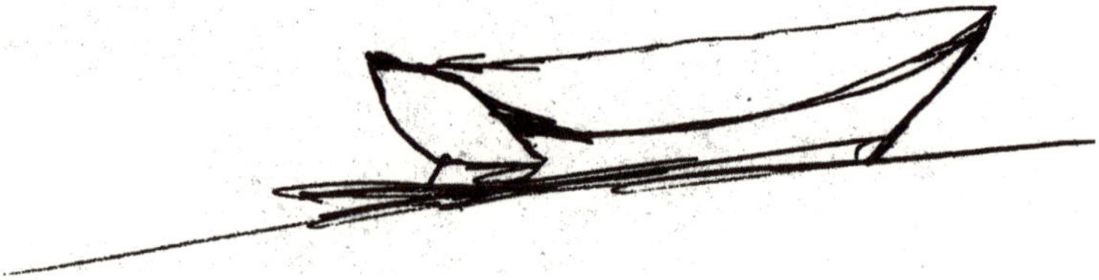

It Often Takes Much Longer
East Pier   Easter Monday

12.05 "That's a cold spot you've got today"

12.09 "Cold isn't it"

12.15 "Very cold today Pat"

12.17 "It's a cold one"

12.20 "It's cold today isn't it"

12.40 "That's a cool breeze Pat"

12.50 "You're in a cold spot there"

12.51 "That's a lovely day"

CHRISTIE

Behind high walls and gates
I remember the wild fire of you
light and crazy crackles
putting so much fright into your poems
that they danced miles high
and scared the living shite
out of anyone who was not dancing.
Your eyes flying out at tangents
and skipping round the doors
before they banged.
Your words coming out so fast
they whizzed white hot
along your wires
and made a mockery of cages.

I will never forget the way
you set our Wednesday nights
on fire.

Mad
they called you trouble.
God I knew the steps.

They put you down the Base
and stopped you dancing.

FÍOCHMHAR

Mo theanga ag léimt istigh leat
tusa rince do theanga i mo bhéal
as do mheabhair
le sleamhain leachtach na seilide
mise meidhreach ar strae
is pléisiúr am' greimt le chéile
go dtagann sruth na lán taoide
le scréach is le h-osna.

What else?

A couple came walking towards me on Castle Avenue
the pair of them wearing grey colours
which would never argue with you
husband and wife swinging their arms
and making brisk steps for better health
wrapped up warm the way you'd never get a cold
a plastic mac each just in case
and him saying to her as they sped past me –
"Yes – that would be the sensible thing alright."

Trains and the sounds of them

Our lives were governed by trains and by the sounds of them
– "Get a move on! There's the ten past eight!"
and there she'd be sure enough
huffing and puffing below in the station
and we'd be rushing up and down the stairs
searching for bags and books
are you going to be all day in the bathroom
there's the half eight God we're rightly late
and if you were still in the kitchen
when the twenty to nine made a huge noise
slipping on the rails with her icy wheels
you may as well sling your bag in under the stairs
and forget about it.

Or the slow creak of a goods backing in off the main line
metal weight making lovely rhythmic clicks
over and over the joints
a long line pushing curving
the driver leaning out looking back
then he does the "Cock-a-doodle-doo"
down the distance to the signal-man
we're in we're clear
moments later the heavy switching of the points.
Soon we'll hear cold metal plinking clinking
tumbling up and down the siding.

Or warm in bed on a winter's night
my brother breathing sleep beside me
I can hear a train coming round the corner
into the echo of the quarry
and down the straight between the steep embankments
the engine crazy with fire flinging up high sparks.
A different sound again
when she has burst under the station bridge
and started to fade away across the water
I'm listening for the rumble very faint
she is passing over the Twelve Arches
and then a slow slow fade away from the estuary
and inland with the saddest call you've ever heard.
Good night late goods....good night.

Mo

Maureen
beautiful friend
for you
the fields are waiting
Maureen
the rabbits want a chase
Maureen
the wind wants to race you
loving again
the gypsy in your face.

Let the branches
touch you with amazement
Let the beauty of the world
ask *you* to dance
Let love light up and laugh
because it knows you
and
Let your friends rejoice
we had the chance.

Mister Rusty Bum

When the crocodile eats you for his lunch
he doesn't know a thing
about
taking out the pips
peeling away the skin
cutting off the rind
or other groovy civilised practices like that
so he wallops into you whole
and includes the lot
– hearing-aids
  dentures
  navel rings
  artificial hips –
if it's in you – he eats it.
Bon appetit.
Metallic fart bubbles.

Incident With "me bollix" **In it** (One)

Pearse Station Northbound platform
13.28 23rd April 1998

The man with the big dark wee-wee stain on his jeans
and much alcohol on his breath
reached down into the green litter bin
and fished out a newspaper
He look at the name on it

– *EXAM BRIEF Supplement* –

"Me bollix" he said
and rammed it back in again.

Beautiful Cracked Eyes

It's the people with the beautiful cracked eyes
who come and speak with me
time after time.
Sometimes they are trembling
and they get their words mixed-up.
Sometimes they do not know what to say at all.
In a world of people afraid
to speak or to listen
it is the ones
who are supposed not to know
time after time
they are the ones
who come and speak with me
time after time
when I am drowning
they are the ones
who come and save me.

The Naomh Sniffle

The Naomh Sniffle Pipe Drum and Spit Band
loved their jigs and reels and hornpipes so much
that they rickedy rattled on and on and on
careering out of one jig and wildly into another
leaving dancers weak, exhausted, dead and dying
inside the parish hall and out strewn across the fields.
"Aw Jaysus" groaned a person. "If they don't give it a rest
I swear to God I'm goin' to split somebody."
And he did.

Nobody shouted "Me life on ye lads!" anymore.
Nobody shouted anything.
They had not got the strength.
The Naomh Sniffle Pipe Drum and Spit Band
rocked roared and rattled through jigs to take the paint off a door,
hornpipes that would drive you up a chimney
and reels to ransack your sanity.
People who toppled into ditches stayed there.
Men climbed tall trees and when they reached the top
kept on climbing.
"For the love and honour of God would some kind person
not machine gun them or something?" asked the parish priest
in his monthly bulletin but nobody had the energy.
They played *'Maggots in the Minnestrone'*, *'Bridie Carey's Gallstones,*
and *'The Knickers of Carmel O'Flaherty.'*
They played *'Pearse McGonnigle's Favourite'*, *'The Horsewhip On
The Dresser'* and *'We'll Have To Eat The Feathers.'*
They played *'The Pimple On Your Arse'*, *Johnny Hargadon's Ricketts'*
and *'If You Didn't Yet You Will Now.'*
People bricked up windows, cemented chimnies and talked to
the wall.
People went under rivers and stayed there.
People sacrificed goats.
But they could still hear them.
'Around The House And Mortify Your Mother', *'Raggle Taggle
Gypsies'*, and *'The Boys Of Killaloo.'*

The Naomh Sniffle Pipe, Drum and Spit Band
offered to do requests and people left the parish.

"Mother o' Jaysus if they don't stop soon
I'll fetch me boot through someone's bodhran"
vowed Muiris Durack but his brother got there before him.

And then
miraculously
the sudden unexpected sweet sound of silence.
All together......they stopped.
For the first time in fifteen years
The Naomh Sniffle Pipe, Drum and Spit Band stopped playing.
All together, they trooped out of the parish hall for a wee-wee.
And then
all together, they trooped back in and started to play again.
"Martha Hahesy's Nightie, "Go Easy On The Buttermilk'
and *'Nora Kilfeather Is Stretched On The Roof.'*

One of them has a weak heart.

Oh Mister Pringle

"What a delight it is for me to see you this morning Mister Pringle. I hope that you have a lovely happy day."

"What?!!"

"I said – 'what a delight...' "

"I know...it's just eh...people don't usually say things like that."

"Oh...right."

He walked on a bit further.

"My goodness me...it's Mister Pringle.
What a gorgeous surprise for me to meet you this morning.
I hope that the rest of your life is filled with the boundless joy you so richly deserve."

"What?! Oh...I see....yes....thank you."

He went in to buy a paper.

"Mister Pringle – I cannot find words to express the glow
which fills me up at the sight of you.
Good health and long life. You deserve it."

"Thank you."

He got into the DART to go into town.
Everyone in the carriage stood and sang
– "For He's A Jolly Good Fellow."

He approached a bus stop to travel home
and said to the man who was standing beside it
– "With respect...before you say anything to me...
will you kindly take each word individually,
encase it in hard pre-cast concrete,
and ram it up your arse."

The man shot him.
Dead.

Oideachas in Aisce

East Pier, Howth 13.40 30-4-98

A man stopped and said to me
– "I can remember the definition of a sonnet
– *'Fourteen lines of Iambic Pentameter.'* "

"God – I can remember the rhyming scheme
– *'abba*
abba
cd
cd
cd'."

But no matter how hard we tried
neither of us could remember a sonnet.

When I'm Big

All the way up the railway bank at Clontarf
yellow flowers growing taller and taller
until they can peep over the top
and watch the trains.
The shorter ones down the bank asking
– "What can you see? What can you see?"

Ná Bris

"Bí cúramach! Ná bris mo h-ionga!"
ar sise.
Lá ina dhiaidh sin
do bhris sí mo chroí.

Overheard at Howth

Sunday 3rd May 1998   evening time

"Do you like them?"

A young trainee hard-man was pointing at a huge new 98D reg car.

"No" said his girl-friend. "I prefer the lovely little ones."

"If I bleedin' called round for you in tha' you'd get into it."

"Yeah...right......but I love the nice little ones."

"There's somethin' bleedin' the matther with you."

LOVELY THINGS THAT HAPPEN

Bank Holiday Monday 4th May 1998
On the East Pier at Howth

LITTLE BOY (racing up to me wide-eyed): "Are you a pirate?"

THIS ONE WASN'T

Myself and another man were chatting together
about great Irish soccer players of the past.

"Do you remember Joe Haverty?" he asked me.

"Joe Haverty! Joe Haverty!!" butted-in an old man
who was shuffling past on his walking stick.
"Sure that fellah couldn't kick a ball!"

At the same Royal Irish Academy For Advanced Begrudgery
you learn that James Joyce couldn't write his name.

THIS ONE WASN'T EITHER

WOMAN (casting her eye over my books): "The man who lives next
door to us.......Henry........
Henry somebody.......he
writes better books than
them."

This is for you

I have taken myself as far away as I can
from the world of literature
because I do not speak the language
nor do I ever want to.
I have taken myself as far away as I can
from the emphasis on sales figures,
number of weeks in the charts,
being nice to arseholes
because it will influence
the figures,
accepting ex-cathedra judgements
from windbags
because a publisher believes
they are important,
being placed on hold
going where you are sent
staying till it is time.

I have taken myself as far away as I can
from the approval of arbiters, editors,
directors, producers
and all the other indicators
who decide what is successful
and what is not.

I have come to a very beautiful
and a very lonely place.
Sometimes it is a scary one.

From time to time
I meet you there.

You make me safe.

Just mind yourself

The wind blew the shadow of a gull off the pier
and whipped it out to sea.
"Come back with that this minute!"
screeched the gull.
"I'm not finished with it!"
"God I'm sorry" said the wind
"I'm always doing that"
and blew it straight back again.

Hinting into hard

When you came and talked with me on the pier
I enjoyed watching the way the woman of you
moved and walked and sat.
You reached and you kept reaching
the man of me.

Slightly

very slightly

damp.

Nothin' Shakin'

When the power went off in my house this morning
Willow and Hoot remained exactly as they were
– Hoot at his dish in mid-lick
– Willow almost up on the window-sill.
They stayed like that
– immobile –
until the power came on again.
I'm going to ask the ESB about it.

RIPSNORTER

Rupert Netherwhisk
unleashed an unmerciful fart
which drilled ferociously
downwards
into the harbour wall
and lifted every person
who was sitting on it
twenty feet up into the air.
Some of them twirled around
as they were rising
and took in the view
on the other side.
"Lovely" they said.

Rupert raised chips
out of the concrete
but
there was no malice in it.
Just wind.

Granny and Golf

My grandmother would regularly tilt the television
as soon as a golfer had played his putt.
"Take that you person of privilege" she would say
as the ball changed direction and curved past the hole.
She was equally good at snooker
but she seldom ever interfered with it.
She had it in for golfers and that was that.
I have absolutely no idea why she did it.
She never said and none of us ever asked.
Possibly something about years ago
– plus-fours, tweedy socks and exclusion.
She could make traffic lights change colour
just by looking at them
but the golf balls were the best.
She spent a lot of time praying to a statue of John Bosco
down in the church.
She used to rub the statue's feet every time she passed.
They were rubbed bare.
Perhaps this is where she got the power.
Perhaps she sacrificed goats.
I don't really know.

Nick Faldo suffered most.

Give Me A Break

The Corporation Information Board
flashed its lights

– "ILAC   254 spaces
  STEPHEN'S GREEN
  SHOPPING CENTRE   372
  EH...PAT...YOUR FLY IS OPEN"

Jesus – there's no need to tell the whole city!
A whisper would do.

London Cold Metal A Bit Of Warm

The lift stopped.
A metallic voice announced.

– *"Second floor –doors opening"* –

The doors opened.
3 people got out and left me on my own.

– *"Doors closing"*

The doors closed.
The lift moved on up.

– *"Third floor – doors opening....eh....Pat."*

I don't know how it knew my name
but it put a bit of crying into me.
Nobody else had spoken it all day.

Fly Away Peter

The little boy came racing down the pier
eyes fixed on the crow
hungry pecking bread
biding its time
knowing it can take flight
any time it chooses
watching with unblinking beady eye
the little boy feet flying
hands outstretched
nearer nearer
any second now
GO!!!
the crow rising up
high and higher
laughing in its caws
the little boy flapping his arms
and rising up after it
fast and strong and sure
putting the fear of God and his feathers
into the crow
chasing it clear across the harbour
out to sea
over the top of Ireland's Eye
into a huge black hole.

Willow Publications Board Meeting

in the hot press

When your fellow directors
are a couple of pussy cats
there is not much point
in fixing a date for board meetings
because they will not come.
It is much much better to wait
until the pair of them are together
in the hot press or on a beanbag.
Then hold the meeting straight away.
Yesterday evening
I stood beside the hot press
and announced into it
– "Willow Publications – up 27"
(not true really but we think it sounds great).
Willow said "meow".
Hoot was asleep.

In the absence of any other sort of record
I suppose that these are the minutes.

Clontarf Road Roulette

Look across at the far platform.
Do you see those people waiting
for a north-bound train?
Look down the line towards Connolly Station.
Do you see that train approaching at speed?
Well it's not going to stop here.
It's going to hurtle on through.
Here it comes.
Now watch closely.
Wait till it's gone
and you can see the platform again.
Hang on.....there's the last carriage.
Now look
......all the people have disappeared.

It's simple.
The driver presses a button.
This activates a big tippex brush
which springs out from the side of the train
and swishes them all away.

He only does it when he's in a bad humour.

I don't know how he brings them back again.

I'm don't even know if he does.

You'll have to ask somebody.

I Remember Kevin

I remember little Kevin P. standing inside the front gate
holding onto the bars with his tiny hands
and asking everybody who passed
– "Where are you goin' ....tih?"
It was a really long pause
every time
it was lovely.
He asked every man, woman and child
who passed up or down our road
– "Where are you goin' ....tih?"
By the time his mother let him out in the morning
all the people who went to work were well gone.
The schoolchildren were gone as well.
So he only ever got two answers.

"Where are you goin' ....tih?"

"Down the village."

"Home"

He kept right on asking.
Little boy
asking
all the time asking.

I would love to know where he is now.
I would love to know.
Kevin – where did you go ....tih?

Shafted By The Bin Fairy

My heart is broken.
The Bin Fairy didn't come today.
I cannot understand it.
I certainly didn't stop believing in him
or anything disloyal like that.
When I put out the rubbish last night
I pinned my letter to a black plastic bag
the way that I always do.
"Dear Bin Fairy –
Thank you for the autographed Jimi Hendrix albums,
the Beatles box-set, the 24" television and video
which you brought me last week.
This week could I please have a signed first edition
of Ulysses with illustrations by Henri Matisse,
the title deeds to a Norman castle somewhere nice
near the sea, a leather football and a surprise.
If you're a bit pushed for space you can forget
about the football.
Lots of love
PAT."

But he never came.

I was bold two or three times during the week.
I have no problem admitting that.
But bloody bin fairies are bold too.
I never asked him about where he gets all his stuff.
The Da Vinci cartoon didn't come out of a Pound Shop.

I don't understand it.
I don't understand it at all.

You knew where you were with Santy.

Slowly Dying Very Very Slowly

They have got nothing else to do
so they sit all day and they stare
– at the man delivering the bread
– at the people walking by
– at the dog scratching itself
– at the crow pecking chips out of the bin
– at anybody who gets off the bus
– at you
– at me

Sometimes they watch one person very closely
then one of them says something to the other
– they both laugh

most of the time they say nothing at all

at night they go home and stare
– at the fire
– at the screen
– at the ceiling

in the morning they start again.

Two Ways You Can Do It

(1) Build billions of strong kitchen chairs.

Give one to every man, woman and child (of jumping age) in the whole world.

Fit each and every person with an extremely heavy pair of clumpy boots.

Let them all jump off the chairs at exactly the same time on the same day and hit the ground with a good thump.

or

(2) Knock the high apples down with a long stick.

GOOD TO HAVE YOU BACK MISTER QUENCH

Wesley Quench's next shop will without doubt be his finest ever.
The man plans to market culottes (made-to-measure)
to be worn during your assumption into heaven.
These are specially designed for holy people
who don't want perverts peeping up their robes.
It will be a winner Wesley.
Put me down for a pair.

Flow Soft

I looked into the Liffey from the Halfpenny Bridge
and saw a dead rainbow floating down towards the sea
– all red and orange and green
moving this way and that.
It must have fallen in and drowned
when nobody was looking.
Or maybe it jumped.
Either way
in ten minutes or so
the water will be lovely and salty
the water will be lovely and deep.

A Car For Rocky A Car For Brad

I've just been looking at the sort of names they put on cars
so that men will buy them

MUSTANG

COLT

RANGE ROVER

SIERRA

When guys feel comfortable driving around in cars called

TWITCHY WHISKERS

DIMPLES

WINKY WOOKEY

I think we'll be getting somewhere.

Sunday Morning On The Dart Looking Out With Delight

Between Sutton and Howth 17-5-98

All the trees have got their summer clothes.
They are trying them for size
– spreading their branches this way and that
– letting the sun warmly admire them
– letting you see such subtleties of green
– letting their shadows wait upon the wind.

HIGH POINT/LOW POINT
Week ending 17th May 1998

HIGH POINT: A young Swiss visitor stopped
and said to me – "I carry three poems with me
everywhere I go."
And she took from her back pocket three poems
written by me and lovingly transcribed in her own hand.
"I treasure these" she said.

LOW POINT: Five young Dublin guys
all suntan lotion, shiny shorts and swagger
walked past me and my books on the pier.
One of them said in a voice loud enough for me to hear
– "NO BEGGARS ALLOWED!!"

Thank you Ireland.
You shite on your own.
You always did.
You always will.

Bloody Blood

Some mornings in bed
I hold my right arm against my ear.
I can hear my blood attempting to clot.
It is gurgling to itself over and over and over
– *"clot clot clot clot clot clot clot clot"* –.

I drop a little soluble aspirin into a glass of warm water
and let it fizzle for a while.
Then I drink it.

Five minutes later I listen to my arm again.
My blood is gurgling to itself over and over
– *"shit shit shit shit shit shit shit shit"* –.

Tough.

extract from "Report Of Steering Committee For The Restoration Of Irish Dance To Its Former Place Amongst Fainnes, Pioneer Pins, Scapulars And Flagillation

"Rigidity and repression being our great historical choreographers,
we feel there is still far too much movement going on.
In future (as sin amach) in addition to locking both arms by their sides,
the dancers shall also fuse their legs together, stiff and tight.
For the purposes of dance, they shall now move their nostrils only.
If they move anything else they shall receive a healthy whacking
with a stout stick.
Pray."

Arriving On The East Pier Beautiful Sunny Morning

Friday 22nd May 1998

THE VERY FIRST WORDS I HEAR TODAY

WOMAN: "You're late!"

WHAT A BEAUTIFUL MOMENT

12.31 sitting beside the barrier

While I am writing on this page
a little sparrow peeps down
from the top of the wall
and makes a lovely shadow
across my words with its head.
I wasn't actually writing about it
but I am now.

Billy Gruff

I was full of ice and cold and not fitting in anywhere
so I went down to the Bull Wall
and approached the wooden bridge
where the water rushes under.
I was making loud noises with my feet
so that the Troll would hear me
– the one who lives under the bridge
and listens for hungry footsteps.
I was making extra-loud ones.
Then I stood on the bridge and waited for him to sing
– "I'm a Troll forty wole
  I'm a Troll forty wole
  I'm a Troll forty wole
  And I'll eat you for my supper."
There was no sound but the water in a wild hurry
and the sharp gulls.
Maybe he was away or something.
Maybe he was up to his eyes eating someone else.

I left a note.

"Dear Troll,
I'm fed-up so I came down for you to eat me.
Sorry I missed you.
PAT."

When I looked back
the wind had blown it away.

How Lovely It Is To Sit Still And Gaze Down

16.45 22nd May 1998 at the edge of the quay

How quietly the tide comes creeping over the harbour floor.
Edging in and around
with easy ripples.
No great rush or hurry.
Knowing that in the good and beautiful scheme of things
it will be high up against the wall
before the fishing boats come.

We Were The Same Queueing For The Pictures

Waves on Claremont Beach come rushing in
shouting back – “Stop pushin’!! Stop pushin’!!”
the next waves shouting – “It’s not us! It’s the ones behind!
Hey youse! Take your time! Stop pushin’!! Stop pushin’!!”
And so on, shouting back all the way to the edge of storm.

AND THE BEST OF LUCK

We would love to pay you but

– our entire accounts department is off sick

– the person who signs the cheques has
  *lost his pen
  *broken his arm
  *gone mad
  *run away to sea

– all of our cheques require two signatures
  both of the signatories have
  *lost their pens
  *shot one another in the head

– our accounts department
  has relocated somewhere
  in the West of England
  and nobody knows the address

– we operate a two month credit system
  and even though we sold
  all your stuff six months ago
  there is still some perfectly valid reason
  why you're not getting paid –
  what exactly is it we don't know
  but if you insist
  we'll come up with a real beauty.

Smiling With His Mouth Dying In His Eyes

"All of my life" he said. "Madness has been three steps in front
– making deep shadows and whispering things.
Madness knows my name."

"It will never catch me again" he said.
"Now it's afraid of **me**.
I point my finger and shout – *'I'm madder than you!!"*

And he laughed.
But it didn't sound like one.

Help me.

I Shouldn't Have Been Listening And It's None Of My Business Anyway But God Almighty

Nassau Street 25-5-98 13.29

I know that it is extremely bad manners
to eavesdrop
on other peoples' conversations
but when I passed the man
in Nassau Street
he was saying the following
into his mobile phone
– "blah blah blah".

You could say that much cheaper.

Anyday Now I'll Be Wild

On the DART to Howth   Friday 29th May 1998

For days I've been waiting
speeding up and down the line to Howth
watching
all the time watching
the bright yellow along the bank
the bright yellow just before Sutton
any amount of flowers
putting their yellow together
to make your breath away.
Waiting
for the very first bit of red
the first blazing brave one
the poppy
that's the one to make me crazy
that's the one to put the cinders in my blood.

Dollymount

near the wooden bridge   30th May 1998
early in the morning

The tide there last night would drown you
but now it's gone
and two men are digging on the floor
with forks and long handles
digging and squelching in the weed
finding ragged worm.

Their dog is sitting on a bum
wet with salt.

When the sea comes again
they will be well gone.

SEACHRÁNACH

Ceo
mar anam damanta
ceo
ag druidim ar strae
cuimhne mí-chinnte éigin
cuimhne mí-chinnte gan lae.
Ceo
gan fuaim coitianta
caillte
go h-uaigneach
ar strae.

Cuairteoirí

Beann Eadair 29-5-98

Daoine daingeana dúnta
a ndorn is a bpóca faoi ghlas
gach pingean ina phríosúnach
a spiorad chomh fuar le sneacht.

Under The Tree

Lazy Sunday 31st May 1998 most of the afternoon

I am sitting under the yellow tree.
It is abundant with blossom.
An infinite number of yellow lanterns
are hanging down.
Bees are buzzing up
into the rich treasure house
believing they are in heaven.
All the grass around rests gently
under petals.
The wind is bringing them down.
Single little bits of yellow
drifting onto me
and onto Hoot
the two of us together on the ground
receiving little sprinkles from the tree
and loving it.

OFF THE ROAD

Many people have got single memories of me.
I know this because from time to time
they tell me.

"I was at your gig in Cork."

"I remember seeing you in a cart
in the parade at Tralee."

"Me and my sister saw you in Andrew's Lane."

Most of the time
all I can remember is crowds of people,
lots of faces and a hotel afterwards.

Sometimes it broke my heart.

A Brief Note To Whoever Is responsible For The Huge Jets Which Fly Over My House And Deafen The Heavens With The Thunder Of Their Engines

If you are not already thoroughly ashamed of yourself
you bloody well should be.
Your obscene din is an insult to nature.

Please go and listen to a sparrow.

Listening Under The Yellow Tree

Sunday afternoon 31st May 1998

I am listening to the wind
making a huge noise in the big trees.
The same sound it was making
long before I was born.
The same sound it will be making
long after.
I feel my father.

Lord And Master

"Darling"
panic-stricken
she said.
"The servants have stampeded all the cattle
drained the lake, burned down the West Wing,
wedged Master Myles headfirst into the rain barrel
and chopped down our prize geraniums."

"It must be something you said during the evictions,
They LOVE me."

"Oh."

That Too I Suppose

lunchtime Tuesday 2nd June 1998

A bad-tempered traffic snarl in Pearse Street.
I am standing beside two school-boys.
We are waiting to cross.
A motorist alongside us is beeping his horn
over and over.
"God – he's a very beepy driver"
I say to one of the boys.
"He's a fuckin' prick" he replies.

Hoot Of Little Faith

Hoot seems to believe that each and every morning
he has to make an eloquent case for his breakfast.
He thinks that unless he puts forward
a powerful series of persuasions he will not get fed.
I walk into the kitchen and he goes
– "Meow meow meow meow meow meow."
I say – "It's alright Hoot – I'm just going to open a tin."
He goes – "Meow meow meow meow meow."
I say – "There's no need to worry Hoot
– I'll never forget you."
He goes – "Meow meow meow meow meow."
I plunge in the tin-opener.
He goes – "Meow meow meow meow meow."
I say – "Any second now Hoot – Chef's Special
– Chicken and Tuna" and I splosh it into his dish.
He eats and purrs at the same time
making a frantic "prrrrrangang....prrrrrangang"
over and over
as if he believes there will never be any more.

Tomorrow morning we'll do it all again.

More Like Them

When I look at the little round plump and busy birds
who hop about my feed on the pier
I think what a pity we put a hard-edged name
like sparrows on them.
It sounds nothing at all like they look.
They deserve a nice soft gentle sort of a name.
Today at 11.54 precisely I propose to give them one.
"PLUMPY BEAG"
There.
Now.
That's it.
"PLUMPY BEAG."
Let the word go out.
As sin amach.

Let the sparrows* rejoice.

•from 11.54 on Friday 5th June 1998 – the plumpy beags.

Transcript Of Evidence Given By Brinsley Buffigan In Connection With The Woodification Of His Grandmother

"For my eighth birthday my grandmother
gave me a 10,000 piece jigsaw of the sea.
There was a small galleon sailing on it
but mostly it was just miles and miles of blue.
I hardly ever went out after that.
I finished it one week before my ninth birthday.

For my ninth birthday my grandmother
gave me a 20,000 piece jigsaw of the Sahara Desert.
There was one little oasis and two camels in it
but mostly it was miles and miles and miles of sand.
I didn't go out at all.
With my brother's help
I got it finished two days before my tenth birthday.

For my tenth birthday my grandmother
gave me a 30,000 piece jigsaw of the North Pole.
There were no polar bears or penguins or anything in it.
Just miles and miles and miles of fucking snow.
With my father's electric sanding tool
I powdered the pieces down
and sprinkled them gently over my grandmother's food.
For the next five years or so
she ate a huge amount of the North Pole puzzle
and eventually floated away out to sea.

I have only got 27 pieces left.
I lost three.
Therefore my grandmother ate 29,970.
It's an awful lot really."

"Now I've Heard It All

12.11 Saturday 6th June 1998

(from a man who has never ever bought one of my books and whose dog pisses on my basket every chance it gets.)

"Why don't you translate them into French?"

Overheard On The East Pier

17.12 Saturday 6th June 1998

two women talking

"D'ye see that? I've got all me breathing things in there. I have to go on a nebuliser twice a day."

"And you look so well!"

12.55 16th June 1998

one woman to another

"Actually she's more competitive than she let's on."

15.25 24-6-98

two women deep in conversation

"I just love parsnips in a stew."

"No....no....I like a parsnip on its own."

An Awful Lot Of Daisies

It's probably my own fault because I started it.
On my way down to the pier from the station this morning
all the daisies were standing up yellow and white
opening their petals to the sun and smiling with the warmth.
So I said it – happy and giddy and gay – "Good morning
daisies!!!"
And by God – how they answered me.
One by one at first – "Morning Pat!" "Morning Pat!"
"Morning Pat!"
Little daisy voices singing and songing all along the line
"Morning Pat!" "Morning Pat!"
Then in twos and threes, four and fives, sixes and sevens
hundreds and thousands of tiny voices rising and falling,
"Morning Pat!" "Morning Pat!" "Morning Pat!"
I have never heard so many dizzy little daisies in my life.
It's a long way down to the pier.
It's an awful lot of daisies.
I think I might get the bus in tomorrow.
The other way.
Over the hill.
They'll be waiting.

LETTING IT SLIP

He never knew that sometimes when she saw him
she would wonder
about the feel of his mouth upon hers.
He thought that she was so untouchably beautiful
she would never notice him.

He never knew that sometimes in her bed alone
she would think
about the pleasure of his fingers touching her
and after a while
she would begin to touch herself.
He though that unreachable girls like her
would never be interested in him.

Sometimes she would imagine the two of them
laughing together.
He would imagine her boyfriend.

From time to time she would wonder
why somebody as successful as her
was so deeply lonely.
He fancied that if he was rich, elegant
and handsome
he would have a chance.

Her name was Alison.

He never knew.

Conversation With A Man Who Has Become Emotional With Drink

Gent's toilet Mahaffy's Pub Pearse Street
Monday lunchtime 8th June 1998

"How are the books goin' Pat?"

"Great thanks – they're flyin' at the moment."

"The wife has three of them."

"I really appreciate that."

"There's nothin' to appreciate – she likes them."

"I'm delighted to hear it."

"You should have been promoted years ago
instead of doin' that."

"But I'm happy – I love it."

"Aw that's alright so...you carry on."

THEY WEREN'T DOING HIM ANY HARM

Across the road from the Pier House   Howth.

This afternoon at 14.05
a man came driving on a little orange tractor
registration number 98 D 31672.
He buzzed up and down
                  up and down the front
till he had chopped the heads off all the daisies.
I couldn't bear to look.
It was horrible.
Little heads were flying in all directions.
Little heads were spewing this way and that
with the wind.
The man on the orange tractor didn't miss one.
A terrible silence fell behind him.

I'm not sure exactly how many he beheaded.
If you write to Fingal Council with the date
they might be able to tell you.   (12th June 1998).
One thing I do know.
It was an awful lot of daisies.

They won't be saying "Morning Pat" again
that's for sure.

Maybe If They Went A Bit Quicker Or Something

God – I feel so terrible when I stand on a snail.
I really do.
There's this awful crunch and I go
– "Aw shit – not another one!"
There's bits of broken shell and squishy stuff
all over the place and I look at all the broken bits
and think – "It took years to make that shell
and I've just smashed it to bits with one clumsy step."
I would love to be able to put it all back together again.
I really would but it just isn't possible.
Sometimes I try and blame the snail
and say he should have looked where he was going.
Sometimes I try and blame God
and say he should have put little blue winkey lights
on top of them.
But it's my fault really.
I'm so sorry.
Every time.
I really am.

WARM TO ICE

I was just talking to the bathwater this afternoon.

"How did you get here?"

"Out of the sky Pat and I'll tell you
– it's a hell of a way down."

I was looking with wonder at the beautiful water
which has risen in vapour and fallen in rain
and come down from the tank in the attic
to hold me safe and warm.

"I think you're lovely. I really do.
Thanks for a brilliant bath."

"You're welcome."

And away it went down the plughole
with the same strangled roar that came from me
when I fell with huge fear down the black vortex
in Jervis Street.
Jesus Christ – suddenly it's not so nice anymore.
There is a nightmare in my stomach.

I'll Smack Your B*M You Polite Newspaper You

I am still trying to track down
the precise edition of the dictionary
which Irish newspapers use.

The one with words like f**k and sh*t in it.

I spell them "fuck" and "shit" myself.

Holy Terror

It's a wild fright to be going faster than your heart is
and you're sitting in a chair
your body pumping wild and wicked speed
waiting for the bloody burst
to blow you broken from the inside out.

And nobody knows the terror in you.

Saying "Peace" to yourself quietly
over and over and over
trying to slow the ructions down.

Nothing looks the same
it's all wild and wicked clear
mad ideas spitting out of everything
running rings around you
and you have to look
'cos if you close your eyes
you'll make it faster still.

For Christ's sake make it stop.

INDEED

in the 130 bus queue   Lower Abbey St.
Friday evening   17.30   18-6-98

When I read the poster from a distance I thought it said

– *IF THERE IS A GOD*
WHY DOESN'T HE SHOOT HIMSELF?

HE HAS!!

When I got closer I saw that it really said

– *WHY DOESN'T HE SHOW HIMSELF?*

I preferred it the first time.

My Dear Friends

My dear buttercups,
I don't know who puts gloss on your petals for you
but it's lovely.
I don't know who told you to hide away in secret places
beside the railway
where you can see the trains
and only the sun can see you
but it's a great idea.
I don't know why I'd rather be a buttercup than a person
but I would.
I don't know why I start to cry when I write this
but I do.

I love you.

Business Procedures Willow Publications

From time to time we receive letters
addressed to

– CREDIT CONTROL MANAGER
WILLOW PUBLICATIONS –

or worse again

– SCHOLARLY BOOKS DIVISION –
WILLOW PUBLICATIONS –

There is nobody here
who answers to names like that
or would ever want to.

I usually write

– HOOT PLEASE DEAL –

on them
and place them beside his dish.

He operates on the basis – "If you
can't eat it, it's not important."

So that takes care of that.

Willow Publications
Report Of Board Meeting Which Happened Under The Yellow Blossom Tree This Afternoon

Tuesday 23rd June 1998

Willow proposed that the comfiest place for board meetings is upstairs in the hot press.
Hoot said would you have a bit of sense it's the middle of the bloody summer Willow we'd all roast up there and anyway Pat doesn't fit.
I proposed that we carry on the way we're going – board meetings continue to happen whenever the three of us are lying down in the grass or resting on beanbags or dozing in front of the fire.
Willow said wasn't it lovely when Blackie used to be here and we passed a resolution that it hasn't been the same since she went away, we miss her terribly and we hope that someday she comes home. Then we all stayed quiet for a while.
Hoot wanted it put on the record that mice are very hard to catch when they go in under things.
Willow fell asleep.
Hoot wandered into the kitchen and stared at the press under the sink.

We'll have another board meeting sometime.

On The Dart To Howth
25th June 1998

Safe in a grassy curve on the railway bank
a young ginger fox is curled up
sleeping.
The wind through the red poppies
and yellow flowers
does not disturb him.
Nor does the flight of butterflies and moths.
As the DART rattles past
he opens one eye and looks at it.

Sleep on ginger fox
sleep on
we'll be gone in a second.

When Your Eyes Look Far Back In Your Ears Are Shut

On the DART to Howth 11.20 25-6-98

When the seat opposite became vacant
the heavy old man stood up
and sat into it.
"I prefer going frontwards as well"
I said to him.
He was so used to people
not talking to him
that he didn't hear me.

Cothrom Me Bollix

Nuair a bhí mé óg
lag Dia mo lámh.
D'fág sé
úsáid an lámh eile.

Nuair a bhí mé óg
lag Dia mo chosa.
Ní raibh fhios agam
nó fhios
ag aon duine eile.
Lag sé iad
mar uisce faoi thalamh.

Anois
fad anonn sna blianta
laige trom
ag teacht chun cinn
diaidh ar ndiaidh
an siúl ag dul asam
diaidh ar ndiadh
Dia ag dul i bfad.

Never Ever Ever Overheard In Conversation

(today or any other day)

POSH WOMAN: "I was just saying to my son, the addict..."

COMBAT with pink flowers

A shaky old lady with a pink flowery hat got onto the bus
and a young man stood up.
"There's a seat for you ma'm' he said.

"It's alright thanks" she said. "I'm only going to the next stop."

"Well that's bloody lovely" he said. "After me standing up
and everything."

"It's bleedin' typical" said another man.
"They're always doin' that to me."

"Aw hang on a minute lads" said another man.
"Let's not go over the top about this –
it's only a bloody bus seat."

"We didn't see you offering yours all the same pal"
said the first man before whacking him.
The other man belted him back.
Another man booted him and very shortly
they were all kicking and thumping and yelling
– *'Hold me back. Hold me back – I'll kill him!!'*

The shaky old lady with the pink flowery hat
got off at the next stop
and poked the driver in the eye with her umbrella.

"What was that for?" he gasped.

"I haven't got the faintest idea" she said.
"You'll have to ask one of the men."

Wrong Room Wrong Room

On one of those sepulchral nights
when dead bodies
stick their fingers up
out of the ground
and generally
scare the living shite out of people
a giant liquorish allsort
lumbered
into Wesley Quench's bedroom
and snarled at him.
So he ate it.

Beware The Primary Colours

The generation who have always bought clothes
which are brown but faintly brown enough
to be no colour at all
grey but faintly grey enough
not to be anything
– this generation chooses its seat on the bus
with care
bland attracting bland
dull safe beside itself

not such a bad day now

red jelly?
No chance.

Back To Bertie Byrnes

The old ladies in the flowery frocks
and holes cut in the sides of their shoes
to let their bunions out
– these old ladies know a way to travel back
to the 1940's
back to Bertie Byrne's shop
the one which sells those lovely hickey sun hats
which they wear up and down the pier
they need old money when they go back
but that's no problem
there's still loads of it in tins
and bottles and jars
and far in underneath the bed.
Sometimes they stop off at Polly Flynn's
for a fizz bag and a Peggy's Leg
but they can't stay too long
because of what happened to Oisín
when he fell off his horse.
Then where would you be?

The Siding I Loved

If you walked up the Back Street
then around by Gilmartin's house
and up the slope
you went in through the back gate
of the station.
A little piece of railway curved in
off the main line
and came to a stop against a wall
with two black and red buffers.

Grass grew up around the brown sleepers.

High white wooden railings
bordered the back.

Every so often
a train would rush past on the main line.
You could see it through the railings.
Otherwise you could hear the bees.
The little siding slept.

Sometimes when you went up to the station
someone had put a goods carriage into it
and left it there
silent
its couplings hanging down.
If you stood beside it on the slope
it towered up huge and heavy
the closest you would ever get.
You could touch a grease box with your finger.
You could smell where cows had been.
You could look in under.

In the morning
it would be gone.

SCARY

When your mother appeared

unexpectedly

you changed

from

beautiful young woman
confident
fascinating
and clear

into

uncertain
hesitant girl
no longer sure
of anything

in a split-second
enchantment cracked
and scattered
little bits on the floor.

Overheard On The East Pier

2nd July 1998   11.55

"Then lo and behold off she went to Wanderers Rugby Club and met this guy...Richard...drop-dead gorgeous........"

Thinking About Mother Nature

There must be an awful lot of green paint in the stores.

My Hero

When I was little
my favourite picture in my Curly Wee album
was Patrick Pig running away from home.
He was sitting on the back buffer of a train
straddling it like a horse
speeding off towards all sorts of adventures.

When I spent all day long on the sofa
with blankets tight around me
and the fire lighting
I would look into that picture
for hours.

When night time came
my father would carry me back up to bed.

On Clontarf Road Station
whenever a train rushes through
I always check the back.

God Save You Polly Flynn

When you opened the door into Polly Flynn's sweetshop
a little bell tinkled over your head
but Polly didn't hear it
because she was profoundly deaf.
She'd be in the back
drinking tea or something.
So we opened the door
and we closed it.
Over and over
tingalingaling
but you'd have needed a cathedral bell
for Polly to feel the vibrations.
Every so often she would peep out from the back
but she never saw us
because the counter was high
and we were too little.

We would leap up and down
and wave our pennies over our heads.
We would stand on tip-toe
and whack our pennies against the counter.
We would lift one another up
and the head which could see over the counter
would yell – "Pennyworth of sweets please Missus Flynn!!"
She would lean forward and look down at us.
"My goodness" she would say. "So many of you.
Why didn't you ring the bell?"

DIG AWAY BOYS

When I put the litter tray in the corner beside the back door
everything was grand.
At night I'd hear Willow or Hoot scratching and digging
and splashing in the distance.
Me upstairs in bed.
Them down on the tray.

One night I heard
the first faint pussy-cat fart in the bedroom.
Over in the far corner.
Then a fusillade and an unmerciful splash.
"Aw shit" I sighed.
Which indeed it was.

I removed the top from my record-player
because it was exactly the right shape.
I filled it full of litter and installed it upstairs in the corner.

From time to time in the night I waken to the sound
of pussy-cats doing what they need to do
upstairs and down.
It does not bother me in the least.
It is Willow and Hoot and I love them.

A woman friend sat up very suddenly in the night beside me
and said – "My God! What was that?!"
Willow or Hoot – it was too dark to see who –
one of them was excavating a pit in the corner.

"It is probably a rat next door....they dig tunnels at night."

"It must be *HUGE*" she said. "Just listen!"

I was afraid to.

I knew what was coming next.

You'd be better off with music.

If He'd Known All This He'd Have Kept His Job In The Garage

At the very first gathering
of The James R. Hickey O'Hare Summer School
Professor R.D. Verity read a paper entitled
– *"The Writings of Hickey O'Hare*
Mythological Celt or Celtological Myth?"
Doctor M. C. McErristy spoke on
– *"Hickey O'Hare – The Almost Complete Absence*
of Upside-down Commas"
and Mervyn Quint asked
– *"Hickey O'Hare – Ah Yes But What If His Mother*
Had Died When He Was Three?"

The following year
Professor Pandit Jaswal from The University of Nepal
delivered his controversial paper
– *"Hickey O'Hare – Supposing It Was Not Snuff At All?"*
The Ambassador of Grovia, Euthan Nint
spoke for two whole days on
– *"Hickey O'Hare – Pan Celtic or Celtic Pant*
& Who Wears Them?"
and a group of American Students* stood and stared
at a house in which it is believed Hickey O'Hare once caught
his finger in a mouse-trap.

For ten consecutive years after that
The James R. Hickey O'Hare Summer School
heard papers such as

– *"Hickey O'Hare – His Intestinal Tract – An Overview"*

– *"Hickey O'Hare – Redefined With An Enema*
And A Bottle Of Tippex"

and

– *"Hickey O'Hare – Are We All Mad Or What?"*

Next year the committee plan to exhume his remains
and see how many four letter words (no plurals or place names)
they can make up with his bones.

You'd be better off eating a box of Smarties.

(*as part of their Liberal Studies)

Sleep Only If You Have To

A huge proportion of the civilised world
still believes there is great merit in
– being busy
– up to your eyes
– run off your feet

– you have your soup first
– then your main course
– then your sweet

– the week is divided in two
– if you have been extremely busy
  during the long bit
  you are entitled to enjoy
  the shorter bit at the end
  providing you intend
  being very busy again
  the moment it is over

Some of the churches
would also have you believe

– if you are utterly miserable here
  there is a reasonable chance
  you will be happy someday
  in a wholesome hymn-singing way
  some place else

Go and eat a bowl of jelly.

The Cormorant's Boat

The black cormorant perches
on the edge of the long brown rowing boat.
Every summer day
holding on tight with his claws
he sits there and thinks about things.
The gentle rise and fall of the water helps him.
The flat calm helps him too.
When a big trawler like the Muriel Grace goes out
big waves roll sideways and rise the boat.
The cormorant rises with it.

The sea-scouts think that they own the boat.

They don't really.

Louder Than The Rest

From my earliest days I can remember "The Mouth."
When we were little outside the picture house
he was the one who boasted loudly

– "Ma'sie Walsh isn't gettin' my fourpence!
You'se can give her yours if you like.
She's not gettin' mine."

When he reached the paybox he'd say

"Fourpence please Missus Walsh."

Wind.

A Letter Written After Looking Into The Middle Of Lots Of Flowers

Monday morning 6th July 1998

Dear Mother Nature,
This isn't a criticism
....O.K......it is really.
Just a very little one.
Why do you put so many yellow middles into your flowers?
They're lovely. They really are.
It's just there are so many of them.
Yellow middles in purple and red and pink.
Daisies even have them.
Like I said – they are very beautiful. Very beautiful indeed.
But......you know......just a bit..........

Oh – one other thing.
All those vivid purple and pink and yellow and white flowers
which you run wild along the railway bank
.....magnificent......it's a riot.
But.....do you know what I miss?
A few blue ones.....just a few here and there.....

Apart from that everything is grand thanks.

P.S. The poppies outside Sutton are your best ever.

God Be With The Days

When I typed just one word
and asked my processor to do a count
it flashed a message on the screen

– *DON'T BE SO BLOODY THICK* –

There's no need for that.

MENU LES AMIS open chicken £3.95

An open chicken
would be a terrible thing indeed.
It would be terrible beyond redemption.

An open chicken
would run around the place
in raw terror of bees flying into it.

Cover it up before the hornets come.

His Brother Was A Priest

"Marty? Marty? Aw...*that* Marty!!
Jesus don't be talkin'.
Rumour had it his fridge was full
of human heads...all lined up
the dates on them an' everything.
They said he oney had the heads
of jockeys but sure you know yourself.
Marty...Marty...don't be talkin'.
They said he tied his grandmother
...the one from Nenagh......
they said he tied his grandmother
to the tracks at Merrion Gates
because she couldn't get her laces open.
"That'll get your boots off for you Granny"
he is reputed to have said.
Marty? MARTY? Marty is right."

He Goes Alone

When Twitching Timmy Gavigan
of the hairy ears and the crooked eye
farts rings around himself into the lift
at Clontarf Road Station
he always asks

– "Anybody else goin' up?"

But nobody ever is.

It must be very lonely
him and his echoes.

BIBLE STUDY YOUR HOMEWORK FOR TONIGHT

(1) What happened to Lot's wife after that?
Did it lash rain or what?

PROMISES

Day after day
I hear mothers warning their children
about falling into the harbour.
But they never do.

We live in hopes.

Tá Súil Mór Agam

Most of my life I have done everything on my own
and it has worked out O.K.
Maybe death will be the same.

The Reason Why A Woman Didn't Buy One Of My Books From Me On The East Pier

(She told me this in Madigan's Pub the following day)

"You were asleep."

THE WAY WE'LL BRING YOU DOWN

told to me on the East Pier

"The first day your-man arrived in for work
with one of them black boxy brief-cases
– the ones with the combination locks
– the gold ones
Vinnie asked him
– 'What have you got on your bread?'"

Tears Roll Down Their Feathers Tears Roll Down Their Fur

Late at night
while we are asleep
all the animals come out
and all the birds land.
They sit together in huge circles
around our "Keep Off The Grass" signs
and piss themselves laughing.

Long Gone Now

Old tall Paddy Dunne
always leaned forward when he walked
the way you'd think he had been fashioned
out of a curved branch
dressed in black
with a bowler hat
and a silver tin to fetch the milk.
He got it free up the Castle
he'd worked there all his life
so why wouldn't they give it to him now
it's little enough they ever gave him
for his work.
Old tall Paddy Dunne
away with him
over the Dublin Road
up the Castle Drive and back
his can creaking empty
then slopping full.

Nine or ten of his cats
underneath the hedge
watching the corner
for when he comes.

HARBOUR MASTER'S REPORT

Caladh Phort Bheag An Bhuidéil Mhóir

At 11.23 on Monday morning the 6th of July 1998
some of the ripples set up the previous day
by a black labrador named Nigel
swimming beside the steps at the South Wall
spilled over the gunwales of a little rowing boat
in Bombay Harbour
which was dangerously overloaded
with television producers
and wet one of their shoes
but apart from the man himself
nobody else ever knew
because the harbour master
misfiled the report.

SUPPER!!!

For about thirty minutes every night while he was sleeping
Pearse St. John Mahaffy O'Toole would change into a fish.
Not all of him.
Just from the waist down.
He would lie there soundly sleeping
and not know a thing about it.
From the waist up he was still Mahaffy O'Toole.
The rest of him had become a haddock called Norman.
All of this began during the summer
which was no problem at all
because his two cats were out for the night.

Yes!!!

16.45 13th July 1998

Today
with the wind loving the sides of it
the hill behind the harbour
is a huge living creature
with green fur rushing
this way
and that way
and everyway.

SAUCE

"No" he said to his maiden aunt.
"The picture is full of birds. It's just that they are hidden.
Keep scrunching your eyes all sorts of ways and you'll see them."

But she never did.

"No" he said to his maiden aunt.
"The painting is full of foxes. It's just that they are hidden.
Twist your eyes. Stare them. Screw them every sort of way
and they will all be there.."

But they never were.

He also lied to her about horses, deer, eagles, goats and donkeys.

"No" she shouted down to her nephew who was struggling for
his life.
"The river is full of lifebelts."

I Thought Those Days Were Gone

Hardly a day passes on the pier just now
that I don't meet somebody
who is running away from home.
Sometimes there are whole groups of them.
It's usually because a cruel stepmother
has been keeping them locked in a tower
and forcing them to eat cabbage.
I generally give them some of the bread
which I'm feeding to the birds
and send them over to Ireland's Eye
with a bottle of milk and a blessing.
When the authorities come down the pier
searching
I point out towards Lambay and say
– "Six of them your honour, and a goat."

You can't do much more.

Go And Eat Your Own

The man with the grey shirt and the brown jacket
and the black trousers
and advanced middle-age overflowing his waist
was standing on the harbour wall
sucking very hard on the butt end of a cigarette
when a young fox forced its head out
through his stomach
snatched the fag from his mouth
and swallowed it.
"Do you know what" the man said.
"That bloody thing thinks I'm made of cigarettes."

An Ghaoth Amuigh

Diaidh ar ndiaidh
bain díot do chuid éadaigh
tá an oíche mórthimpeall
an tine ag rince dearg
is ocras fiáin i mo bhéal

Another Day He'll Tidy The Bathroom Or Something Like That

Vincent P. Thwack reached far over one night
from Clontarf Road Railway Station.
He reached as far over
as the curved side of The Dublin Mountains
and began to click fields sideways,
moving them up, sliding them down
clicking them into new configuration
this way, that way
up down and sideways.

"Now" he said when he was all finished.
"Let them sort that out."

And home he went to bed.

ANYDAY NOW HUNTING TIME

For a while the sparrows were lovely and organised.
The little ones chirping and pleading
and fluffing up their *feathersjjjjuj
each one appealing to its mother.
Now it's mid-July
and they don't seem to give a shit about
who their mothers are anymore.
It's just – "Gimme a bit of bread"
to any bird who will feed them.
It's the same with the crows.
Big black ones
big enough to feed themselves
rawing and cawing
beaks wide open
"gimme a bit of bread
anyone
I don't care who you are
just gimme a bit of bread."
The mothers have had enough.
Huge big seagull chicks
still pestering the life out of them
haunting them round the harbour
"Gimme a bit of fish
gimme a bit of fish."

Anyday now it'll be
– "Fuck off and get it yourself!"
Then we'll all get a bit of peace.

(*Although it may look very much like it, this word is not in fact mis-spelled. After I had written the word feathers, Willow stepped on the keyboard and joined another bit onto it. It is most probably pussy-cat language so himself and Hoot know precisely what it means. I can only guess.)

Cut Yourself And See A Primary Colour

When we went into Clerys with our parents years ago
there were rows and rows and rows of grey coats.
So faintly grey
they were nearly no colour at all.
Brown jackets
fawn trousers
grey jumpers
black gaberdines.
A whole department
dedicated to the bland
    the dull
    the understated.

No other colours need apply.

An entire shoe department offered a choice
between black or brown.
Priests went one better.
Their shop offered them
no choice at all.

Oh happy day.

Grey Very Very Grey

Howth 17-7-98 Friday midday

There is a grey-haired man sitting on his own
in the harbour today.
That is all he is doing.
He is not looking at anything.
He is not watching anyone.
He is just sitting.
For ages.
For ages.
For ages.
Very very occasionally
he puts his tongue out a bit
and moves his mouth.
Nothing else.
He just sits there
for ages
and ages
and ages.

His eyes look like nothing at all.

TRADING FIGURES WILLOW PUBLICATIONS

for the day Friday 17-7-98

OUTGOINGS:

Dart to Howth	–	£1.10	
Rasher and tomato roll	–	£1.80	
Cup of coffee	–	.50	(Joe let me off 30p).
TOTAL EXPENDITURE:	–	**£3.40**	

INCOME:

Found on the road at Fairview	2p	
2 copies of Half A Hug	£14.89	(Should have been £15.98 but I gave the wrong change)
TOTAL INCOME:	**£14.91**	

PROFIT FOR THE DAY: **£11.51**

INVESTMENT PLANS:

Next week's bus ticket
3 tins of the really expensive cat food........brilliant!

We still need four thousand pounds
to publish next year's collection
but there's four or five months to go yet.
Someone might die and leave us something.

Who Likes Butter?

Not a laugh or a smile or the memory of one.
Sitting on the bus. Slumped.
Brown jacket like my father
                    my grandfather
                    everybody's father used to wear.
Left hand under chin.
Elbow on the window.
Eyes straight ahead.
Not looking out.
Not looking in.
Not looking.
Hair cut short
          severe
          up behind his ears
high up the back of his neck.

He does not know the 1960's, 70's, 80's, 90's.
They do not know him.

A book is in his jacket pocket.
A book with a white cover and two words in blue.

"RERUM NOVARUM"

If you tickled him
I think that something terrible would happen.

You Can't Go Out

Most of us made baskets and trays and wallets.
Someday we would be well enough
to go out again.
Old Mister Jennings had been there so long
there was no point in him making anything.
He was in for good.
So he stood in the hall
beside the grandfather clock
and he watched the door.
Always he wore his hat and coat and scarf.
It was his job to watch the people going out.
He knew who was well enough.
He knew who was not.
His grey mustache would never forgive you.

When I was wearing my clothes again
and my parent's visit was over
I walked with them down to the hall.
Mister Jennings stepped forward.
It was alright for my parents
to walk out the door
and over to their car.
They were well enough.
It was alright for me to watch them go.
From the inside out.
The front door and no further.

In my life nothing has ever hurt me more.

Nothing.

THIS MORNING

17-7-98 Howth

Somebody has chained a telegraph pole to a dustbin.
Who in the name of God would want to rob a pole?

FASHIONS GOT THE POWER

If your mother had said to you years ago
"You are not leaving this house
unless you wear your cap back to front"
you'd have said –
"Ah ma.....that's mental cruelty...
that's not fair......you're weird."

People had their heads shaved
either because –
(1) they'd been collaborating
    with the enemy
(2) they had ringworm

At present
couples are not walking around
inside the same pair of jeans.

Give it time.

Oh My God

The best way of instantly recalling
each and every place you have been
is to discover
you've paraded around Dublin
for the last hour with your fly open.

The Reason Why I Didn't Leave School At Eleven But Stayed On A Year Later

(Told to me by a taxi-driver in Dun Laoghaire)

"The masther says to me *'why are you leavin school?"*
an' I says to him *'cos I can't sing'*
an' he says *'can you eat?'*
an' I says *'yes'*
so he says *'eat that'*
an' he gives me his sandwiches
'I can't bring them home agen' he says
'I'm goin' down to Mooneys for me lunch.'
So I didn't leave, I stayed on
an' that was me job every day
to eat the masther's sandwiches."

Pearse

Two little boys holding hands along Great Brunswick Street.
"Hurry Padraig hurry" says one. "We'll miss the trains."
"I'm coming" says his brother.
And they stand together
looking up at the bridge into Westland Row
looking up at the trains.

In the breakers' yard at Kilmainham
they were shot

apart.

There is a much prouder name on the station now.
There is a much prouder name on the street.

Uaigneach

Mo leaba fuar is folamh
mo leaba folamh is fuar
me bhean áluinn imithe

tosnaíonn mo ghorta mór.

Go Go Go!!!

When the wild hunger comes on you
crooked looks perfectly straight
rough appears smooth
port in a storm
and you cannot fucking well wait.

Put Your Hand Around Behind It

I love being able to see in 3D
without having to put on
red and green glasses.

Two Old Women Talking Very Loud And Very Posh

130 bus 14-7-98 11.25

"He's at college....he's a bit of a........eh......like......he likes a good time."

Overheard On The 31 Bus

18.25 25-7-98

YOUNG WOMAN: "I think she's a bit weird to be honest witcha."

Overheard In Omni Park

16.20 27-7-98

WOMAN: "I'm kinda tired pleasing them all."

Agony An tSamhraidh
(on the DART)

Guth garbh aici
ar nós
bosca scian
caint
gan stad
go tapaidh
im' cheann
go h-uafásach
an pian.

Cormorant skates
All the
24/7/98
EAST PIER

UNDER THE YELLOW BLOSSOM TREE

Sunday 26-7-1998 16.25 all the blossom is gone now

All the buttercups are standing up and saying
– "aren't we lovely...we've got yellow petals
and they're all glossy."
All the daisies are saying
– "aren't we lovely...we've got white petals
with yellow in the middle."
And I'm saying – "My God but you're right."
The magpie is making a high rattle in the tree.
Any minute now I think it's going to rain.

STYLE

The wood pigeon who lives around here somewhere
hidden
loves saying –

"Whuh hoo hoo
 huh huh
 Whuh hoo hoo
 huh huh
 Whuh hoo hoo
 huh huh....(little pause)....huh."

That's the best bit.

A Shame Really

Having put matching woolly hats
onto the heads of all his golf clubs
he snuggled them up
side by side
in a soft pink duvet,
rested each head
on a green flowery pillow
and sang a gentle lullaby
about rabbits, burrows
and broccoli
until they were sound asleep.

His hurling stick
is outside
in the shed.

Away Went Cockneberry the Third

“No......I’ll tell you what I can’t stand about here”
Marcus Cocknberry the Third said to his wife.
“There are not sufficient moments in a day
so that you can ask with a very loud voice
   – *‘WHAT TIME IS IT?’*
and everybody on the entire planet
tells you differently.”

“Well why don’t you fuck off
and live someplace else then?” she asked.

So he did.

The Man Who Farted Golden Eagles

The Rt. Honorable Virgo Thormanby-Thrust
simply could not help farting golden eagles.
It was not something that he ever did on purpose.
The Rt. Honorable Virgo Thormanby-Thrust
never said – "I know what I'll do – I think I'll fart
a golden eagle."
It just happened.
He farted
and an eagle soared away – majestic – to the skies.
Not every time.
Some of the time.
The suspense and the uncertainty
made him twitch.

It always seemed to happen
the times he wore his trousers.
Or perhaps he was becoming cynical.

You get that way
when
intermittently
you fart
majestic soaring mountain birds.

PITFUL EXTREMELY PITIFUL
Howth East Pier 12.25 1-8-98

Somebody
a young man
with a big loud voice
is roaring my name
from a hidden place

– "PAH!! PAH!!
HEY PAH INGOLDSBY!!"

Roaring it over and over
and whistling at me.

I do not
like the kind of person
he sounds.
I do not like it at all.

Now from his hiding place
on top of the green hill
he is yelling

– "PAH! YOU PRICK!!
PAH! YOU BASTARD!!
PAH! YOU BOLLOX!!

His companions
hidden with him
are laughing.

There is a deep hurt
in my stomach.

Ten minutes later
away further down the pier
much much further
I am listening to the sparrows.

They are beautiful.

CONFESSION

Sometimes
instead of simply not buying a book
because they don't want to
people try to cover their retreat
by saying
– "Oh...they're all poems are they?"
At which point I do the bastard
and produce
– 'LAUGH WITHOUT PREJUDICE' –

which isn't.

Heh heh heh.

Got It In One

A man approached me this morning
on my way down to the pier
and told me what exactly it was
that he didn't suppose.

"I don't suppose
you've got a pound on you
by any chance."

He was right.

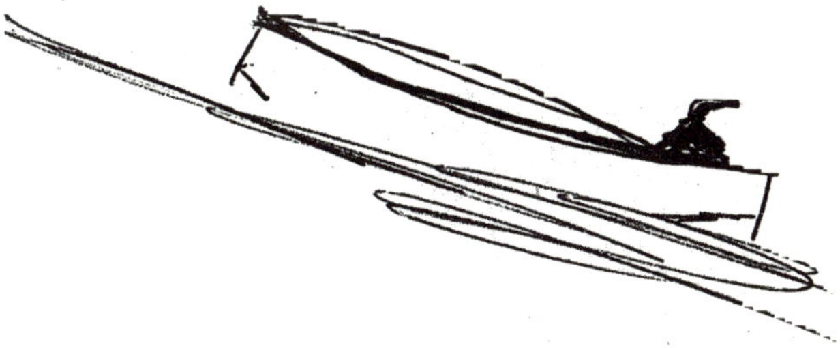

AN HONEST AD

For lovely long years while we enjoyed a complete monopoly
we suited ourselves and frequently treated you like shit
because we knew that you had to take it.
You had nowhere else to go.
We though it would last for ever. We really did.
Now, what with monopolies boards and open competition
and all that other shit we are well and truly fucked.
We can't get away with it anymore.
We don't blame you if you give all your custom
to our competitors from now on.
It's our own fault.
We have brought this upon ourselves.

P.S.
We'd do it all again
if we thought we'd get away with it.

Wanting To Hit Someone

friday 7-8-98 16.15

A man of years enough
not to be wearing
the same grey
as his parents
is walking
the slow walk
between them.
His hands
are hanging at his side.
He is opening them
and
closing them
making tight fists
over
and over
and over.

God It's So Lovely

east pier   howth   16.40   7-8-98

I am watching a gull walking
into the harbour shallows
at the edges of grey water
walking very slowly
with no hurry on him at all
the water deepening up around his legs
till a beautiful fusion of water and weight
floats him on it.
The sun comes out
and blinds silver lights across the tide.
The gull is moving one way.
The lights are going the other.

Through The Roof

When we bought our house she said
it cost six hundred pounds
mind you that wasn't today or yesterday
if we sold it now
the man from the estate agent's said
he could get us two hundred thousand for it
mind you my husband said
what would you buy for that nowadays
and the man from the estate agent's said
– a vow of silence from him
about all the bodies buried out the back –
so we said fair enough
carpets and curtains not included.

The Other Side Of Your Face

If you laugh a lot
people will think that you are nice
and they will like you
so she laughed her suspenders off
at everything the other old lady
said to her on the bus.

"I said, not for sixpence ha'penny you won't."

"Ha ha ha ha ho ho ho ho."

*"Of course when he fell out of the tree
Martin thought he was a squirrel."*

"Ho ho ho ho he he he he."

"Madeline Gilhooley, the pick of the bunch."

"He he he he he ha ha ha ha."

"Cancer of the colon."

"Haaa haaa haaa hooo hooo hooo."

"It buried my poor husband."

"Ho ho ho ho ho ho ho ho."

"What the hell are you laughing at?"

"Ha ha ha ha he he he."

*"I've got a little pistol in my bag
– if you don't stop laughing
I propose to blow your brains out."*

"He he he he he he he."

BANG!

*"Actually it wasn't cancer at all.
He fell under a bus.
Heh heh heh."*

Sweet

MacDara Heevey's pigeons dropped shite
onto everything I owned.
White shite on my roof
all over my clothes
white shite on my wall
all over my shed.
MacDara Heevey's pigeons
cooed and pooed
pooed and cooed
and utterly wrecked my head

I never said anything.
You don't like to complain.

MacDara Heevey's dog barked and barked
day and night
loud and long
I never said a word.
MacDara was big, strong and aggressive.
You don't like to complain.

Dog barking
pigeon shiting
on and on and on.

I shot MacDara Heevey
chopped him up into little squares
nicely diced
cooked
marinated
with a lemon sauce
fed him to the pigeons.
They delished him
bit by bit
then shit him on my roof
my clothes
my shed.

I never said a word.
Complain.
You don't like to.

Homesick At Home

I would like to live in a country
like I used to.
Where the people are warm,
open, friendly and welcoming.
Not tight, closed, inhibited
and suspicious.
It's gone.

Good Night And God Bless

The quayside has got a most exquisite memory.
At night
when heavy trawlers rest lightly at the wall
the quay remembers
each and every seabird's shadow
which moved across it in the day.

Hot Smoke

On August 25th at 11.23 precisely
1,879 different people
telephoned my number
at exactly the same moment
and
unknown to themselves
blew the fuse in my immersion
re-charged my hearing-aid
de-flea-d Hoot and Willow
dry-cleaned everything
in the laundry basket
hot-rodded the car next door
and de-frosted my fridge.

You couldn't depend on it thought.

HEARING AID

50% of my hearing
is hiding somewhere in the house.
I put it down two weeks ago
but I can't for the life of me
remember where it was.
I suspect it's lying hidden
under a pile of books
behind a radiator
down the back of the mattress
or some other sneaky place
like that –
half of my hearing
acting the bollix
leaving me high and dry
with cold silence
all on my left.
When it turns up again
it owes me the other half
of the past fortnight.
So far
I've only had it in mono.

If it's under the mattress
I shudder to think
what it's been hearing.

Verbatim Transcript Of Shouted Conversation In The Harbour

13.50   8th August 1998   Howth

ME (shouting out to three men on board *'SCAMP'*): "IS THAT YOUR BOAT!!!?"

MAN (shouting back): "WE CAN'T HEAR YOU!!!?"

(he switches off the engine)

ME (shouting again): "IS THAT YOUR BOAT!!!?"

MAN (shouting back): "YEAH – IT IS!!!?"

ME (shouting): "I HAVE A COMPLAINT ABOUT IT!!!"

MAN (shouting): "WHAT?!!!"

ME (shouting): "EVERY TIME I TRY TO DRAW YOUR BOAT, IT MOVES!!!"

MAN (shouting): "WELL WAIT TILL THE TIDE GOES OUT?!!!"

Go hObann

Killiney Hill   14.06   10-8-98

Suddenly
one of the little purple flowers is moving
because a bee has just left it.

I Was Travelling Home On The Dart When Some Bloody Ejit Hit Butt Bridge With A Lorry – That's The Second Time This Year So I Got Off At Pearse Station And Walked Into Town Along The Quays Giving Out Hell Under My Breath About Thick Ejits Who Bump Into Bridges When I'm Trying To Get Home – My Legs Gave Out Near Matt Talbot Bridge So I Sat Down And Wrote The Following Poem Which Has Got No Connection Whatsoever With Anything Which Has Gone Before But You Could Well And Truly Wreck Your Head In A Summer School With A Title Like This If You Really Wanted To

Evening time 10-8-98 sitting near Matt Talbot Bridge

A butterfly flew into the railway tunnel at Killiney
heading north because he probably thought
it was a short-cut to somewhere.
Seconds later he hurtled back out again
with a DART chasing along behind him.
He was going at an unmerciful speed
with such white-hot terror
that he burst into flames
burned to a cinder
and whacked against the windscreen
of another DART
which was going in the opposite direction.
He got as far as Bayside before he fell off.
It was a very extreme way indeed
of travelling that distance for nothing
but he didn't do it on purpose.
No butterfly would.

Sitting On Clontarf Road Station
11.25 Wednesday 29th October 1997*

Over behind the city
the mountains are faint and blue.
A rushing noise is coming up
from the sea.
Trees as making noises too.
A DART behind me in the siding
does a little squeak
but it's not going anywhere.
It's wheels are creaking slow.
A hawk hovers high
out behind the far platform.
It is watching something.
The late October sun
still knows how to warm my face.
I don't care if my train never comes.
It is lovely here.

*I know it's a last year poem but I lost it
and only found it again today – 10-8-98
so welcome home.

No Holes Without Bits

Should you ever go into Bartholomew Pringle's Emporium
to get your ears pierced
be sure to ask for the bits back when he is finished.
Say –
"Bartholomew Pringle – can I have the bits back please?"
He will tell you that there are none.
He will swear it to you.
He will say
– "Bits? What bits?"
I don't know what bits you are talking about?"
Do not take any of his nonsense.
Insist.
Say
– "Bartholomew Pringle. I am not a fool. Don't lie to me.
You have just made little holes in my ears.
Of course there are bits. Little bits of me. They are mine.
Hand them over this minute
or you will force me to take the hedge clippers to you."

Otherwise he will put them into a tin box with all the others
– the ones he is saving to build a huge big gigantic ear
capable of listening to kittens
whispering secrets to one another
in the wardrobe after dark
when they think that nobody can hear them.

Get your bits back. Save the kittens.

Forever Trev

Unwittingly
Cedric Winthrop Jessington the Third
stumbled upon a tiny part
of the great feline secret.
On waking one morning
he spied the black tip of his pussy-cat's tail
peeping at him over the pillow
and jiggling around in such a way
that it seemed to have a life of its own.
So he greeted it
with a cheery – "Good morning Trevor."
The tail spoke back to him with amazement.
"Far-out...cool...you know my name....awesome
.....that is awesome."
What he did not know
is that every cat in the world
gives exactly the same name to its tail.
Trevor, Trev, or occasionally Trevvy Toots.
Never anything else.
Perhaps it's the same
for dogs and cows and horses.
You could always ask someone.

Balbriggan Harbour Revisited
13.28 17-8-98

A black pool left deep
between sloping sand
and high walls
the tide gone out so far
you can't float anything
but seabirds
you can't smell anything
but fish stink
you can't hear anything
but gulls going to hell
and lamenting
the eternity of it.

XMAS MORN D426
black rusty brown
sits high and dry
rotting
on her knife edge.

The sun is hot to bake you.

Balbriggan Railway Station Revisited
15.50 17-8-98

I am sitting on the platform
resting my back against the wall
down at the same level
as the flat trucks of the goods train
stopped
after a slow creak into the station
stopped
waiting for a flag or a light
or something.
I think about straddling a flat truck
in wind and rain
whooshing down the line
finishing up in East Wall
North Wall
one of the yards.

The driver is out of his cab
on the platform
looking up the line
the porter is out of his office
on the other platform
looking down.

Maybe it will go soon.

Sitting On The Custom House Steps

Friday evening 14th August 1998

Cool
My word
you have no idea
poised in her perfection
sitting on the steps
young blonde beautiful
tight black velvet trousers
red shirt
black jacket
she kicked off her shoes
and showed blue polish
on her toes
sunglasses on top of her head
A mobile in her pocket
saying
answer me
then she read EAST OF EDEN.
I was thinking
WOW
WOW
WOW
until a pigeon hopped over
all plump, friendly
and ready for some food
she shooed it hard, heavy
and fuck-offy with her foot.

You won't be staying in my house.

Pity

15.50 17-8-98

Written on the wall of the shelter
northbound platform
Balbriggan Railway Station

WE ARE OUT OF THE MIND
ON DRUGS AND DRINK

Happy Day

18-8-98

It seems like every time I got out right now
I am finding money.
Yesterday I found a lovely new penny
on O'Connell Bridge.
This morning it was a twopenny piece
on Killester Station

Another 2p
and I can buy a drumstick.

Harmonstown Station

Tuesday morning 18-8-98

The most beautiful purple flower I have ever seen
is luscious and open.
A bee goes in headfirst and disappears completely.
The sun is putting such bright light through a leaf
I can see the way its veins are going.
A bee flies against the back of it
and makes a dark blob.

MacDunna – There's Your Reason

See,
the look-out fish
swims up and down near the shore
and keeps a close eye on the DART
as it trundles into Howth.
The merest hint of a fishing rod
peeping up at the window
and MACDUNNA (the look-out fish)
is gone like a shot
hightailing it towards the lighthouse
– "Leg it boys and girls...fast....
there's a rod coming!"

That's the reason.
It's got nothing to do with seals or trawlers
or all that other scientific stuff.
It's MACDUNNA – the look-out fish.
(Macker for short)

15.20
18/8/98

I Miss You Beaky

I loved Beaky specially because he had a growth
on the side of his neck.
Every time the sparrows landed beside me
I looked out for him.
Always I hoped he would be there.
I felt happy when he was.
Once or twice he arrived near me
on his own
and I felt extra-special.
Once he didn't come at all for two weeks
and I got sadder and sadder.
I thought he might be dead.
Then one day he came back.
"God Beaky – I thought you were dead."
But he wasn't.

I tried to see if his lump was getting any bigger
but I couldn't really tell.

All through last winter
Christmas
all the way down to the end of June
Beaky hopped and pecked and made me happy.

I have not seen him since.

Today it is the 18th of August 1998.

I miss you Beaky.

Conversation With A Girl In Yellow Shorts

14.50 18-8-98 East Pier

"'Scuse me – are you Pat Ingoldsby?"

"Yes – I am."

"No...but – really?"

"Yes – really."

"It's funny 'cos you don't look like him."

MY FAVOURITE PEOPLE

golden advice on the pier tuesday 18-8-98

Money is so scarce at the moment
that I asked Geoff, Lucy and their friend Shane
– "I need four thousand pounds before next February
to publish my next book. Where am I going to get it?"
I asked them because they are children
and their vision is bright.
Lucy said make a shop in your garden.
We sold Jaffa Cakes and Mi-Wadi for 10p each

or a drink *and* and a Jaffa Cake for 15p
Geoff said.
We made two pounds twenty p.

A bank
Geoff said
a lottery ticket........ if you win
........save up
get a job

After a long think
Lucy said
Get a hot dog thing and make hot dogs

Ali said nothing because she was asleep.
Shane said nothing 'cos he didn't want to.

I feel much better now.
I am very happy that they came along.
They made my life beautiful again.
They always do.

Poor Pennies

When I am giving you your change
please don't say to me – "It's alright.
Keep the penny"
'cos I'll only have to counsel it afterwards.
Long after you are gone
the penny will ask me
– "Why didn't that person want me?"
and I'll be saying
– "Aw – they did really."
"Well why didn't they take me then?"

Sometimes at night
I have got a houseful of pennies
all standing around chanting
– "I AM a good and worthwhile penny."

When I am giving you your change
please don't say to me – "It's alright.
Keep the penny"
'cos it's not really.

A Sure Way To Discover Which Big Unwieldy Yokes Your Neighbours Don't Want Anymore

Put an empty skip outside your house.

Get Out Your Gallic Wars

Don't ever say to me that he didn't
because he most certainly did.
He gave Latin classes
to 28 very carefully selected
cornflakes
The verb *"to be"* was no trouble at all.
"Sum es est" – they flew through it.
Every morning
lined up in rows of four on the dresser
28 eager little flakes chanted
"mensa mensa mensam
  dominus domini domino
  bellum bellum bellum"
till one day he was killed at work.
A blackboard fell on him
– sideways and spinning.

That was the end of it really.

His daughter cleared out the house
and ate the cornflakes.
She did not know they were learning Latin.
They did not know how to tell her.

Arise My Good Good Man

Brought up that way

– *what-a-good-boy*
what a good girl am I –

look teacher
look at my copybook
isn't it spotless
isn't it clean
all my sums right

look at me now
washing my hands
I've just been to the toilet

distinction
nomination
gold medal this
that and the other

up to my eyes
haven't a minute
rushed off my feet
good and busy
busy busy busy
good good good

living a competition
they never stop
one of her poems was shortlisted
for the British Eminence Awards
in 1996

first second third
if places are that important
they'd be happier breeding greyhounds.

Dear Seamus
26-8-98

Today at 13.25
as I was walking along Nassau Street
I had to look very hard
through the leaves and the branches
to make out the words
– FINNS HOTEL –
high up on red brick.
The wooden cut-out of yourself
at BOOK'S UPSTAIRS
could do with a lick of paint.
Apart from that
everything is grand.

Health Warning
Dunnes Stores, Nth. Earl Street

If you're wearing sandals
don't go down to the far corner
to get a yoghurt
or your feet will freeze.

Ah Me

A man
trying
to look
like
he is not
trying
to look
up
a woman's
skirt
looks
just that.

Slán Abhaile

On the days when we were small and had nothing to do
we would go up to the library and look at the sand.
A little pile of it on a yellowing piece of curled-up paper.
Inside a glass case.
A card beside it said – *SAND FROM THE SAHARA DESERT.*
We would stand on tip-toe and stare at it for hours.
We would lift the littlest ones up so that they could stare too.
Some of them thought that it was *the* SAHARA.

I do not know how a very small piece of such a gigantic desert
came to be in the Carnegie Library in Malahide, County Dublin.
I do not know if what Bernie the Milkman said was true.
That it really came from the Velvet Strand in Portmarnock.
It it was so, he should not have said it.
If it was a lie, he should have been taken out and shot.

Years and years later when they modernised the library
and put computers into it
the glass case disappeared.
So did what was inside.
Nobody can remember where it went.

I would like to think it found its way home again.
I would like to think that camels walked on it.
We only had Howard's cattle
and they were up in the Hill Field.

But Of Course

Iarnród Eireann's new ticket
expires
before you reach your destination
unless
you run along ahead of the train
and get there before it.
Buy a return
and you stand where you are.

Cycles For Sale

All together
the motor-bikes manacled outside the shop
said – "VROOOOOOOOMMMM!!!!
and the owner came rushing out.

They were still there
manacled together
in a line
innocent
as always.

"Fuck yez anyway" he said.

About Poetry

A workshop to me is a place
where you fix cars
                make boxes
                or solder bits back
                onto spectacle frames.

Why anybody would wish
to jack a poem up over a pit
and shine torches under it
is something I would rather
you did not explain to me.
I prefer not to know.

Crumble

The little boy down on the strand at Balbriggan
was shouting up

– "MA!! I'm diggin' a HOE-WELL!!
"MAAAA!! I'm diggin' a HOE-WELL!!"

His Ma was up to her eyes
sharing out the sandwiches
pouring lemonade
tying laces
and breaking up fights.

The little boy shouted and shouted

"MAAAA!! I'm diggin' A HOE-WELL!!"

until the tide came in
and made a mockery of it.

Then he started to cry.

Hard Labour

The little old man with the cap and the crinkly grey hair
sat on the pier and sweltered.
For an hour or so.
Then he loosened his tie a little
and opened the top button of his shirt.

The little old man with the tie loosened a little
and the top button open
sat on the pier and sweltered.
For an hour or so.
Then he untied his laces and widened his shoes.
Just a bit.

The little old man with the tie loosened a little
and the top button open
the laces untied and the shoes widened just a bit
sat on the pier and sweltered.
For an hour or so.
Then he took his shoes off completely and placed them
one beside each foot.
He thought for a while about rolling up his sleeves.
But he didn't.

There he sat
underneath his cap
beside his shoes
in the sun
on the pier
for ages and ages and ages
and the sweat ran down him.

At seven o'clock or so it was time.
He began to button and close and tighten things.
He began to put things back on.

"Jaysus" he said to himself. "You'd be wrecked."

It All Depends Who Is Doing it

Whenever he whacked his ball towards the green
it whizzed through the air
just like any other.
Then it dropped down towards its target.
When it was two feet above the ground
it stopped.
And stayed.
There.

When he came to putt
he stood upon a box to reach it.
The ball ran along two feet clear above bumps
and twigs
and dips until it was directly over the hole.
And in it dropped.

"Aw c'mon – that's not right.
No offence but we call that cheating"
said the ones who could not do it.

Patrick Gerard

Da

I loved the way you laughed.
Good deep and hearty mirth
far down inside you.

I loved the special feeling
when you asked me
to fetch something from your pocket
– the faint smell of tobacco
– the rubber feel of your pouch
– the cold of your keys and coins
– the shivers when I touched your hanky.

I loved the way you put soccer into my heart
the way it was in yours
– kicking a little ball with me in the back yard
– heading a balloon in the hall
– push-ha'penny on the table
– Subutteo on the floor
– standing beside me in Dalymount
  cracking Cadbury's in half
– feeling my heart break when Dayo ran rings
  around you out the back and suddenly I knew
  that you couldn't really play at all.

I loved the way we sat near the fire.

I loved the sound of you downstairs
when I was small in bed.

I loved the angle of your finger sticking out
the crazy way you held your cup.

I loved the sizzle when your tobacco spit hit the hearth.

I loved the sounds of you down the garden
metal whanging earth and stones
breaking up the clay.

I loved the way your mother had seen Uachtarán na h-Eireann
putting out his bin.

I loved the way you always put us first.

I miss you terribly.

I really do.

ÁIREAMH

Should you ever turn to Question 26 on Page 331
of – A SCHOOL ARITHMETIC

HALL & STEPHENS

you will read

*"If 4 men with 9 boys can do a piece of work in $1\frac{1}{5}$ days
and 3 men with 6 boys can do the same work in $1\frac{5}{7}$ days
how long would one boy or one man working alone
take to do the work?"*

Is it any wonder we turned out the way we did?

(*The solution is provided on page xxix of the same book.
I will not insult your intelligence by printing it here.
You have probably worked it out in your head already.)

WESLEY ARÍS

You would spend the rest of your life
trying to – *"SPOT THE DIFFERENCES"* –
in Wesley Quench's pictures.
He loves setting them out.
He really does.
Nonetheless
he is finding more and more difficulty
in placing them with magazines
and newspapers.

EXAMPLE

TWO PICTURES OF THE SAME RABBIT
NOT WEARING A HAT

the differences according to quench

PICTURE A: He's not wearing a hat
because he cannot afford to buy one.

PICTURE B: He's not wearing a hat
because a wind blew it into the canal
and it sank.

Rock on Wesley
– when they catch up with you
you'll be long gone.

And These Thy Gifts

When Jeremiah Thrust
fell into the Irish Sea
from the car ferry
all the fishes gave thanks
to their God
for the bounty from the land.
Then they picked him clean.

MUTE

on the train from Crewe to Glasgow Tues 1-9-98

It is a never-ending source of astonishment to me
how so many people can travel together
on so many trains
without speaking a word to one another.

Bursting into a carriage singing –

"All together now!!!
3,000 green bottles standing on the wall!!!!"

would almost certainly get you shot.

Then they would look at one another and nod.
Not the sort of nod which leads to conversation.
More of a – "serves him right. Now let's get back
to lovely lanquid torpor."

ANNOUNCEMENT
"We are now approaching Glasgow Central......"
(that much is allowed.)

SECRET SIDINGS

9.45 on the train from Glasgow to Oban 2-9-98

I love the secret little places
which railway tracks wander into
rusting slowly under weeds and grass
and other straggly things
stopping forever against crumbling brick
which knew the feel of steam
but has long since forgotten.

Foxes play there.

Moments Of Magic

McCaig's Tearoom Heritage Wharf Oban
12.30 2-9-98

ME: "Fresh local haddock with chips and vegetables please."

HERSELF: "Would you like a drink with it?"

ME: "A glass of water please."

HERSELF (loudly): "*JUST* tap water."

Fucking magic.

Mo Chlann Féin

Oban 2-9-98 13.33

When McCaig's Tea Rooms put tables and chairs out on the pier
stray people who wander around the town
and up and down sharp edges in their minds
sometimes put themselves near you.
They sit in silence
and let you fall into the holes of their eyes.
They ask you questions about yourself
or tell you things – "The sea is over the ground
because if it was underneath it, the fishes would bump their
heads."
I am familiar with the things that they see.
I recognise them well.
Sometimes they will not let me sleep.
Sometimes they make me afraid to be awake.

Hold hands with me.

Conversation With A Little Boy Through A Wire Fence

Castlebay thu evening 2-9-98

“Excuse me” he shouted. “Will you give to me your name?”

“Will you give me yours?”

“No.”

“Then I won’t give you mine.”

LONG PAUSE

“My name is Kenneth.”

“Mine is Pat.”

“I told you a lie.”

“I told you one was well.”

I stuck out my tongue at him.

He stuck out his tongue at me.

Lovely.

Brón Ar An mBooze

The drunken man
is watching the two young girls
and the bums of them.
He is making a clumsy grab
as they go past.
They skip around him
the way he is not there
and talk about something else.

They know him well.

He is looking down
at the ground
and telling it things
which are very broken.

Bagh Siar
Thu 3-9-98

The sea does not remember you.
As soon as you are gone
it washes your footprints away.

An Sean Scéal Céanna

Craigard Hotel 17.30 Thu 3-9-98

An t-uaigneas atá orm inniu
an t-uaigneas céanna a bhí orm
ag an mbord seo
'san ostán seo
ar an oileán seo
bliain díreach ó shoin.

Aithníom é.

Cheap mé go mbeadh diofríocht.

Níl.

The Blessed Remains

near the old schoolhouse   Vatersay   3-9-98

A pair of cormorants are diving in the wide water.
If they stop and listen
they will hear the drills of the men
working against the old school
making it into a place
which is not a school any longer.
Down the rocks near the sea
lie the rusted remains
of what looks like a train crash.
"How in the name of God
did a train get down there?"
I asked one of the men.

"The underground" he says.

It must be true.
The cormorants told me the same thing later.

September Evening In Nasc
The Ferry Coming In

Sept 1998 the happiest time ever I spent in Barra

A black and white dog is barking bloody hell at me
through the back window of a red van.
Green and blue fishing nets are tangling with the grass.
A mother duck huddles her brood away fast
saying – *"Heed the dog – that's an Irish poet – he'll eat you."*
"Have a bit of sense woman" says the cat.
A sheep rubs her back to and fro and up against the fence
with an –*"Ohhhhhhh that's lovely and the price of wool."*
Peacock and peahen together on top of a caravan
are looking down the bay
the way you'd think they're watching Clansman
rounding the bend.
They're not.
They'll never tell you what.
Of barking dogs and lumps out of your leg
an old man says – *"They don't know the difference*
between a bloody rabbit and a sheep."
They'd want to.

Later That Night

Sheep snuggled up against the wheel of a van.
Sheep snuggled up against the grass and the rocks.
Settling down for the night.
The cool of the dew is touching them.
I wished them – *"Sámh codladh chugat"*
and walked on up the hill.

Giddy Boots

Giddy giddy giddy boots
do you know it seems to me
hopping skipping dizzy jigs
a ripsnorting place to be
kicking up the fiddler's dust
chasing a rabbit or three
clapping hands at sparrows
and always late for your tea.

Conversation With Two Boys Late At Night

23.10 fri 4-9-98 near the Heritage Centre, Castlebay

"Hiya boys – will you be dancing in the competition tomorrow.?"

LOOK OF DISBELIEF, PAINED AMUSEMENT
AND WHERE HAS THIS MOST MISGUIDED
OF PERSONS COME FROM?

"*That's* for pussies!!!"

NEXT DAY

CONVERSATION WITH A GROUP OF BOYS OUTSIDE THE SCHOOL WHERE YOU CAN HEAR THE SOUND OF THE PIPES AND PEOPLE CLAPPING

"Are you going in to watch the dancing?"

LOOK OF 'OH GOD HE'S BACK'

"Naw! *That's* for pussies!!"
"Michael Jackson dancing.....that's what I do."

"How about Michael Flatley? He's made millions out of dancing?"

"He's a pussy."

"He's one of the fittest men in the world."

"He's still a pussy."

O.K. boys.
O.K.

Mise Amháin

Ag trampáil
suas is síos na trá
gan rian ar an ngainneamh
ach rian mo chosa féin.

Cur do lámh orm.

To A Green Fishing Net In The Middle Of A Crooked Field

Sunday 6th September 1998 12.15
walking with the wind towards St. Brendan's Graveyard

There's no fishes out here you dozy net.
There's only the wind
and that'll go straight through you.
Do you prefer the grass or what?
I saw what's left of a rabbit over there
stretched out sideways
he thinks he's still running.
The birds have had his eyes.
His right front leg is gone as well.
He still has his tail
white
the wind is moving it.
Do you prefer the grass or what?
I'd sooner be in the sea.

Nearly At The Graveyard

Sunday 6th September 1998

Out among the wild grasses
sideways to the sea
I came unexpectedly
upon
a giant yellow rusting JCB
with a little rabbit minding it
although it wasn't really.
It ran away and hid for fear of me.
Don't worry little sentry.
I won't rob your JCB.
I can hardly carry what I have.
Anyway
I'll be gone in a minute.

I'm Not That Bad And I Know It

St. Brendan's Graveyard, Borv
Sunday 6th September 1998

I came up out of the rain towards the graveyard wall
singing at the top of my voice to the sheep

– *"I just dropped in to see yez all*
I'll only stay a while
I'd like to know how you're getting on
I'd like to see you smile!!!!"

They did not know what I was.
For a second or two
They did not believe that I was happening.
Then with wild-eyed scatter
they ran like hell.

I Was Just About To Leave The Isle Of Barra Hotel When Three Young Girls Began To Dance And Suddenly My Pen Was Rushing These Words Across A Page

(very specially for Marion MacNeil, Rosemarie Maclean and Anna Agnew – it was you)

My oh my
the dignity and the pride
you have riches
beyond
the widest world of wonder.
I watch you dance
a thousand years
your quiet beauty
takes my heart asunder.

Monday night 7-9-98

Green Smithereens

When you sat upon a rock
and made green music
with your harp
I looked for a second
between the strings

a wave broke
and filled the sky
with smithereens

the dandelions were listening

the grass was too

and the sea was putting
its deep noise
at the back of me
and
at the back of you.

9-9-98 Seal Bay

MEMO

My dear Wallace Arnold,

On the back of your huge big tour bus
parked outside The Isle of Barra Hotel
you say

– *WE THINK OF EVERYTHING* –

You didn't think of my birthday

Taking Bits Of Me With Them

Quite by accident
I stumbled upon a crucial secret
regarding the midges who swarm darkly
and bite lumps out of you over on Vatersay.
They have all got exactly the same name,
understand biblical language
and respond instantly to fearful threats.
When I yelled

– "BEGONE CELESTINE OR I'LL TAKE
THE LIQUIDISER TO YOU!!"

they were gone like a shot.

WHACK
with dried flowers

Macanhenry Wimple went down to the library
borrowed a hardback edition of
– THE DRIED FLOWER BOOK –
went home
and beat his wife's brains out with it.
He returned the book
before a fine came upon him.

It Hurts Me

I would like to play the uillean pipes
till every rock and stone and hard thing
cracks inside myself
and barbed wire tangles out in torrents.
Then the tears will come.

Bo Peep – You Don't Lose Them Far Enough

One sheep is thinking to himself

– "If I don't get away from here
they will wreck my head entirely."

And he is moving up to a corner
where none of them ever goes.

"Hey....c'mon...he's going up there....
c'mon quick or we'll miss something!!"

He slowed a bit.
They did too.

He stopped.
They did too.

He started again.
They did too.

BAAAAAAA!!!!!!

You'd be demented.

Cool

The daddy-long-legs of Barra
suction themselves to your window
and let the wind do all their moving
for them.
They'd be dead for months
and not know it till the wind drops.

Lonely

Lick me moo-cow
please lick me too
a sick sadness
is making lumps in me
I do not know what to do.

Lines Written Leaning Against A Gate Looking Over Towards Seal Bay

The disappearings of a rainbow
at opposite ends of the sky
up and in through heaven
and out the other side.

Skitthery Diddle – then he does it backways

If ever a rabbit feels the slightest fear of being bored
he takes off suddenly towards the top of the hill
to see can be pass himself out
and run all the way back
before he gets there.

Pat's Dream

I would like some organisation
with more money than sense
to buy me a cottage with an open fireplace
a hen
and the window slanted in such a way
that the sun off the sea doesn't blind me.
And the only thing they want from me in return
is the occasional egg
and that I am happy.

15/9/98
16.01
You
across
a table
from
me.

Oh Shit

The last time that I pushed down
one of those KENCO COFFEE plungers
3 railway stations
an abandoned mine
Ronald Hegarty's bonded warehouse
and 10 guinea-pigs
all exploded.

I am not sure what else it is they do.

Overheard On The 130 Bus

Saturday afternoon 15-8-98

ONE GIRL TO ANOTHER: "She's so attractive she's disgusting."

Overheard In Fairview

Late September 1998

WOMAN: "She's in an awful state. They took the leg off an' it didn't make any difference."

Overheard On Vernon Avenue

Late September 1998

WOMAN: "Then the other fellah went in after him an' he was drowned as well.
It was fuckin' terrible really."

ADJUSTMENTS

Lim Terese eased her bus ticket in
the machine pushed it out
she eased it in
the machine pushed it out
in out in out
suddenly it sucked the ticket
and Lim
her hand to the wrist
her arm to the elbow
herself to the neck
and said beep beep beep.
When it spat her out again
she was 30 years older
and eligible for the free travel.
"Feck it" said the driver.
"I wish it'd stop doing that."

On The Harbour Wall

14.34 21-9-98 east pier

Old women with beards
who chew imaginary food
and
sit with arms locked tight
trying hard to remember
what it was
some man said
before
Martha
went away
to
that
other place
and
did not come back

or something.

Harmonstown Dart Station
overheard further down the carriage behind me
22-9-98 11.15

"There's your man that used to do Pat's Cha'.
d'ye remember him?"

"Buy a book off him."

"Bollox!"

Late Effects Of Polio

They introduced a round lemon pipe
up through the ball of her foot
and channel trains ran through it
as far as her knee
which was O.K. with her
because that's as far
as she was going.

Something Lovely To Do The Next Time It Rains

Go out and get wet.

Ar Aghaidh Libh

Sat 26-9-98   near Naas

All the people on the Waterford train
including myself
just sitting here
not using our feet at all.
What a pity
we can't send them
out for a gallop somewhere
racing sheep across the fields
chasing behind the train
hopping up and down on the roof.

If I was a foot I'd love it.

Boomisacomin'

When people are feeling frightened
they often become aggressive
when people are feeling fed-up
pissed-off and at their wit's-end
they often become aggressive
when people are feeling
like they are not being appreciated
they often become aggressive
when people are feeling
like they do not count at all
they often become aggressive.

Dublin '99 – keep your head down
well down.
For Christ's sake
don't light a match.

Rince

The leaves are down
life with the tree is done
lying on the ground
they're waiting for the wind
to twirl and dance and run
into empty air
and giddy kingdom come.

Where Indeed?

"You have to have a hobby" he said.
"Somethin' to get you out of the house.
Otherwise you'd go mental.

I smash windows.

Plate glass
stained glass
it's all the one to me.
I go out everyday
with a song in me heart
and a bagful of bricks.
Oney for that
I'd be in the house all day
talkin' to meself
– then where would I be?"

Dublin 1998

The trouble with feeling
that we've all stopped talking
to one another
is that unless you're very careful
you'll stop talking to people too.

Heuston Station   Sat   26-9-98   13.20
The only person who has spoken to me
since I arrived one hour ago
is the man who told me I need 20p
for the toilet.

He Loves It

All the supermarket trolleys
are keeping still and quiet.
Here comes the man
oh God here he comes
looking at the wheels
Jesus
he's looking at the wheels
walking on
he's walking on
oh God
he's writing something

the music stops.

"O.K." he says.
"Let's see
if you can swim."

What Makes Dublin Great

1-10-98   at the 13 bus terminus   off Merrion Square

The bus was standing at the stop
with the front door open
so I went to get on.
The driver looked at me.
"I'm not goin' anywhere" he said.

Ah Come On Lads

Sunday 15.30 4-10-98 East Pier Howth.

The cormorant is sitting black and happy
clutching the side of his boat
the way he wants it
half-way along
exactly in the middle
making black diamonds with his wings
hanging them with the wind.
A seagull
not knowing the arrangements
lands on the up curve at the bow
and settles
nice and comfy
resting with the tide.
HAH!
The cormorant tightropes sideways
skips and jumps
fast as he can
broken scampers
right away to the end.

The pair of them
sitting stiff and still
the pair of them
looking away
from each other.

High on the harbour wall
the raggy crow is making
his laughing noises.

Slowing Down

Be not deceived by the elderly man
who shuffles down the pier
with a white pillow.
Holding it against his bum
he leaps from the wall
yelling “whooooooo!!!”,
lands on his bottom
and bounces back the way he came.
He used to set fire to himself
and do a somersault
but he’s older now.

CREIDIM

I believe the day will come
when people will realise
that television and newspapers
simply
cannot mind their own business.

In the meanwhile
– *"Mavis Netherby*
first noticed this appalling rash
in July 1987.
Mavis....tell us exacttly how......."

Wrong Night

The cow who jumped over the moon
desperately wanted to be noticed
but everybody was asleep.

OH GRANNY G.

"Don't worry – it might never happen"
said Granny Gavigan.
Next morning
in the tiny community of Whisk
the ambulance crashed
the lifeboat sank
and the family doctor died.

You're absolutely right Granny
It mightn't.
Then again
– it might.

Don't Mind Me
I'm Only Selling My Poems

11-10-98 East Pier

Two wrinkly old ladies
pulled in at 15.30 this afternoon
and talked in their car for a long time
about how to dispose of the body in the boot,
how to rob another bank
without generating any more bodies
and how to keep me from finishing this
without stuffing me into the boot as well.

After a decade of the rosary
they chugged away up the hill.

Tortoise High Tortoise Down

A tortoise is moving along flat and heavy and round
opening its beak and swallowing full gusts of wind.
Now it is creating little holes underneath itself
and firing tiny jets of air down
with force and lift and power
raising itself higher and higher above the ground.
Now it is turning the pillars of air
into rigid poles of ice
and shaping them and carving them
into fantastic totems
which hold it high and beautiful for an age.

When it is time
the fireflies come
breathing gently on the bottom of each pole
melting the totems down.

once again
a tortoise is moving along flat and heavy and round.

Yahoo Happy Yahoo Happy Skip And Dance And Jig

14.10 11-10-98 on the East Pier in Howth

BEAKY CAME BACK!!!!!!!!!

Fíor Chara

My goodness me
my good Mother Sea
so many shells to be washed
waves to be made
and swooshes to sound.
So much seaweed to dance
gravel to shift
amd rocks to surround.

You give me peace
and I love you.

Shut In Shut Out

All around me in my native city
I see people closed up tight
afraid to talk to one another.
Whatever it is
they are frightened of
it must be very terrible indeed.

ENCOUNTER WITH *"Me Bollix"* IN IT (TWO)

9-10-98 11.50

at the bus stop outside The Belgrove Stores

"I'm goin' out to Malahide to walk in the woods."

"Fair play to you – it's freezin'."

"Yeah but I love the noise the wind makes in the trees."

"Me bollix!"

Bing Bong

I would like to meet the polite courteous man
who apologises to me
every time I attempt to travel Northwards
on the DART.
He tells me –
"all Northbound DARTS
are running 20 minutes late."
He tells me –
"all Northbound DARTS
have lost the will to live
and will probably never work again."
He tells me
"all Northbound DARTS
have been returned by sea to the makers."

In the meanwhile
all Southbound DARTS are running so regularly
you would think that someone
had pumped prune juice up their pipes.

I would like to meet the polite courteous man
who apologises to me
every time I attempt to travel Northwards
on the DART.

I would like to ask him is it only me?

FOUR TO GO

I would like tubes of toothpaste to be marked
– suitable for 4-9 teeth
10-14 teeth
and so on.
I bought a large tube for my four today.
I hope to God they appreciate it.

Up Your Aperture

An overseas tourist was taking a dishonest photo
of my friend selling her newspapers
in the city centre
taking it in such a way
that she thought Jessie wouldn't know.
Some chance.
Jessie erupted from behind her stand
and read the riot act
and fair bloody play to her.
The tourist looked at her partner
and threw her eyes up to heaven.

"Don't be throwin' your eyes up to heaven"
said Jessie. "Or you'll go up after them!!"

HIS ALARM IS WORSE

When you press the doorbell
of the man
who works the pneumatic drill
you set off such a roaring racket
that dividing walls shudder
floors collapse inwards
and crockery falls off the dresser.

He's not in.

Peg-Legs Eye-Patches And Stripey Jumpers

When the remote-controlled model yachts
are put back in the box
and the lid is shut........
that's when the little sailors come out.

Mister Crow I Hope You Live Forever

Half-cracked raggy baggy
raucous Mister Crow
meandering all over the pier
giving out hell about everything
hands in his pockets
black cloak in tatters
looks like he slept in it
nothing is right
the sparrows the starlings
the sun the clouds the tide
everything wrecks his head
giving out hell
arguing blue blazes with himself
stating the case with crazy caws
refuting it with whistles
giving out the livelong day
he loves the merry hell of it
me, I love to listen.

Said To Me By A Man
(in the Flowing Tide)

"We used to watch you when we'd no cable."

Said To Me By A Woman
(in Howth)

"You were much nicer before shock treatment."

I Remember The Men I Remember The Horses And Carts

The horses and carts left in the dark of hungry morning.
Each one in isolation
down a lane
and onto the cold road to Dublin.
A man harnessed the horse
and looked into the silence of empty fields.
All the town's asleep.
A man blew into his hands and rubbed them hard.
A man walked along beside the horse
or sat high against a piled-up pyramid of cabbage.
Little sounds are wakening in the hedgerows
little stones are making powder with the wheels.
A long walk.
A long walk to market.

Have you see the cart?
Red heavy and square
two high wheels wooden spoked
metal cold and banded round each rim
a hub you could step on and haul yourself up.

All the way to Dublin on a hard and frosty morning
each man holding short the rein and walking at the head
each hoof stepped into sack and tied
against the time you'd slide and slip
and maybe snap a leg.
A long walk.
A long way to market.
Man and horse
breathing clouds together
measuring the miles.

Down The High Rock

Late at night
while I lie in bed
awake
the sea is still moving
in and out
beyond The High Rock
where this afternoon
I sat with the stones
and the gravel
moving in and out
without me
and I lie here
alone.

Some Days You'd Be Better Off In Your Bed

Saturday 24-10-98   13.47.   Killester

I came into The Beachcomber today
because I was starving.
My *favourite* was on the menu.
Roast tuffed turkey.
"YES!!!"
I'll have that.
"YES!!"

As I was approaching the Carvery
the chef took up his cloth
and very carefully wiped the words
Roast stuffed turkey
away from his blackboard.
He left every thing else.

It has been that sort of a fucking week.

LATER

sitting upstairs in The Beachcomber
looking out through the huge skylight
at drifts of cloud moving east
I was thinking
– "God – it's just like being on a ship.
You'd think *we* were moving."

When I went outside again
I had to get the bus home from Kimmage.

YOUR SHOES'D BE RUINED

Gregory Grooth fell asleep
on the DART
and woke up in his bath.
He was giving out hell
because he didn't have time
to get his clothes off.

For A Beautiful Young Woman Called Che Who I Hardly Met At All

I love the way you look at things
they delight you
and you dance!

LISTENING TO A LEAF COMING DOWN

OW
OW
OW
OW
OW
OW
OW
OW
OW
OW
OW
OW
OW
OW
OW
OW

OW!!

The Most Beautiful Reason For Learning To Read That I Have Ever Heard

told to me by Willie

"I taught myself to read 'cos I wanted to laugh at the comics like the others."

ABOUT SOMETHING ELSE
SOMEBODY SAID

"For life being short....it's very long."

I have found deep wisdom and agonising beauty
with the men and the women
who have got no home
little hope
and pain which never goes away.
Always they remind me
one meal
one coat
one bed
one me
one you.

Laugh
and it goes away

till you stop.

DANCE

Shadows are saying

"Please don't blow out the candle
please don't switch off the light.

Please.

Don't."

The candle will burn down.

White Flying Shite Machine

I would never have written a title like that
if a gull had not done what it did.
I would have written something lyrical
like – *gull*
        hang yourself high upon the wind
        and let it take you sideways –

I would never have written a title like that
if a gull had not splattered
a spectacular two-tone shite
against the cover of IF YOU DON'T TELL
ANYBODY I WON'T
two-tone with a heart of hydrochloric acid
which burned a hole through my hat.

What in the name of God are you eating?

A Single Back To There Please

God be with the times
when railway stations
were alive with porters
who swarmed all over the place
pushing trolleys and clipping tickets,
sweeping up tiny cardboard triangles
and slamming heavy doors shut,
all the trains were late
and nobody seemed to give a fuck.

Now
you can't find a porter to save your life,
a machine validates your ticket,
you carry your own baggage,
a machine sells you coffee,
you pay twenty pence to do a wee wee,
and nobody seems to give a fuck.

I preferred it the first time.

Forever Bunnies

The man with the plasticine leg
fashioned it
into a legion of hungry rabbits
who hopped all over the table
and finished off his dinner.
Then
with a wilful wildness
which surprised him utterly
they finished him off as well.

No matter how hard they tried
they could not fashion
themselves
back into a leg.

WORSE THAN THAT

When we were little
somebody
was doing evil things to children
up Townyard Lane.
My mother wanted to make sure
that none of us went up there
so she told us the most fearful thing
she could think of.

"Don't go up Townyard Lane.
There are communists up there."

We used to run past the end of it.

There was a comic in our house.
A different sort of a comic.
It was called – IS THIS TOMORROW?
All our other comics were full of Biffo the Bear,
Corky the Cat, Desperate Dan,
Buck Jones, Kit Carson.
This one was full of Communists
doing terrible things to Catholics.
There was a picture of a priest.
He was preaching in the pulpit,
waving his finger and warning us all to watch out.
But he wasn't watching out himself
because two men were standing down the back
in long leather coats with the collars turned up,
black hats with the brims pulled down
and little thin lines where their mouths should be.
In the next picture
the priest was sitting in the back of a car
between the two men.
In the next one, the priest's hat was lying in a ditch.
His broken glasses lay beside it.
The car was speeding away.

That was 1947.

I haven't been up Townyard Lane since.

Not Looking Not Laughing

He was one of my father's uncles.
We were told not to look at his ears.
Luxuriant bushes of hair
were bursting out of each one.
He would ask us questions
about what class we were in
and what we were best at
and we would try to answer him
without looking at his ears.
He was one of my father's uncles
so we were told to sit on the sofa
and not to laugh
when he took off around the room
with long lunges of his legs
dancing to imaginary music
played by himself upon an invisible flute.
We sat there
in an agony of not laughing
as he careered around the furniture
making "dee idle dee didle" noises
and terrifying the cats.

Before he left
he always reached into his pockets
and took out loads and loads of statues
of St. John Bosco.
I do not know how he was able to dance
with the weight of them.
Each of us got one and said
– "Thank you very much"
without looking at his ears.

Then he was gone.
My mother would stare down the road
after him and say
– "Another couple of years."
My father would say
– "With the help of God."
And we would all say
– "Amen."

Taking It Out On Me

The tree who wanted to climb
to the top of itself
fell on a dolphin instead.
It was easier.

Ow

I remember the bite of your mouth
the way I liked it
little nips
that pinched my flesh and skin

the lick
    lick
    lick

the touch of the hungry tip
flicking in.

Oh come
    come
    come again

    come

and let us begin.

One Per Cent

The lovely little young and friendly man
simply could not stop talking to people
and asking them things
and telling them things
so many little detailed things
that he made them scream inside their heads.

"Hello – what's your name? Where do you live?
That's a nice pair of shoes. I have an Uncle Brendan
and an Auntie Kitty. Canaries can fly but I can't.
Last Monday got lost. I don't know where.
It never came. I hope I meet you every day.
Can you fly? I can't."

The lovely little young and friendly man
made people scream inside their heads so loud
that they bumped into things.
He made people scream inside their heads so loud
that they fell into canals and stayed there.
He made people scream inside their heads so loud
that they flapped their arms and flew over hedges.

Then he toddled down to the day-care centre
and made gorgeous little wallets.

The Pensioner's Lament

– 12 FOR THE PRICE OF 9!!! –

"I only want one."

Loops To Be Looped

A concertina
of crazy flying sea-birds
doubling and halving
up and down the sky
feather-close
skimming o'er the water
tight tucked under
keeping both feet dry.

We Were Robbed

This morning as I was sitting on the toilet
with the seat up
because it's lovely and cold that way
the following question came into my mind

– *If you could change one thing so far*
in the course of history, what would it be?

The answer whizzed into my mind immediately.

"John Ateyo would have hit the bar in 1957."

It's not great really.
Nothing about famines or wars or awful things like that.
But there you are.
Hopefully your answer will take care of them.
As long as John Ateyo hits the bar
you can change anything you like.

EVERYTHING ELSE WASN'T

The man with the artificial-rest-of-his-body
had a pain in his arse with people saying
– "That's not <u>really</u> hair sure it's not"

Because it <u>was</u>.

Proper Order

The little boy was watching the artist at work
staring at every single brush stroke
bombarding him with questions
utterly wrecking his head

"Would you ever fuck off
and stop annoying me!" he said.

The little boy raced over to his Mammy
and protested with much injured innocence
– "But I only.........I only said.........
I didn't do anything!"

So the mother whipped out a pistol
and shot the artist dead.

The little boy's eyes were wide.

"But I only told you because......
I was only saying.........I never meant........"

So she shot him as well.

Mammy There's Fishes Swimming Over The Sky

East Pier 28-8-98

The wide water
is taking huge white clouds
and blue sky
deep into itself
turning it upside down
and letting mullet swim high
over it.

OVERHEARD ON THE EAST PIER

19.04 18-8-98

GIRL (On mobile): "I'm missin' you loads....yeah......fuckin' heaps."

14.50 29-8-98

YOUNG MAN (to his girlfriend): "I'll be a nervous fucking wreck."

God I Was So Tired

I wandered into Easons
and went up the escalator to the top floor
where all the paintings are.
When I was certain that nobody was looking
I ducked into Vincent Van Gogh's bedroom,
hopped into his bed
and was sound asleep in seconds.

Down on the ground floor
a lady sometimes wanders up and down
beside the magazines
saying – "No reading."

As yet she does not realise
that people sleep upstairs
in the paintings.

Keep your head well under the covers
and you'll be grand.

The Whistle's Bewitched

In the park before the game starts
all the men run around elegantly
with deliberate exaggerated style
kicking, passing, posing, stetching,
the way the professionals do.

Then the referee blows his whistle
and changes them all into
headless chickens.

Questions That I Hate

(East Pier   Saturday 22nd August 1998)

"Will you be here next Tuesday?"

Signs

The sign in a shop window in Dawson Street
said – *"ONE CAR IS STOLEN EVERY MINUTE"* –

Your-man must be wrecked.

The sign on the train said
– *ALARM – Pull this handle to stop train.*
Penalty for improper use.

You'd nearly do it
to find out what the penalty is.

Nearly Never Killed

The man inserted his ticket in the machine.
"Beep beep beep beep beep beep beep!!!"

"This ticket is invalid"
announced the bus driver.
"What will we do with him?"

"Chop his head off!" yelled an old lady
who had always wanted to see that
done to somebody in real life
and not just in the films.

"Lie him down in front of the bus
and drive over him" suggested a man
who wished to get his own back
on someone
for the way next-door's dog
always dumped its hefties
on his lawn.

"Tie him up between two horses
and make them gallop
in opposite directions"
said Martha Birtles who hated everybody.

"Hang on a second" said the driver.
"It's the ticket machine.
It's not working properly."

"Does this mean
we won't be chopping his head off?"

"Or driving over him?"

"Or stretching him between the horses?"

"Yes – I'm afraid it does?"

"Fuck it. Everything was grand
till yez got rid of the conductors."

OUL ONES ON THE BUS

"The old man was dying from cancer
bad he was
blind in one eye one of his lungs gone
he was up in Glasnevin
tidying his wife's grave
these three lads hopped on him
gave him a terrible hiding
robbed his two pounds fifty all he had
such a terrible hiding
he died slowly under a hedge
they didn't find him for six weeks
his cat had died by then
locked up in the house
all by itself
waiting for himself to come back."

"Aw the poor cat."

On The Ha'penny Bridge

18-11-98

Would you look down there
Would you look at the leaves
in the Liffey
speeding along
with the flood
green ones
brown ones
in the rich river
swollen
with winter and mud
under the bridges
down to the ocean
melting
like salty mulch should

Put This In Your Ads And Say It

All of the banks
who

– *"put you first"* –
– *"want what's best for you"* –
– *"are at your service"* –

would
still be closing
at three o'clock
in the afternoon
if they thought
they'd get away with it.

A Very Scary Thought Indeed

If I was to accept sponsorship
from a certain source
I would spend
the rest of my life
writing
McPoems.

For Mrs. Pete

Ever since I bought a pair of red woollen gloves
and lost the right hand one
I cannot believe the number of people I see
who have bought a pair of red woollen gloves
and lost the left one.
Any amount of people going around
with a red woollen glove on their right hand
and nothing on their left.
And me going around
with a red woollen glove on my left hand
and nothing on my right.

I saw Pete Cummin's Ma on the Pier last Sunday.
She has the pair of them and long life to her.

It Doesn't Add Up

Jimmy Tuairisc peeped into the future
and saw
the half of the world's population
which was under thirty
worked in Gateworks International
and was on duty
telephoning
the other half of the world's population
which was under thirty
worked in Gateworks International
and was not
and attempting to sell them
even more computers
while
the other half of the world's population
looked on and said
"Thanks be to Jaysus I don't work there."

Your Number Is Up

I have just bought a new phone
a lovely yellow one
an extremely brilliant one
because
for the first time ever
it enables me
to *put people on hold*

My God it is so brilliant.
Now I can ring up
all those cold-voiced
disinterested people
who ever said to me
– "Just a moment
I'm putting you on hold"
and I do it to them.
Holy shit
the excitement is killing me.
I can play tinny music to them
breaking in every so often to say
"thank you for holding"
and wham the music back in
before they get the chance
to say a word.
If you have never done this to me
YOU HAVE NOTHING TO FEAR.
It's lovely lovely revenge
that I'm after
and with tinny music
it is very sweet indeed.

If I Had A Brown Boat I'd Give It To You

Saturday morning 21st November 1998 East Pier

The cormorant is wandering around the harbour
like a lost soul

popping his head up and down

wondering

"where the fuck is me boat gone?"

The lovely long brown one.
The one he sits on in the summer.

The sea-scouts think it is theirs.
They have taken it up for the winter.

He has nowhere to sit
and dry his wings.

I think it is the pits myself
but I do not know cormorant language
so I cannot tell him.

I grieve with him instead.

A Day Out

There are people
who drive to the harbour of Howth
on Sundays
and park their cars
directly facing the high grey wall
and sit there for hours
staring straight ahead.
Sometimes they park
directly facing the bit of the wall
with a sign on it
– a white rectangle
containing a red circle.
Inside the circle stands a dog
looking away to his left.
Behind him on the ground
is a tiny pile of his shite
with criss-cross red lines
drawn through it
and black steam rising.
Underneath are the words

– MAXIMUM PENALTY £100
– CLEAN IT UP –

Sometimes
they stare at that.

So There

In these days of correctness
sexist awareness
chauvinism
and all the other pitfalls
for the spontaneous
I write the following words

– I was looking at your face
much of the time
while you were sleeping
on the Galway train
because
you are one
of the most beautiful women
I have ever seen.
You really are.

Yours
with deep admiration

me.

Vernon Avenue To Fairview Whooshed

Vulcan whooshed his finger
directly pointing down the line of cars
jammed up for a mile along the road.
With furious heat
he fused them all together
into one rigid length
full of people giving out hell.
Then
firmly grasping one end
he began to raise them up high
getting higher to the sky.
The man at the far end shouted
– "Jesus – I can see the coast of Wales!!"
and Vulcan saying – "Isn't it better
than being stuck in the rush hour."

Gotcha

Perhaps it's a historical thing.
I don't know
but Irish people have mastered the art
of pretending not to be looking.
Jesus – an Irish pair of eyes
will whizz away a split second
before you catch them.
You'd want to be fast
I tell you.
You'd want to be bloody fast indeed.

We never look
never ever ever.

We never miss a thing.

LINES TO SALUTE THE LAUNCH OF

– "A COMIC TOUCH" –

(a drug prevention initiative by *Thomas Larkin & The Art Group*)
at *The Merchant's Quay Project)*
Friday morning 20th November 1998

There's nothing glamorous
about running out of veins.
There's nothing glamorous
about matchstick legs.
There's nothing glamorous
about young voices slurring and slow.
There's nothing glamorous
about tying rubber around your arm.
There's nothing glamorous
about young men and young women
puncturing 40, 50, 60 years
with a dirty tip.

There's nothing cool
about burying your friends.
There's nothing cool
about being forgotten.
There's nothing cool
about blood you'd be afraid to touch.
There's nothing cool
about a very short life.

I salute your new and very brave comic.
I salute the very brave men and women
who have put it together
and made the words
and made the pictures.
You know what this scourge is really about.
You have paid a high and a painful price
for your knowing.
I salute you for putting the real story
into these words and these pictures.
Some of them are so good
they make my blood run cold.
Congratulations.
I wish you huge success.

You Make The Difference

This is the only time in my life
that I feel safe.
The only time.
Writing words.
These words.
Making a message.
Sending it to you.
A message to say
if you need to feel safe
come inside these words
come inside these words
with me
now.

You are in here.
I know you are
because
now
I feel safer.
Now
I do.

How It Was

If any roundy blue plaques
are put up about me
after I die
I would like one
in North Earl Street
which simply says

– PAT INGOLDSBY
WAS MOVED
FROM HERE –

Thank you.

Piddle

When the flames had died
and his house was gutted
when suffocating smoke
had cleared
and firemen rolled away
their hoses
when crooked black timbers
showed where his staircase
used to be
and the sky looked down
into his sitting room
his black and white cat
picked her way through
the bricks and the rubble
to where her litter tray
still sat in the corner
she perched upon it
and did a huge big
spraying wee-wee.

"A bit late for that now"
he said.

I Just Wondered

Did you ever sit upstairs
shouting
– "MAAAAA..........
THERE'S NO PAAAAPER!!!"

Me too.

I Don't Know Either

I like to walk up to a real Dublin girl and say
in heavily accented tones
– "Yah-ten bah-ten
    hah-ten shah-ten
    skunthen dah-ten."

She will usually reply
– "WHA!!!?"

From time to time I enjoy
the following conversation
with strangers

– "Yah-ten bah-ten
    hah-ten shah-ten
    skunthen dah-ten."

"WHA!!!?"

Deadly.

All Other Places

There were lots of stamps on it already
when the man asked in the post office
"Is there enough on this for Germany?"
and the counter clerk said
– "That'll be another 5p" –
and gave him two more stamps
which he stuck on
and the counter clerk said
– "With those two stamps
you're over the weight
– that'll be another 5p" –
and he gave him two more stamps
but the man said
– "You can stick these up your arse –
I'm delivering it by hand" –
and he got the bus, the car ferry
and the train.

Otherwise They'll All Be At It

My right hand shot up high above my head
and said – "I'm never coming down."

I tried being gentle.

"My dear beloved hand
come down to me now who loves you
and I will dizzy you with kisses."

"Go and barnacle your bum!"

"O.K. Listen to me carefully
listen to me good.
Be down here in 30 seconds
or I'll cut you off at the elbow."

"Eh...how about if I send down
a finger and thumb?"

"Come down in any order you like
but if all the bits
don't join up the right way
you're going into the mincer."

"Hold your fire
– we're sending down a thumb."

It's the only way to talk to them.

Toit Is Tine

Sinn sínte i ndearg na tine
fliuch is crua is bog
do theanga ag fliceáil i ngach áit
is mil ag driopáil ar an rug.

SERVICE 12½%

"Would you like tea or coffee with that sir?"

"Coffee please."

"Oh."

"Where would you like to sit madam?"

"A corner table please with windscreen wipers."

"I'm sorry – all our corner tables are gone."

"Are you ready to order dessert sir?"

"Eh yes....I'll have the banana split please."

"The chef hates making those!"

As Go Bráth Leat.....Man

Guys come along in the depths of winter
with bottles of cider and tattoos
dressed for living on a beach
in Thailand
and clutch my hand very forcefully
with that weird emotional
thumb-twisting torture
which signifies –
"I too have suffered man
and within the deep intensity
of my pain
I recognise a fellow pilgrim."

If it's all the one to you lads
will you kindly fuck off
and break someone else's thumb
while you're at it.
Or else just go and live in Thailand.

Now It's All Pink Sacks And Calculators

Huge big coalmen with caps on back to front
and an empty sack as a shoulder-cloak
would come pounding down your entry
bent under the weight
and defying your mother or anyone
to count the number of bags.

"I ordered six....you only brought me five."

And they would count out six empties
on your step.

My father stayed out of sight
and counted the crashes.

AN FHÍRINNE

The Holy Spirit
and
Dublin Taximen
know
EVERYTHING.

I Was There Once And I'm Never Going Back

A gigantic broadcasting mirror
surrounded by hungry egos
who need to keep appearing in it
and hear the voices of people
whom they otherwise dismiss
as ordinary Joe Soaps
telling them
how wonderful
their reflection still is.

Thickly coated chocolate
eating itself.

A Cousin Died Of It

Without any regard whatsoever
for my years
or my physical condition
the young woman
unapologetically
ragged, bagged,
licked, flicked,
chewed, screwed,
rucked, fucked,
humped
and bumped me
until
I was scarcely able
to stand.

What in the name of God
has become
of respect
for age?
Is it ever coming back?

Kindly stop that groaning
agus
tabhair dom mo bhata.

Words To The Virus Which Found Me In 1942

How did you find me?
How in the name of God did you find me?
A tiny infant....down in the heart of old Malahide village
.....in the little house beside the coal yard.
Were you looking for me or what?
Jesus – I was only just born.
I knew nothing at all.
The name they put on you was bigger than me.
Infantile Paralysis.
That is what they called you.
You went around the place paralysing babies.
And you found me down there in that beautiful place
beside the village green
looking over to the island
looking out to the sea.
You came into me and you made me cry.
What time did you come in?
What day was it because I don't know?
What brought you down that way in 1942?
Which way did you come?
Was it down New Street past Bertie Boyle's
or did you come round from the Back Strand
under the railway arch past Lesley Riley's?
Was the tide full in?
I don't know that either.
Why did you go past Annie Daniel's house
and come into ours?
There was me and Michael, Ma and Da.
Why did you come into me?
Did I make a sound as you were going past?
Was that the way it happened?
Was the wireless on?
What could you hear?
I think it was a cowardly thing that you did.
To take the life away from my arm
and put a hidden slow wasting in my legs

was a cowardly thing indeed.
Fifty years on
I am still finding the damage of you.
Fifty years on
you find new tears in me.

Are you still alive?
Are you still doing your virus stuff?
Are you still alive
or did something wipe you out?
I would like to know your name?
Mine is Pat.
I played soccer in spite of you
and I was good.

The Way She Might

"You're right" he said.
"You are absolutely right."

"What?" she said.

"I was wrong" he said.
"Wrong completely.
It was all my fault."

"Hold on" she said.
"Hold on.
Perhaps
I was a little bit
to blame myself."

"A little bit!" he said.
"A little bit!
Is that the most
you can say?!
It was ALL
your bloody fault!"

Saved

The old nun was hitch-hiking
in the bicycle lane at Fairview
trudging slowly to town.
Suddenly
with a deep lunge
of her still powerful hips
she sprang upwards
and plumped onto the crossbar
of a very startled bike.

"Town!"
she commanded.
"And don't spare the horses!"

"Hang onto your beads"
he replied
and away they went together
whooshing
down the road to salvation.

BLAZE UP SPIN DOWN

I came upon a hubcap at the side of the road.
It was dying.
I didn't know what to do for the best.
The last time I found one as weak as this
I rushed it to the nearest garage.

They shot it.

I took it up to the top of the highest hill
and let it die
spinning
– *"wheeeeeeeeeeeeeeeeeeeeeeeeee"* –
all the way to the bottom.

Dubh Linn

This town is dying slowly.
The humanity is dying.
The fun is dying.
The ease is dying.
The profits are getting huge.

WHUHOOOOOOO!!!!!!
3-12-98

Today
I was watching the worst shoplifter
in the world
going about her business
in the flower shop.
She was robbing little teddy bears
and she did not give a shit.
She was stuffing them into her pockets.
She was stuffing them into her bags.
She was dropping them all over the floor.
The staff did not give a hoot
so she took her time.
When she was quite finished
she picked the teddies up off the floor
and sat them all back in a comfy row
on the shelf.
Then she walked out of the shop
with a look of pure virtue on her face
and her pockets stuffed with little bears.

A Lovely Encounter In The Irish Life Mall Talbot Street

11.05 1-12-98

"Would you like to go halves in a wee-wee?"
I asked the man who was just about
to put his 10p in the door.

"Aw Jaysus Pat.....you can have this one on me."

EXCHANGE

Irish Life Mall   Talbot Street
11.15   5-12-98

The man approached me in desperation
earnestly knotting his legs
and holding a 20 pence piece.

"Would you have two tens Pat?
I'm burstin'."

"God I'm sorry. I haven't.
Do you know what?
I've just written a poem about this.
Would you like to hear it?"

"Aw fuck no Pat – I'm burstin'"
and he hurtled away down the Mall.

Moments later he shot past me again
going the other way.

"Did you get one?" I yelled.

"Yeah but it cost me the twenty p!!!"

Value

Irish Life Mall   Talbot Street
5-12-98

A little girl rushed around the corner.

"Martin wants to do a wee-wee
an' we can't get in!"

I rushed back with her.

One of her little friends
was hoisting another little one
as high up in the air as she could
– trying to get ten p into the slot.
Martin was holding himself
and wailing –
– "Hurry up...hurry up
it's comin'...it's comin'!!!"

I opened the door
and they were in like a shot.

Three others nipped in behind them.

Half The Time They Don't Listen

Talbot Street 12.25 9-12-98

"There y'are Pat – how's business?"

"Shite."

"That's great."

WORDS OF HOPE

later on same day

"How's it goin'?
You won't sell many of *those.*"

TORPEDOED IN TALBOT STREET
13-12-98

After a week standing on the steps
outside the Department of Education
I finally worked out
what the faded words say
– the two faded words
high up on the yellow wall opposite
high up between the two windows

– *BOYS* –

that bit was easy.
I got it right away.
The other word took me the week.
When I finally cracked it
I felt great.

I was telling an American woman about it
this morning.
I pointed over at the words.

– "BOYS" – she said.
She got it right away.

Then she made out a "T" in the other word.

She has just set me back a week.

From The 130

14-12-98 Clontarf Seafront

I don't know if the dog
was having visions
or what.
Out in the middle of nowhere
no trees or bushes or anything
he lifted his leg
and pissed so powerfully
sideways
against nothing at all
that I almost saw the tree myself.

A Man With Much Drink On His Breath Telling Me About His Walking Stick

13.52 20-12-98 Talbot Street

"He was a big man.....look at the bend on that
........look at the bend on that....
he was a big man alright
I got the stick off of him when he died
his wife gave it to me.....
seventy years old that stick.
Jesus look at the bend
....he was big alright....he was big."

Conversation With A Small Boy

Later on the same day

"Where will I get the lend of a chair?"

"Just go into a pub and whip a bleedin' stool."

Aisling

"I'm still waiting for a beautiful woman
to come along and say
– *"I live very near here.....*
the fire is lighting in my apartment....
come there with me now.
stretch out on the rug
and write two or three of your poems
upon my inner thigh."

Perhaps she has said it to me already.
In French or Spanish.
I wouldn't know.

Things You'd Be Thinking About

Selling my books in Talbot Street
Christmas week 1998

In the morning when the shutters go up
outside HALL'S CAMERAS LTD
and H. HOPKIN & SONS
the beds directly above them
in The Talbot
go up as well
and whoever is asleep in them
gets shoved against the ceiling.

You could always have an early breakfast.

Another Thing

People who are testing binoculars
stand outside HALL'S CAMERAS LTD
and spy down towards Connolly Station.
I would like to arrange for a Spanish Armada
to come surging up the street under full sail.

They'd buy them then.

My Prayer

All I ever received
from the love of you
was
massive doses of guilt
migraine headaches
and psychiatric care.

If it's all the one to you
kindly fuck off
and love someone else.

I See

I asked Hoot this morning

– "How come
I'm not a world name
with my poems?"

And he said – "Meow."

So now I know.

God Bless You

The man in black with glasses
who twitches frequently
on both sides of his face
– whenever he sneezes
he sounds exactly the same
as a Spanish sailor
removing with a penknife
the appendix of a tortoise
with force.

Whenever he sneezes
on a crowded train
someone will always exclaim

– "My God
there is a Spanish sailor
on board
removing with a penknife
the appendix of a tortoise
with force."

But they can never find him
which is just as well
because it would go
very hard
with him.

To My Parents After Visiting Their Grave

St. Stephen's Day 1998   Malahide

It's as well that you're lying here.
The old house
is part of McGowan's Carpets now.
They've taken away the front porch.
They store things in your bedroom.
There are tenants in the new house.
The front lawn and the railings are gone.
You wouldn't like it.

It's as well that you're lying here.
You can hear the crows high up
and the children at play in The Reds.
You can hear the trains over the estuary.
You can hear the Malahide that you knew.

Rest in peace.